IS CONSCIOUSNESS EVERYWHERE?

Essays on Panpsychism

Edited by
Philip Goff and
Alex Moran

Imprint-academic.com

Copyright © Imprint Academic Ltd., 2022

The moral rights of the authors have been asserted.
No part of this publication may be reproduced in any form
without permission, except for the quotation of brief passages
in criticism and discussion.

Published in the UK by
Imprint Academic, PO Box 200, Exeter EX5 5YX, UK

Distributed in the USA by
Ingram Book Company,
One Ingram Blvd., La Vergne, TN 37086, USA

ISBN 9781788360876

A CIP catalogue record for this book is available from the
British Library and US Library of Congress

Contents

- 5   About Authors
- 9   Is Consciousness Everywhere?   *Philip Goff & Essays on Panpsychism*   *Alex Moran*

1. The Scientists

- 16   Consciousness and the Laws of Physics   *Sean Carroll*
- 32   Relations and Panpsychism   *Carlo Rovelli*
- 36   Physics, Time, and Qualia   *Marina Cortês, Lee Smolin & Clelia Verde*
- 52   The Real Problem(s) with Panpsychism   *Anil K. Seth*
- 65   Reflections of a Natural Scientist on Panpsychism   *Christof Koch*
- 76   Autism and Panpsychism: Putting Process in Mind   *Jonathan Delafield-Butt*
- 91   Dr Goff, Tear Down This Wall! The Interface Theory of Perception and the Science of Consciousness   *Robert Prentner*
- 104   What is a Theory of Consciousness for?   *Chris Fields*

2. The Philosophers

- 116   Is Panpsychism at Odds with Science?   *Luke Roelofs*
- 129   A Solution to the Combination Problem and the Future of Panpsychism   *Annaka Harris*
- 141   Galileo's Real Error   *Keith Frankish*
- 147   Qualities and the Galilean View   *Michelle Liu*
- 163   Grounding the Qualitative: A New Challenge for Panpsychism   *Alex Moran*
- 181   Panpsychism and the Limits of Physical Science   *Alyssa Ney*
- 194   Missing Entities: Has Panpsychism Lost the Physical World?   *Damian Aleksiev*

212	Can a Post-Galilean Science of Consciousness Avoid Substance Dualism?	*Ralph Stefan Weir*
229	'Oh You Materialist!'	*Galen Strawson*

3. The Theologians

250	Why a Panpsychist Should Adopt Theism: God, Galileo, and Goff	*Joanna Leidenhag*
268	Panpsychism and Spiritual Flourishing: Constructive Engagement with the New Science of Psychedelics	*Sarah Lane Ritchie*

4. Replies from Philip Goff

289	Putting Consciousness First: Replies to Critics	*Philip Goff*

ABOUT AUTHORS

Damian Aleksiev holds a PhD in philosophy from the Central European University. He works primarily on the metaphysics of consciousness and questions of fundamental ontology. He is currently working on a project that explores the use of explanatory gaps as a guide to reality.

Sean Carroll is an American theoretical physicist who specializes in quantum mechanics, gravity, and cosmology. He is a Research Professor in the Walter Burke Institute for Theoretical Physics in the California Institute of Technology Department of Physics and an External Professor at the Santa Fe Institute.

Marina Cortês is a physicist who works at Berkeley Lab in California. She completed her PhD at the University of Sussex. Research interests include physics of the early universe, Bayesian statistics, dark energy reconstruction and phenomenology, and gravity of discrete space times.

Jonathan Delafield-Butt is Professor of Child Development and Director of the Laboratory for Innovation in Autism at the University of Strathclyde. His work examines the origins of conscious experience and the embodied and emotional foundations of psychological development. He took his PhD in Neurobiology at the University of Edinburgh, extended to Developmental Psychology at the Universities of Edinburgh and Copenhagen. He has trained pre-clinically in psychoanalytic psychotherapy.

Chris Fields is an independent scientist working at the intersection of quantum information theory, evo-devo biology, and cognitive neuroscience. His primary interest is in how systems at all scales identify objects in their environments, represent state changes over time, and develop communicative conventions that enable cooperative and/or competitive activity. Fields has published over 190 refereed papers in various ares of physics, cognitive science, and biology.

Keith Frankish is Honorary Reader in Philosophy at the University of Sheffield, Visiting Research Fellow at the Open University, and Adjunct Professor with the Brain and Mind Programme in Neurosciences at the University of Crete. Keith has wide research interests in philosophy of mind and cognitive science, and he has published numerous articles, books, and edited collections, including *Illusionism as a Theory of Consciousness* (Imprint Academic 2017).

Philip Goff is an Associate Professor of Philosophy at Durham University. He defends panpsychism as the best account of how consciousness fits into our overall theory of reality. Goff has authored an academic book, *Consciousness and Fundamental Reality* (OUP), and a book aimed at a general audience, *Galileo's Error: Foundations for a New Science of Consciousness*. He has published 45 articles as well as writing extensively for newspapers and magazines.

Annaka Harris is the *New York Times* best-selling author of *Conscious: A Brief Guide to the Fundamental Mystery of the Mind*. She is an editor and consultant for science writers, specializing in neuroscience and physics, and her work has appeared in the *New York Times*. Annaka is the author of the children's book *I Wonder*.

Christof Koch is a neuroscientist best known for his writings on the basis of consciousness, starting with the molecular biologist Francis Crick over 25 years ago. Trained as a physicist, Christof was for 27 years a professor of biology and engineering at the Caltech. In 2015, he became the President of the Allen Institute for Brain Science in Seattle. He's now the Chief Scientist of the MindScope Program at the Allen Institute. He's also the Chief Scientist of the Tiny Blue Dot Foundation, with its focus on understanding consciousness.

Joanna Leidenhag is Lecturer of Theology and Liberal Arts at the University of Leeds. She earned her PhD from the University of Edinburgh in 2018. Her first monograph, *Minding Creation: Theological Panpsychism and the Doctrine of Creation*, was published in 2021 by T&T Clark. Her research includes theological evaluations of panpsychism, as well as theological engagement with autism and mental health.

Michelle Liu is a Leverhulme Early Career Fellow at the University of Hertfordshire where she also worked as a lecturer in philosophy (2019–21). She completed her DPhil at Oxford in 2019. She works in philosophy of mind, philosophy of language, and aesthetics, and is particularly interested in phenomenal experience, pain, polysemy, and the expressiveness of art.

Alex Moran is a Leverhulme Early Career Fellow at the University of Oxford. His PhD was completed at the University of Cambridge (Queens' College). His primary research interests are in metaphysics and the philosophy of mind. Moran also has research interests in metaethics, early analytic philosophy, and early modern philosophy.

Alyssa Ney is Professor of Philosophy at the University of California, Davis. She is the author of *The World in the Wave Function* (OUP 2021), *Metaphysics: An Introduction* (Routledge 2014), and editor, with David Albert, of *The Wave Function: Essays in the Metaphysics of Quantum Mechanics* (OUP 2013). She works primarily on the interpretation of quantum mechanics and the unity of science.

Robert Prentner completed a PhD in physical chemistry and, later, in philosophy at ETH Zürich. He did his postdoc studies on the interface theory of perception at UC Irvine, and worked as Senior Research Fellow at the Center for the Future Mind at Florida Atlantic University. His research pertains to the philosophical foundations in the sciences of mind, in particular mathematical consciousness science. He is co-editor of the journal *Mind and Matter* and co-organizer of the 'Models of Consciousness' conference series. He is currently based at the Munich Center for Mathematical Philosophy at LMU Munich.

Sarah Lane Ritchie is Lecturer in Theology and Science at the University of Edinburgh. Her research is in science and religion, with a particular focus on the intersection of cognitive science of religion, psychology, theology, philosophy of mind, and metaphysics.

Luke Roelofs is a postdoc at NYU's Centre for Mind, Brain, and Consciousness, where they work on issues in the metaphysics of consciousness and the philosophy of imagination. Their first book, *Combining Minds*, seeks to defend panpsychism against its much-discussed 'combination problem'.

Carlo Rovelli is a theoretical physicist known for his work in quantum gravity. He founded the Quantum Gravity group of Aix-Marseille University. He is member of the Institute Universitaire de France, Honorary Professor of the Beijing Normal University, Honoris Causa Laureate of the Universidad de San Martin, Buenos Aires, and member of the Académie Internationale de Philosophie des Sciences. He is currently Adjunct Professor of Philosophy at the Western University and Distinguished Visiting Researcher at the Perimeter Institute in Canada.

Anil K. Seth is Professor of Cognitive and Computational Neuroscience at the University of Sussex, Co-Director of the Canadian Institute for Advanced Research (CIFAR) Program on Brain, Mind and Consciousness, an ERC Advanced Investigator, Editor-in-Chief of the journal *Neuroscience of Consciousness*, and a Wellcome Trust

Engagement Fellow. He has published more than 180 papers. His TED talk has been viewed more than twelve million times, and his new book, *Being You: A New Science of Consciousness*, is an instant *Sunday Times* Top 10 Bestseller and *Guardian* book of the week.

Lee Smolin is a theoretical physicist who has done most of his work on quantum gravity. He has co-founded a number of approaches including loop quantum gravity and relative locality. He has broad interests and has contributed new ideas to cosmology (cosmological natural selection), foundations of quantum mechanics (non-local hidden variables theories, energetic causal sets, etc.). He is the author of five semi-popular books that explore the philosophical ramifications of the big open questions in physics. Since 2001 he is a founding and senior faculty member of Perimeter Institute.

Galen Strawson holds the President's Chair of Philosophy at the University of Texas at Austin. His books include *Mental Reality* (1994) and *Selves* (2009). His most recent books are *The Subject of Experience* (2017) and *Things That Bother Me* (2018).

Clelia Verde is the product development manager at an innovative insurance company based in Milan. Trained as an engineer (MSc: Risk Management, MIB School, Trieste, Italy), she has over the years consistently pursued her interests in advanced physics and the literary arts. She is a contributing science editor for an independent Italian general-interest news site (Gli Stati Generali) and an award-winning poet (Bookcity Milano 2019).

Ralph Stefan Weir is Lecturer in Philosophy at the University of Lincoln and Associate Member of the Faculty of Theology and Religion at the University of Oxford. He received his PhD in philosophy from the University of Cambridge in 2020. His recent publications include 'Does Idealism Solve the Problem of Consciousness?' in *The Routledge Handbook of Idealism and Immaterialism*, 'Christian Physicalism and the Biblical Argument for Dualism' in *International Journal for Philosophy of Religion*, 'Bring Back Substances!' in *Review of Metaphysics*, and, with Benedikt Paul Göcke, *From Existentialism to Metaphysics: The Philosophy of Stephen Priest* (Peter Lang Publishers).

Philip Goff[1]
and Alex Moran[2]

Is Consciousness Everywhere?

Essays on Panpsychism

1. Introduction

The mind–body problem, broadly speaking, is the challenge of understanding how the conscious mind relates to the physical world. On the one hand, it seems that there is nothing more familiar to us, nothing we know better, than the phenomenon of consciousness. On the other hand, there is much about consciousness that has defied our understanding. In particular, there seems to be a special difficulty when it comes to reconciling the *subjective* and *qualitative* aspects of consciousness with our *objective* and *quantitative* scientific picture of the physical world. One major task for science and philosophy, therefore, is to find a way to bridge this gap, thereby explicating the place of consciousness in nature.

Historically, two main positions have been defended in answer to the mind–body problem. One is dualism, the view that consciousness lies outside the physical domain, and therefore has a wholly non-physical nature. The other is materialism (or physicalism), the doctrine that, since reality is wholly physical in nature, consciousness must ultimately be thought of as a part of the material world. The

Correspondence:
Email: philip.a.goff@durham.ac.uk; alexander.moran@philosophy.ox.ac.uk

[1] Durham University, UK.
[2] University of Oxford, UK.

present volume, however, focuses on an alternative theory; namely, *panpsychism*. Like the dualist, the panpsychist claims that consciousness is irreducible and fundamental, and hence cannot be understood in other, more basic, terms. Like the materialist, however, the panpsychist also thinks that consciousness does not lie outside of the rest of nature, but is rather firmly located within the material world. Indeed, the panpsychist makes the radical claim that consciousness is *ubiquitous* in nature, in so far as it is a property instantiated, not just by humans and some animals, but even by the most fundamental constituents of physical reality. The panpsychist thus takes nature to be permeated by consciousness, rather than viewing consciousness as a derivative phenomenon that only emerges at higher levels.

One central thought motivating the panpsychist position is an observation about our understanding of the physical world itself. The key idea is that physics only describes the relational or dispositional properties of matter, not its intrinsic nature: what matter *does* rather than what it *is*. This then raises a question about what the intrinsic nature of matter actually is. According to the panpsychist, the intrinsic nature of basic matter is constituted by rudimentary kinds of consciousness. This then allows us — or so it is hoped — to explicate the more familiar phenomenon of human consciousness in terms of the more basic kind of consciousness instantiated by fundamental physical things.

Proponents of panpsychism argue that the theory enables us to make significant headway with the traditional mind–body problem, while avoiding the well-known problems that dualists and materialists face. Its detractors, meanwhile, urge that the theory faces substantial problems of its own. The papers in this special issue explore these and related issues, from the perspectives of science, philosophy, and theology. Some papers focus on further motivating and developing the panpsychist position. Others explore various challenges that the panpsychist faces. Collectively, they shed new and important light not only on panpsychism, but on the fundamental question of the place of consciousness in nature more generally.

As stalking horse, many of the papers focus on Philip Goff's book *Galileo's Error: Foundations for a New Science of Consciousness*, which offers an important and accessible defence of the panpsychist view. The special issue also includes a response piece from Goff to these various articles.

2. The Scientists

Our first three papers are by theoretical physicists. Carlo Rovelli has previously defended a relational interpretation of quantum mechanics, according to which quantum mechanics concerns not how physical entities are in themselves, but how they are *in relation to one another*. In his paper, Rovelli suggests that this relational conception of physics is itself a — very mild — form of panpsychism; and moreover that it may help address the intuitions underlying the 'hard problem' of consciousness, as the differences between the mental and the physical are now less stark than they might previously have appeared.

The next two papers share a common focus. Panpsychists believe that consciousness is a fundamental feature of reality. Both of the next two papers (the first by Sean Carroll, the second by Marina Cortês, Lee Smolin, and Clelia Verde) agree that this conviction requires rewriting current physics, as the understanding of the basic physics in brains given to us by the 'Core Theory' (the standard model of particle physics combined with the weak-field limit of general relativity) does not make reference to consciousness. From this point of agreement, they go in different directions. Carroll infers that consciousness is not a fundamental feature of reality, but rather an emergent property of certain complex systems. Cortês, Smolin and Verde, in contrast, present a new interpretation of fundamental physics in which qualia — as well as the passage of time — play a fundamental role.

The next two papers are by neuroscientists, one pro panpsychism and one opposed. Anil Seth believes that, rather than focusing on the 'hard problem' of where consciousness comes from in the first place, it is more profitable to focus on the 'real problem' of explaining, predicting, and controlling the various properties of consciousness in terms of physical processes in the body and brain. Moreover, he rejects panpsychism as an untestable and therefore unfruitful hypothesis. Whilst Christof Koch is much more sympathetic to panpsychism, like Seth, he emphasizes the importance of experimental science and testable predictions. Koch outlines and defends the integrated information theory of consciousness, a theory which entails that consciousness is ubiquitous in the physical world and hence can be seen as a form of panpsychism. Koch also expresses disagreement with Philip Goff's exegesis of Galileo in *Galileo's Error*.

We often forget that there are philosophical assumptions lying at the bedrock of any scientific worldview. In his work as an experimental psychologist, Jonathan Delafield-Butt has found that dropping the philosophical assumptions of materialism, and adopting in their place panpsychist assumptions, affords deeper insights into the nature and character of autism. In his paper, Delafield-Butt lays out the case for this, with reference to the panpsychist framework of Alfred North Whitehead.

In our next paper, Robert Prentner defends the *interface theory*, which he has developed in collaboration with Donald Hoffman and others. The interface theory has much in common with panpsychism. Both take fundamental reality to be made up of consciousness. However, whereas panpsychism holds that the physical world is also fundamental (because the physical world is made up of consciousness), the interface theory — or 'idealism' as we philosophers have called it for a couple of hundred years — holds that there is a more fundamental (mental) reality *underlying* the physical world. Prentner pits panpsychism against the interface theory, arguing that the latter offers a more robust, less dualistic theory of consciousness.

Chris Fields has developed a detailed form of panpsychism: minimal physicalism. In his paper in this special issue, however, Fields raises some challenges to Goff's conception of the science of consciousness. Fields is sceptical that science can or should be in the business of accounting for the specific character of conscious states, e.g. the redness of a red experience, as opposed merely to accounting for why those conscious experiences exist at all (it is interesting that this approach seems to be the precise opposite of the 'real problem' approach Seth defends in his paper). Fields also expresses his suspicions that the real motivation for the 'consciousness war' between dualists, materialists, and panpsychists may be a yearning, on the part of some opponents of materialism, for human exceptionalism.

3. The Philosophers

The special issue includes nine papers by academic philosophers (excluding the response piece from Goff). Two of these papers defend and elaborate the panpsychist position. Luke Roelofs explores the question as to whether, and to what extent, panpsychism classifies as a scientific worldview. In addition, the paper offers a nuanced discussion of what exactly it means for a philosophical thesis such as panpsychism to count as scientific in the relevant way. In the following paper, Annika Harris explores a fundamental problem that panpsychists face, which has gained much attention in the recent literature, known as the 'combination problem'. Harris diagnoses the problem as resulting from the conviction that, for any given conscious experience, there exists a 'subject' or a 'self' that has the experience. Once we realize that subjects don't really exist in the first place, she argues, the problem goes away.

Five further papers focus on raising challenges for panpsychism; a couple also promote an alternative, physicalist view. Damian Aleksiev argues that, while they focus on explaining consciousness, panpsychists face problems when it comes to accounting for the physical world. In particular, the argument is that if we conceive of the intrinsic nature of matter in terms of consciousness then several facts about physical reality become difficult to explain. Alyssa Ney, meanwhile, argues that panpsychism is not, contrary to what its defenders maintain, sufficiently well-motivated by its underlying claims concerning physics and the nature of matter. She also argues that physicalists are in just as good a position as panpsychists when it comes to accounting for free will, objective value, and meaning.

The remaining three essays from this grouping challenge panpsychism from a different angle. Again, panpsychists are primarily concerned with explaining consciousness. However, as these authors point out, there is a related explanatory task, concerning, not consciousness, but the sensory qualities of external things (such as the redness of a rose, or the distinctive smell of coffee). Both Keith Frankish and Michelle Liu argue that panpsychists repeat the 'Galilean mistake' of supposing that the sensory qualities are instantiated 'in the mind' rather than by external objects. Frankish then argues that, instead, we should deny that *anything* instantiates such qualities, which, he claims, then makes it that much easier to embrace a reductive materialist position. Michelle Liu, by contrast, defends sensory quality realism, i.e. the view that external things really possess the full range of sensory qualities they seem to, and criticizes the panpsychist for not respecting this position. In a similar vein, Moran presupposes precisely the kind of sensory quality realism that Liu defends, and then maintains that, on this assumption, there is much about the physical world that panpsychists are not in a position to explain. He also claims that, while certain nearby positions to panpsychism may be able to meet the challenge he articulates, there is a certain kind of non-reductive physicalism that fares just as well.

The two final papers by philosophers are concerned with connections between panpsychism and the more familiar materialist and dualist positions. While sympathetic to Goff's proposal for a 'post-Galilean' science of consciousness, in which consciousness is taken to be a fundamental feature of reality, Ralph Weir argues that post-Galileans end up being committed to a radical form of dualism known as substance dualism. Pushing in a contrary direction, Galen Strawson urges that panpsychism in fact classifies as a form of materialism, and indeed arguably the most defensible form. Strawson also provides a helpful, detailed discussion of what the doctrine of materialism actually amounts to.

4. The Theologians

Panpsychism is often assumed to be a spiritual doctrine. However, many contemporary proponents of panpsychism are resolutely secular. They may not believe in a transcendent reality but they do believe that people have experiences — they feel pain, they see colour — and this mundane and everyday reality needs to be accounted for in our overall theory of reality.

Nonetheless, some have argued that there are important connections between panpsychism and spiritual convictions of some form or another. Joanna Leidenhag, in her paper, argues that the motivations that lie behind the arguments put forth in support of panpsychism, if applied consistently, lead to belief in God. The panpsychist demands an intelligible account of how consciousness emerges. But is it consistent to demand an explanation of consciousness without also demanding an explanation of the existence of the universe itself? Leidenhag thinks not, and suggests this line of reasoning can be satisfactorily concluded only by a commitment to theism.

In the final essay of the volume, Sara Lane Ritchie explores the connections between panpsychism, psychedelic experiences, and spiritual flourishing. It is common for the person undergoing a psychedelic experience to feel a connection to *ultimate reality*, however that is conceived (Ritchie examines a number of options, including panentheism, the view that 'the entire natural world exists within God, but also that God is, in some sense, more than the natural world'). Presumably a materialist must reject these experiences as delusional. However, Ritchie argues that the worldview of a panpsychist may be consistent with the verdicality, i.e. the non-delusional nature, of these kinds of psychedelic experiences.

5. Replies from Philip Goff

In the final article, Philip Goff responds to the essays of the volume. He also explores some ideas on what a 'post-Galilean' science of consciousness — one which takes consciousness to be a fundamental feature of reality — might look like.

Sean Carroll[1]

Consciousness and the Laws of Physics

Abstract: We have a much better understanding of the dynamics and ontology of physics than we do of consciousness. I consider ways in which intrinsically mental aspects of fundamental ontology might induce modifications of the known laws of physics, or whether they could be relevant to accounting for consciousness if no such modifications exist. I suggest that our current knowledge of physics should make us sceptical of hypothetical modifications of the known rules, and that without such modifications it's hard to imagine how intrinsically mental aspects could play a useful explanatory role.

1. Introduction

We don't fully understand consciousness. That's hardly surprising. The human brain, which is at least somewhat involved in consciousness, contains roughly 100 billion neurons and 700 trillion synaptic connections. It is arguably the most complex structure in the known universe. Even as neuroscience makes impressive advances in understanding the brain, it seems prudent to anticipate that we have a number of conceptual and technical breakthroughs yet to come that could bear in important ways on the question of consciousness.

We do, on the other hand, understand the basic laws of physics governing the stuff of which brains are made. They take the form of an effective quantum field theory describing a particular collection of matter particles interacting via force fields. There is certainly much of

Correspondence:
Email: seancarroll@gmail.com

[1] California Institute of Technology and Santa Fe Institute, USA.

physics remaining to be discovered, but in the specific regime covering the particles and forces that make up human beings and their environments, we have good reason to think that all of the ingredients and their dynamics are understood to extremely high precision (Carroll, 2021a). Modern physics, in other words, provides evidence for what philosophers call 'causal closure of the physical': physical events have purely physical causes (Loewer, 1995; Papineau, 1995), at least in the regime relevant to human life. Without dramatically upending our understanding of quantum field theory, there is no room for any new influences that could bear on the problem of consciousness.

Given this situation, it might seem surprising to a disinterested observer to learn that anyone would argue that the best route toward understanding consciousness involves augmenting or altering the ontology suggested by fundamental physics. To start with the least-well-understood aspects of reality and draw sweeping conclusions about the best-understood aspects is arguably the tail wagging the dog. When we can't remember where we put our car keys, we don't typically respond by going out and buying a new car.

Nevertheless, a prominent strain in the philosophy of consciousness proposes to do just that (Chalmers, 1996; Goff, 2017; 2019). The justification for such a radical move is that there will be something *qualitatively* missing in any account of consciousness based purely on physical ontology as we currently understand it. This perspective arises from a conviction that physics can explain behaviour, but not the first-person experiences characteristic of human consciousness; that physics may account for the dynamics of the stuff in the universe, but it doesn't illuminate the intrinsic nature of that stuff.

In this paper I support the idea that physics is in such good shape that the most promising strategy for trying to understand consciousness is as a (weakly) emergent phenomenon that leaves physical ontology untouched, rather than trying to extend or elaborate that ontology with specifically mental aspects (*cf.* Moran, this issue; for a contrary view see Cortês, Smolin and Verde, this issue). After reviewing the 'Core Theory' and our reasons for being confident in its accuracy, I will discuss what it would mean to modify it, either directly in the dynamics or by adding additional ontological features. It is always possible that contemporary physics is inadequate and in need of modification, but a close examination highlights the difficulty of doing so in a rigorous and convincing way.

2. The Physics Underlying Everyday Life

Science often employs multiple vocabularies or theories for describing the same physical situation, often at different degrees of focus or coarse-graining. These are often called 'levels', although strictly speaking they need not be arranged hierarchically. Within any level, we can specify the domain of circumstances in which a particular theory is applicable. The claim here is that there is one level of description — that of effective quantum field theory — and a well-defined regime — interaction energies below certain thresholds, broad enough to include every situation encountered in ordinary human life — where we have very good reasons to believe we know precisely what is going on.

The fundamental ontology of any quantum theory is specified by a 'quantum state' or 'wave function', expressed mathematically as a vector in an abstract Hilbert space (Carroll, 2021b). In a quantum field theory, that state can be thought of as being constructed from possible configurations of fields that take on values at each point in space-time.

Fortunately, the details of this formalism are not necessary for our present purposes. Once we quantize the fields, appropriate configurations — essentially, low-lying energy states — can be interpreted as collections of interacting particles. These circumstances are more than broad enough to encompass human beings and their environments. Thus, we can think of people and the objects around them as configurations of certain particles. In particular, human beings are made of atoms; those atoms are made of protons, neutrons, and electrons; the protons and neutrons are made of quarks and gluons. These particles interact through gravitation, electromagnetism, and the nuclear forces, and get mass from a background Higgs field.

The dynamics of these particles and forces are governed by an effective quantum field theory known as the 'Core Theory', consisting of both the standard model of particle physics and the weak-field limit of general relativity (Wilczek, 2015). This theory is not the ultimate theory of everything, nor is it intended to be. The world might not be described by a quantum field theory at the deepest level; that description might emerge from a more fundamental set of degrees of freedom and dynamical laws. And the Core Theory is certainly not supposed to cover every circumstance — dark matter and the Big Bang, to name some obvious examples, are not included. But we have excellent reasons to believe that the entirety of the 'everyday life regime' supervenes on the ontology and dynamics of this theory

(Carroll, 2021a). If there is a more fundamental level, its properties are irrelevant to the autonomous dynamics of the Core Theory. And if there are additional particles and forces, they interact too weakly with the known fields to exert any influence on human behaviour; otherwise they would have already been detected in experiments.

Our confidence in this picture derives from the fact that quantum field theories are the practically unique way to satisfy the general principles of quantum mechanics and relativity; from symmetries ensuring that any unobserved fields must be too weakly-interacting with ordinary matter to be relevant for everyday-life dynamics; and the property of effective field theories that the dynamics themselves are fully determined in terms of a very small number of parameters. We can't know for certain that the Core Theory suffices to correctly describe the behaviour of the particles and fields making up human beings, no matter how good our arguments become, but any proposed modification of this theory should be held to a very high standard indeed. Just as with any hypothetical new physical model, it should be quantitative and precise, detailing exactly how the explicit dynamics of the Core Theory are meant to be modified, and how such modifications are consistent (or not) with features such as unitarity, locality, symmetries, and conservation laws, not to mention experiments.

3. Domains of Applicability

In the context of the relationship between consciousness and the laws of physics, it is worth being a bit more explicit about how we specify the 'domain of applicability' of a theory (Carroll, 2016). The general idea is that there is a set of physical situations in which the predictions of the theory are meant to be accurate, with no claims being made for situations outside that set.

The empirical foundation of the Core Theory has been established through a line of experimental and observational results stretching back to Faraday, Rutherford, and many others. But the most precise constraints come from modern-day particle colliders, which typically measure the results of scattering individual particles off of each other. One might sensibly wonder whether results from such a paradigmatically reductionist setting can be straightforwardly extrapolated to something as complex as a human brain, which contains roughly 10^{27} particles. Perhaps brains are just not within the domain of applicability of the Core Theory.

If we accept the basic framework of effective quantum field theory, this concern is unfounded; everything that happens inside biological organisms here on Earth is unambiguously within the purview of the Core Theory. Its domain of applicability is bounded by two criteria. The first is that gravity must be weak, so that we can treat the gravitational field as an ordinary quantum field, sidestepping subtleties of horizons and Hawking radiation. 'Weak' is a relative term, and in this case means 'the gravitational potential is much smaller than one'. In practice, this means 'we are nowhere near a black hole'. This criterion is easily met by everything we know of in the Solar System, human brains included.

The other criterion comes from effective field theory. The modifier 'effective' indicates that the domain of applicability of the theory is specified in terms of energies — in particular, the amount of energy transferred between particles when they interact. An effective field theory is meant to be accurate when energy transfers remain lower than some explicit cut-off. In the case of the Core Theory, experiments have established its accuracy at energy transfers of up to 10^{11} electron volts. Electrochemical reactions inside biological organisms, meanwhile, happen at less than 10^2 eV. Shrinking the domain of applicability of the Core Theory while remaining within the framework of effective quantum field theory requires a mistake in our current understanding by a factor of over a billion, which seems implausible.

The effective field theory paradigm also features very specific properties of the field dynamics: they are local (interacting only with other fields at the same space-time point), and governed by a simple and inflexible set of equations. So below, when I refer to 'within the effective field theory paradigm', this is what is meant: a theory of quantum fields, evolving under the appropriate simple dynamical equations, applicable in circumstances where gravity is weak and interactions feature energy transfers below the cut-off.

Within its domain of applicability, the Core Theory is what we might label *causally comprehensive*. If we give a complete specification of the quantum state of the Core Theory fields within that regime, there is a specific equation that unambiguously predicts how it will evolve over time. This equation is sufficient to describe everything human beings generally do, unless they jump into a black hole or stick their hand inside the beam of a high-energy particle accelerator. There are no ambiguities or loose ends. The fact that brains are big, complex things is irrelevant. The Core Theory makes specific predictions for how any particular brain will behave; our choice is to either accept that prediction, or modify the theory in some way. There is no third alternative (Aristotle, 2002).

4. Ontology and Dynamics

Despite the extraordinary empirical success of the Core Theory, and the fact that human beings and their brains are made out of particles interacting within its domain of applicability, there is a lingering worry that no physicalist picture is up to the task of accounting for consciousness, even as some higher-level weakly-emergent phenomenon. There are various ways of expressing this concern: conscious experiences are inherently first-personal and subjective; merely physical objects cannot feel what it is to be like something; describing the behaviour and functions of objects does not explain their intrinsic nature; and others. See Goff (2019) for an overview.

One common reaction to these concerns is to contemplate modifications of the underlying ontology suggested by modern physics: to suggest that a quantum state built upon interacting fields obeying strict equations of motion is incapable in principle of accounting for consciousness, and that we instead need to add specifically mental aspects to our description of reality. We may contemplate ontological modifications as dramatic as substance dualism, in which an immaterial mind is distinct from the physical body but interacts with it, or idealism, in which the physical world is a kind of projection of a fundamentally mental reality. I will focus on more subtle approaches, in which mental aspects or properties are related to, but augment, the basic physical reality. Approaches under this umbrella include property dualism, which posits distinct mental properties in addition to physical properties (Chalmers, 2003); Russellian monism, which posits both physical and mental aspects belonging to a single underlying set of properties (Russell, 1927; Chalmers, 1996; Strawson, 2006; Goff, 2017); and other forms of panpsychism, epiphenomenalism, and related approaches (Papineau, 2020). For convenience I will refer to any new ontological features as 'mental aspects', which is meant to include potentially autonomous properties as well as intrinsic qualities that might supervene on the physical situation.

Any such approach must specify whether, and how, it modifies the dynamics of the theory as well as the ontology. In our conventional understanding, consciousness exerts an important influence on behaviour: I can have a conscious experience and talk about it.

(Admittedly, highly trained philosophers are reported to be able to imagine removing consciousness from a being without affecting its behaviour in any way.) Observed human behaviour can be traced to electrical and chemical signals in our brains and nervous systems. Explicitly mental aspects of ontology could affect this behaviour by, for example, influencing the rates of chemical reactions, or the strength of electromagnetic forces, or the probability of certain quantum outcomes.

In what follows we will examine different kinds of relationship that a theory of consciousness might have to the physical dynamics of the Core Theory, as well as the possibility that there is no relationship at all. Our goal is not to comprehensively catalogue the possibilities, but just to highlight some of the challenges faced by any approach that aspires to explain consciousness by adding mental aspects to the fundamental ontology of the world.

5. Consciousness and Quantum Mechanics

Quantum field theory is a subset of quantum mechanics. Like any quantum theory (and in contrast with classical theories), the dynamics of the Core Theory come in two parts. There is a law of evolution that describes how an undisturbed quantum state evolves deterministically over time, referred to as the *unitary* dynamics. The other part takes the form of a probabilistic algorithm expressing how the wave function responds to being measured. Operationally, a measured wave function 'collapses' onto a state with a definite value of the quantity being measured, so we label this the *collapse* dynamics. Collapse introduces a stochastic element, with the probability of different outcomes being related to the original wave function by the Born rule. There are therefore two broad strategies one could contemplate for modifying the dynamics of the Core Theory: altering its unitary dynamics, or its collapse dynamics.

The questions of what precisely constitutes a quantum measurement, what happens when one is performed, and what is the correct ontology describing quantum systems have remained controversial. We don't need to distinguish between these competing theories for our present purposes; see overviews by Norsen (2017) and Maudlin (2019), and *cf.* Rovelli (this issue).

According to textbook quantum mechanics, when one measures a quantum observable such as position or momentum or spin, only certain specified outcomes can be obtained. To each possible outcome, the quantum state assigns a complex number, the amplitude. The Born rule states that the probability of obtaining that outcome is the modulus-squared of the corresponding amplitude. Importantly, there is no hidden structure within this rule; once we know the amplitudes, experimental outcomes are truly randomly chosen from the appropriate probability distribution.

The Born rule has thus far passed experimental tests (Jin *et al.*, 2017), but the fact that both consciousness and quantum measurement remain mysterious makes it tempting to imagine that there is a connection. What we are interested in here is not the prospect that consciousness causes wave-function collapse (Wigner, 1961; Stapp, 2001; Chalmers and McQueen, 2014), but that somehow wave functions collapse in just the right way to account for consciousness. Penrose and Hameroff have developed an approach in which wave functions collapse when certain physical criteria are met, which they argue can explain aspects of human cognition (Penrose, 1989; 2014; Penrose and Hameroff, 2011). However, although this programme is often described as an approach to 'consciousness', it does not attempt to answer the qualitative questions of first-person experience any differently than any other purely physical account. Similarly, quantum entanglement may play a role in cognition (Fisher, 2015), but this is a matter of information processing, without any special connection to qualitative experience.

If one were interested in allowing mental aspects to affect the probability of quantum measurement outcomes, presumably that could be done. The Born rule states that the probability of obtaining an outcome a is given by $p(a) = |\psi_a|^2$, where ψ_a is the component of the wave function corresponding to that outcome. We could imagine a new rule

$$p(a) = f(\psi_a, M_a), \qquad (1)$$

where M_a represents some novel mental aspect of the situation. This modified Born rule might affect the rate of certain chemical reactions inside a human brain, thereby allowing mental aspects of consciousness to influence our physical behaviour, without showing up in experiments performed with non-conscious equipment.

Of course, such a rule for wave-function collapse represents a wild modification of conventional physics, not merely a loophole within it. A respectable theory along these lines would include a specification of what the mental aspects M_a are, an understanding of their independent dynamics, and an explicit form of the new rule (1). All of these are possible to contemplate, but they remind us of the high standards to which any modified laws of fundamental physics should be held.

Furthermore, if one were convinced that purely physical ontologies are incapable in principle of accounting for the qualitative features of consciousness, the process of wave-function collapse does not offer any unique opportunities. Regardless of when and how wave functions collapse, at the end of the day they are still wave functions. If we think in terms of novel mental properties affecting chemical reaction rates, there would be no relevant difference between modifying the collapse dynamics and the unitary dynamics.

6. Consciousness and Quantum Field Theory

We turn next to the unitary dynamics. As discussed above, if we stay entirely within the effective field theory paradigm, both for the Core Theory fields and potentially new dynamical elements, there is no room for modifying the dynamics in ways that would be relevant for human behaviour while remaining compatible with experimental constraints. Any new fields that would be relevant for what goes on in the human brain would have been discovered long ago.

We can nevertheless imagine that new mental aspects influence the quantum fields of the Core Theory, without themselves obeying the rules of quantum field theory. To see how that might work, it is useful to look at one part of the Core Theory: quantum electrodynamics, the theory of charged particles (including electrons, protons, and even atomic nuclei) with electromagnetic fields. To the extent that gravity, nuclear reactions, and radioactive decays can be ignored, this is enough to include all of the physics relevant for human biology. The unitary dynamics can be summarized in a one-line equation:

$$A = \int_{k<\Lambda} [DA][D\psi] \exp\left\{ i \int d^4x \left[\sum_n \bar{\psi}_n(i\gamma^\mu \partial_\mu + q_n\gamma^\mu A_\mu - m_n)\psi_n - \frac{1}{4}F_{\mu\nu}F^{\mu\nu} \right] \right\} \quad (2)$$

We don't need to dive into this equation in detail, but a few points are worth highlighting. The expression tells us how a quantum state (describing, for example, a human brain) consisting of charged particles ψ_n and electromagnetic fields $F_{\mu\nu}$ evolves from a given initial state to a final one. It is entirely deterministic and causally comprehensive; indeterminism only comes from the non-unitary collapse dynamics. The expression in square brackets is the *Lagrangian*, which encodes the properties of different kinds of particles. The notation $\int d^4x$ indicates that the Lagrangian is integrated over space-time. This reflects the locality of the unitary dynamics: fields only interact with other fields (and with themselves) at the same point in space-time. The notation $k<\Lambda$ indicates that this is meant to be an effective theory, applicable below the cut-off energy Λ. See (Carroll, 2016) for further elaboration. Any given approach to consciousness will either modify this equation, or it won't.

The properties specified by the Lagrangian include the masses of the particles m_n, as well as the parameters characterizing their interactions, such as electric charges q_n. The strength and rate of electrochemical processes, including those in human brains and bodies, are calculable in terms of these parameters.

An obvious way that mental aspects could modify physical dynamics is for them to affect the values of these parameters, which would in turn affect the rate of processes in the brain. Given some physical/mental situation **S**, we could imagine context-dependent changes in the values of the physical constants that govern Core Theory dynamics, of the form

$$m_n \to m_n(\mathbf{S}), q_n \to q_n(\mathbf{S}). \tag{3}$$

If **S** included mental aspects of our ontology, this would be a mechanism by which those aspects could affect human behaviour, such as our testimony concerning our introspective experiences.

As in the case of the Born rule, we are welcome to contemplate mentally-induced modifications of particle-physics parameters, but a number of questions present themselves. What precisely is meant by the situation **S**, and what kind of dynamics does it have? Naïvely, changes in the masses or charges of particles would lead directly to violations of conservation of energy and momentum. Are these compensated by transfers of energy between conventional matter and a 'mental sector'? These questions are potentially answerable, but they highlight the challenges faced by any proposed theory of this form.

7. Mental Degrees of Freedom

Panpsychists sometimes analogize consciousness to electric charge, as a property that inheres in appropriate fundamental particles. There are at least two severe limitations to this analogy. First, electric charge is a paradigmatic example of a property with dynamical consequences; placed in an electric field, particles with opposite charges move in opposite directions. In the case where we imagine that the properties associated with consciousness have no dynamical consequences, it is not clear what the analogy is supposed to illuminate. Second, charge is conserved. An elementary particle has a single, unchanging value of charge throughout its existence, whereas it is generally supposed that conscious states can take on different values.

We should therefore distinguish between two alternatives: that hypothetical new mental aspects of our ontology supervene on the physical situation, or that there are independent 'mental degrees of freedom' that are not determined by the physical situation. In panpsychist terms, these correspond to the possibility that any particular electron has a definite value of consciousness, versus the idea that any given electron might have multiple conscious states.

If the new mental aspects of our ontology supervene on the physical aspects, as far as physical behaviour is concerned this is indistinguishable from not introducing mental states at all. Consider some causal chain $P_i \rightarrow M_i \rightarrow B_i$, where P_i is some physical state, M_i is some mental state, B_i is some behaviour, and '$\rightarrow$' stands for 'inevitably leads to'. If we think of mental aspects as primary, but nevertheless supervening on the physical situation, we could write $M_i(P_i) \rightarrow B_i$. Either case is functionally equivalent to the shorter chain $P_i \rightarrow B_i$. For all intents and purposes this is equivalent to positing that mental aspects have no effect on physical behaviour.

Turn instead to the alternative where the same physical configuration might be associated with different mental degrees of freedom. These aspects would be roughly analogous to the spin of an electron, which is some combination of 'spin-up' and 'spin-down' for any given particle, but can be specified independently of the particle's position.

The idea of new independent mental degrees of freedom runs into immediate trouble. In conventional field theory, the existence of new degrees of freedom quantitatively affects processes that rely on quantum fluctuations, in which each property value represents a separate contribution that should be added together (e.g. we 'sum over spins' in a scattering calculation). But we know empirically how many degrees of freedom actual electrons have — two spin states for the electron, and another two for its antiparticle, the positron. If electrons could also be found in both 'happy' and 'sad' states, it would have an unmistakable impact on their scattering rates, in flagrant contradiction with experiment.

Our allowed alternatives are to posit that all electrons have the same conscious state, in which case there is effectively no dynamical impact, or that mental degrees of freedom are somehow not like physical ones. In the latter case, we are left with the question of what mental degrees of freedom *are* like, if they are not like physical ones. We can avoid conflict with what we know, but only at the expense of pushing our ideas further away from clarity and tangibility.

8. Strong Emergence

One approach to the relationship between consciousness and physics is to appeal to *strong emergence* — the idea that legitimately new behaviours arise in collective phenomena that cannot be derived in terms of the individual behaviours of constituent parts of the system. Strong emergence is sometimes invoked as a way to allow for specifically mental causal powers (O'Connor and Wong, 2005). It is worth spending a moment on the relationship between strong emergence and the underlying framework of effective quantum field theory — namely, they are entirely incompatible.

As discussed above, the Core Theory provides a comprehensive specification of the quantum-field dynamics within its domain of applicability, which includes any processes between known particles with energy transfers less than a hundred billion electron volts. The field equations are precisely local: the unitary dynamics of each field at any one space-time point are influenced only on the values and derivatives of the other fields at the same point, and not directly by what is happening elsewhere. Electrons and other particles obey the same equations whether they are inside a rock or inside a human brain.

The strong emergentist must therefore deviate from the paradigm of effective field theory entirely, while maintaining the empirical successes of the Core Theory. The most direct way to do this would be to postulate a new restriction on the domain of applicability that is not given in terms of energy transfers in particle interactions, but on some explicitly macroscopic criterion. For example, one could hypothesize that quantum field theory breaks down when the number of particle excitations in a region surpasses a certain number, or when the configuration of such particles reaches a certain quantifiable degree of complexity or information processing capacity. The effective masses and couplings of elementary particles might, for example, be modified as in equation (3), where the situation **S** could involve a quantitative measure of consciousness from an approach such as integrated information theory (Tononi *et al.*, 2016).

One is, of course, free to contemplate whatever extravagant deviations from contemporary physics one likes. Particle-physics experiments typically examine the interactions of just a few particles at a time, so new physical laws that only kick in for complex agglomerations of particles are not necessarily ruled out by data we currently have. It's worth noting, however, how profound a departure such laws would represent. The most fundamental principle of quantum field theory is locality: fields at any one point in space-time are only influenced by the values and derivatives of other fields at that same point, not the behaviour of fields at other points. Modifying the dynamical equations in ways that were sensitive to the complexity of a configuration of surrounding particles would represent a dramatic overthrow of this principle.

Moreover, based on purely physical grounds rather than consciousness-based motivations, our expectation that the laws of quantum field theory might break down in biological organisms would be very low indeed. To we macroscopic people, the 10^{27} particles in a human brain seems like a lot, certainly far greater than the number physicists typically collide in high-energy accelerators. But the density of those particles is very low by particle-physics standards. To be conservative, we might take as a standard length scale the Compton wavelength of the electron, $\lambda_e = 2 \times 10^{-10}$ cm. The volume of a human brain is 1,260 cubic centimetres, or about 10^{32} cubic Compton wavelengths. The number of particles is therefore less than 10^{-5} per standard volume. From the point of view of particle physics, a brain is not a densely packed system; indeed, it's practically empty space. There is no physical rationale for expecting the dynamics of the Core Theory to break down in such an environment, regardless of how complex the overall situation is. For any particular electron or nucleus, almost all of the rest of the brain is so far away as to be essentially irrelevant.

This is not to say that the concept of strong emergence might not be relevant in other contexts, where the 'micro' theory is something other than elementary particles. If a complex system consists of a collection of smaller systems that are themselves complex, a purely local theory might not suffice, and the best description of the overall dynamics could conceivably involve microphysical dynamics that depend on macrophysical contexts in interesting ways (Flack, 2017). In the phenomenon of 'quorum sensing', for example, gene expression in bacteria is affected by their overall population density (Miller and Bassler, 2001). In such cases, the subsystems themselves are extended objects with non-trivial internal dynamics, so the criterion of locality is significantly less severe. Quantum field theory is a very different situation, where subsystems (elementary particles) have no internal structure. In that case, the locality of interactions is exact; what matters to the dynamics of a field at each point is only the other fields at that same point, not anything elsewhere. Any new context-dependent behaviour departing from the predictions of equation (2) would be a violation of our expectations from effective quantum field theory, not a supplement to them.

9. Conclusions

The temptation to augment the ontology of the world with specifically mental aspects stems from a conviction that describing the mere behaviour or function of matter cannot be sufficient to account for consciousness or innate nature. As characterized by Levine (1983), there seems to be an 'explanatory gap' between physical states and conscious experiences. Physicalism posits that a conscious experience is an emergent phenomenon that arises in higher-level models of the same underlying processes described by physics. To a panpsychist, as Goff (2019) says about the brute identity theory, this 'is very unsatisfying'. Arguably it is this 'satisfaction gap', more than any explanatory or ontological gap, that prompts the introduction of intrinsically mental concepts and categories into fundamental ontology.

Any discussion of mental aspects of ontology must specify one of two alternatives: changing the known laws of physics, or positing that these aspects exert no causal influence over physical behaviour. We cannot rule out the first option either through pure thought or by appeal to existing experimental data, but we can ask that any modification of the Core Theory be held to the same standards of rigour and specificity that physics itself is held to. The point of expressions like (1) and (3) is not that mentally-induced modifications of physical parameters are impossible, but that a promising theory of consciousness should be specific about how they are to be implemented.

We don't know everything there is to know about the laws of physics, and there is always the possibility of a surprise. But the solidity of our confidence in the Core Theory within its domain of applicability stands in stark contrast with our fuzzy grasp of the nature of consciousness. The most promising route to understanding consciousness is likely to involve further neuroscientific insights and a more refined philosophical understanding of weak emergence, rather than rethinking the fundamental nature of reality.

Acknowledgments

It is a pleasure to thank Philip Goff for stimulating this work and for enlightening discussions. I'd also like to thank Jenann Ismael, Barry Loewer, Alex Moran, David Papineau, Alex Rosenberg, and Eric Schwitzgebel for helpful comments and pointers to references. This work is supported in part by the Foundational Questions Institute.

References

Aristotle (2002) *Metaphysics*, Book 3, 996b, 2nd ed., Sachs, J. (trans.), Santa Fe, NM: Green Lion Press.

Carroll, S.M. (2016) *The Big Picture: On the Origins of Life, Meaning, and the Universe Itself*, New York: Dutton.

Carroll, S.M. (2021a) The quantum field theory on which the everyday world supervenes, submitted to Shenker, O., Hemmo, M., Iannidis, S. & Vishne, G. (eds.) *Levels of Reality: A Scientific and Metaphysical Investigation* (Jerusalem Studies in Philosophy and History of Science), [Online], https://arxiv.org/abs/2101.07884.

Carroll, S.M. (2021b) Reality as a vector in Hilbert space, submitted to Allori, V. (ed.) *Quantum Mechanics and Fundamentality: Naturalizing Quantum Theory Between Scientific Realism and Ontological Indeterminacy*, [Online], https://arxiv.org/abs/2103.09780.

Chalmers, D.J. (1996) *The Conscious Mind: In Search of a Fundamental Theory*, New York: Oxford University Press.

Chalmers, D.J. (2003) Consciousness and its place in nature, in Stich, S.P. & Warfield, T.A. (eds.) *The Blackwell Guide to Philosophy of Mind*, Malden, MA: Blackwell.

Chalmers, D.J. & McQueen, K.J. (2014) Consciousness and the collapse of the wave function, in Gao, S. (ed.) *Consciousness and Quantum Mechanics*, Oxford: Oxford University Press.

Fisher, M.P.A. (2015) Quantum cognition: The possibility of processing with nuclear spins in the brain, *Annals of Physics*, **362**, pp. 593–602.

Flack, J.C. (2017) Coarse-graining as a downward causation mechanism, *Philosophical Transactions of the Royal Society A: Mathematical, Physical and Engineering Sciences*, **375** (2109), 20160338.

Goff, P. (2017) *Consciousness and Fundamental Reality*, New York: Oxford University Press.

Goff, P. (2019) *Galileo's Error: Foundations for a New Science of Consciousness*, New York: Vintage Books.

Jin, F., Liu, Y., Geng, J., Huang, P., Ma, W., Shi, M., Duan, C.-K., Shi, F., Rong, X. & Du, J. (2017) Experimental test of Born's rule by inspecting third-order quantum interference on a single spin in solids, *Physical Review A*, **95** (1), 012107.

Levine, J. (1983) Materialism and qualia: The explanatory gap, *Pacific Philosophical Quarterly*, **64**, pp. 354–361.

Loewer, B. (1995) An argument for strong supervenience, in Savellos, E. & Yalcin, U. (eds.) *Supervenience: New Essays*, Cambridge: Cambridge University Press.

Maudlin, T. (2019) *Philosophy of Physics: Quantum Theory*, Princeton, NJ: Princeton University Press.
Miller, M.B. & Bassler, B.L. (2001) Quorum sensing in bacteria, *Annual Review of Microbiology*, **55**, pp. 165–199.
Moran, A. (this issue) Grounding the qualitative: A new challenge for panpsychism, *Journal of Consciousness Studies*, **28** (9-10).
Norsen, T. (2017) *Foundations of Quantum Mechanics: An Exploration of the Physical Meaning of Quantum Theory*, Berlin: Springer.
O'Connor, T. & Wong, H.Y. (2005) The metaphysics of emergence, *Noûs*, **39** (4), pp. 658–678.
Papineau, D. (1995) Arguments for supervenience and physical realization, in Savellos, E. & Yalcin, U. (eds.) *Supervenience: New Essays*, Cambridge: Cambridge University Press.
Papineau, D. (2020) The problem of consciousness, in Kriegel, U. (ed.) *The Oxford Handbook of the Philosophy of Consciousness*, Oxford: Oxford University Press.
Penrose, R. (1989) *The Emperor's New Mind: Concerning Computers, Minds and The Laws of Physics*, Oxford: Oxford University Press.
Penrose, R. (2014) On the gravitization of quantum mechanics 1: Quantum state reduction, *Foundations of Physics*, **44**, pp. 557–575.
Penrose, R. & Hameroff, S. (2011) Consciousness in the universe: Neuroscience, quantum space-time geometry and Orch OR Theory, *Journal of Cosmology*, **14**.
Rovelli, C. (this issue) Relations and panpsychism, *Journal of Consciousness Studies*, **28** (9–10).
Russell, B. (1927) *The Analysis of Matter*, London: Kegan Paul.
Smolin, L. & Verde, C. (this issue) Physics, views, and qualia, *Journal of Consciousness Studies*, **28** (9–10).
Stapp, H. (2001) Quantum theory and the role of mind in nature, *Foundations of Physics*, **31** (10), pp. 1465–1499.
Strawson, G. (2006) Realistic monism: Why physicalism entails panpsychism, *Journal of Consciousness Studies*, **13** (10–11), pp. 3–31.
Tononi, G., Boly, M., Massimini, M. & Koch, C. (2016) Integrated information theory: From consciousness to its physical substrate, *Nature Reviews Neuroscience*, **17** (7), pp. 450–461.
Wigner, E.P. (1961) Remarks on the mind–body question, in Good, I.J. (ed.) *The Scientist Speculates*, Portsmouth, NH: Heinemann.
Wilczek, F. (2015) *A Beautiful Question: Finding Nature's Deep Design*, New York: Penguin.

Carlo Rovelli[1]

Relations and Panpsychism

Abstract: *Twentieth-century physics has revealed a pervasive relational aspect of the physical world. This fact is relevant in view of some of the motivations for panpsychism. In fact, it may be seen as a vindication of the panpsychist idea of a monist continuity where some aspects of consciousness's perspectivalism are universal. On the other hand, this same fact may undermine some of the motivations for more marked forms of panpsychism.*

If some aspects of mind are universal, which ones are so? I point out in this note that twentieth-century physics has *already* vindicated a — very mild — form of panpsychism. This is because of the profoundly relational aspect of physics, manifest in general relativity, but especially in quantum mechanics. Twentieth-century physics is not about how individual entities *are* by themselves. It is about how entities *manifest* themselves to one another. It is about relations.

This is particularly evident for quantum theory, especially (but not uniquely) if one reads it in terms of its relational interpretation (Rovelli, 1996; Laudisa and Rovelli, 2021). Niels Bohr expressed this 'contextual' aspect of quantum theory by saying that a physical system can only be described taking into account the systems it is interacting with. The current account of the world in fundamental physics is therefore always the account of how a system affects another system.

This implies that the most effective way of thinking about the world is not in terms of entities with properties, but rather in terms of systems that have properties in relation to other systems. In turn, this

Correspondence:
Email: rovelli.carlo@gmail.com

[1] Centre de Physique Théorique de Luminy, Aix-Marseille University, France.

implies that any physical description of a system is necessarily perspectival: relative to another system.

In the textbook Copenhagen formulation of quantum theory, the physical system with respect to which properties take value is variously and a bit obscurely interpreted as 'the apparatus', 'the observer', 'the macroscopic world', and similar. In relational quantum mechanics, the properties of a system **S** are defined with respect to *any* other physical system **O** with which **S** interacts, and are relative to **O** (Laudisa and Rovelli, 2021). Similarly, in the many-worlds interpretation of quantum mechanics, variables have values only in relation to (components in a branch of) other systems when the two have got entangled (Saunders *et al.*, 2010).

A direct consequence is that our physics today is not a physics of the world seen from the outside. It is a physics of the world always seen from the perspective of a physical system.

This is perhaps not panpsychism, because there is nothing specifically psychic or mental in the relational properties of a system with respect to another system. But there is definitely something in common with panpsychism, because the world is not described from the outside: it is always described relative to a physical system (Dorato, 2016). So, physical reality is, in our current physics, perspectival reality (Rovelli, 2021).

On the other hand, this very relationalism may suffice to resolve the very problems that motivated panpsychism in the first place: do we really need elementary physics to include more aspects in common with the mental world than this? Which ones? All phenomena of which I am aware that are more related to psyche or mind are connected to a brain, a neural system, sensory organs, feedback loops, or the like. I cannot imagine anything even vaguely more like psyche or mind, without some structure like those.

A motivation for panpsychism is the idea that there is a 'hard problem' of consciousness (Chalmers, 1995). This is the expectation that, even if we had figured out how our body works in terms of our current science, there would still be something mysterious about 'consciousness'. This expectation is based upon an intuition: that subjective experience, the 'first-person perspective', *must* be incompatible with the world as we describe it now in physics. For instance, Chalmers argues that it is *conceivable* to imagine a body with the same physics as my body, but with no subjective experience; hence subjective experience must be over and above this physics (*ibid.*).

I do not know how Chalmers can 'conceive' this: I can't. The point is that what we can 'conceive' depends on the conceptual structure we have, and this keeps changing and includes a great deal of presuppositions, sometimes wrong. The history of science should warn us that trusting 'intuition' blindly can freeze us into wrong ideas. We do not yet fully understand the physics of thunderstorms but trusting the compelling 'intuition' that lightning and thunder are evidence for the rage of Odin may not be very wise.

More specifically, the intuition of a tension between mental and physical is based on two assumptions. The first is that the subject of experience is an irreducible entity. The second is that the material world is formed by substances having properties. There is indeed a tension between these two concepts that appear radically distinct. In particular, mental phenomena are intentional: they are relative to something else. Material phenomena at first may seem incompatible with this.

Both these assumptions seem to me to be wrong. There is a vast philosophical literature about the first and I will not go into it. Rather, I focus here on the second. The intuition of the 'material world' as formed by substance with properties, for instance just particles moving in space, is based on eighteenth-century physics. Today we have a much more subtle view of the physical world. In particular, as I mentioned above, physics today is about relations between systems, about how systems affect one another. In a precise sense, quantum mechanics, in particular, has undermined the idea of an elementary level where components' matter can be described independently from anything else.

If our basic understanding of the physical world is in terms of more or less complex systems that interact with one another and affect one another, the discrepancy between the mental and the physical seems much less dramatic, and I do not see the need for stronger versions of panpsychism.

This obviously is not a solution to the 'easy problems' of consciousness — figuring out the physics of the body when mental processes happen. But it definitely undercuts, it seems to me, the intuition underpinning the belief that there is any 'hard' problem of consciousness. Mental phenomena are like other complex phenomena, and the existence of a subjective perspective is precisely the *generic* situation in physics: how systems 'appear to one another'.

So, relationalism can be seen as a very mild form of panpsychism. That is, there is something in common between mind and matter: they are ways that the reality of physical systems manifests itself to other physical systems, which is precisely what our physics currently describes.

The world as we know it today in physics is neither the material world of eighteenth-century mechanical philosophy, nor embodied with any peculiar proto-consciousness. It is a world where physical systems — simple and complex — manifest themselves to other systems — single and complex — in a way that our physics describes. I see no reason to believe that this should not be sufficient to account for stones, thunderstorms, and thoughts.

References

Chalmers, D.J. (1995) Facing up to the problem of consciousness, *Journal of Consciousness Studies*, **2** (3), pp. 200–219.

Dorato, M. (2016) Rovelli's relational quantum mechanics, monism and quantum becoming, in Marmodpro, A. & Yates, A. (eds.) *The Metaphysics of Relations*, pp. 290–324, Oxford: Oxford University Press, arXiv:1309.0132.

Laudisa, F. & Rovelli, C. (2021) Relational quantum mechanics, in Zalta, E.N. (ed.) *The Stanford Encyclopedia of Philosophy*, [Online], https://plato.stanford.edu/entries/qm-relational/.

Rovelli, C. (1996) Relational quantum mechanics, *International Journal of Theoretical Physics*, **35**, pp. 1637–1678, arXiv:9609002 [quant-ph].

Rovelli, C. (2021) *Helgoland*, London: Penguin.

Saunders, S., Barrett, J., Kent, A. & Wallace, D. (2010) *Many Worlds? Everett, Quantum Theory, and Realism*, Oxford: Oxford University Press.

Marina Cortês,[1] Lee Smolin[2] and Clelia Verde[3]

Physics, Time, and Qualia

Abstract: *We suggest that four of the deepest problems in science are closely related and may share a common resolution. These are (1) the foundational problems in quantum theory, (2) the problem of quantum gravity, (3) the role of qualia and conscious awareness in nature, (4) the nature of time. We begin by proposing an answer to the question of what a quantum event is: an event is a process in which an aspect of the world which has been indefinite becomes definite. We build from this an architecture of the world in which qualia are real and consequential and time is active, fundamental, and irreversible.*

> *This formulation makes it clear that the uncertainty relation does not refer to the past. (Heisenberg, 1949)*

1. Introduction

The core of Galileo's new science was the idea that all motion could be represented mathematically while all change could be rendered as motion. To make this credible not only was part of the world discarded but memory of it erased. Sensations, colours, and thoughts were not part of the mathematical universe and that came to be thought of as the only universe there was. This splitting of the world

Correspondence:
Email: lsmolin@perimeterinstitute.ca; clelia.verde@gmail.com

---

[1] Instituto de Astrofísica e Ciências do Espaço, Universidade de Lisboa, Faculdade de Ciências, Campo Grande, PT1749-016 Lisboa, Portugal.
[2] Perimeter Institute for Theoretical Physics, Waterloo, Canada.
[3] Corso Concordia, 20129 — Milano, Italy.

had many implications for our understanding of the world itself. Philip Goff's book, *Galileo's Error* (2019), is focused on one of those implications which is our understanding of qualia and consciousness. But there is another set of implications which is not discussed as often, and that is the understanding of time.

The aim of this paper is to show how Galileo's original error made possible an even more serious error, this one having to do with our understanding of time. Furthermore, the two issues — the place of qualia in a physical universe and the nature of time — influence each other. The hard problem of consciousness is a very different problem depending on one's conception of time. Tied up in this confluence of questions there are two more: the problems in the foundations of quantum mechanics and the problem of quantum gravity. To us, all four issues — the nature of time, the place of qualia in a physical universe, the foundational problems of quantum mechanics, and quantum gravity — are all tangled up. They have a common solution. The solutions we adopt to each of them alter the way we understand the others.

2. The Hard Problem of Time

Galileo's error in removing qualities, sensations, and awareness from the world, leaving only a universe of quantities and of quantifiable relationships, led to an equally profound error about the nature of time. From the first to the second error is only a few steps. The success of the science of motion led to the hypothesis that the motions and forces of the world — from atoms up to stars — were a system that was causally closed, as all explanations of motion pointed to more motion.

Meanwhile, the success of the 'universal laws' posited by Newton made it plausible that all of the motions could be described in mathematical terms — that is, in terms of mathematical figures, such as parabolas and hyperbolas, that were quickly found to be solutions to equations that expressed a few simple principles. The achievements of this straightforward methodology were and are stunning. It begins to seem as if nature is intrinsically 'mathematical'. This can be expressed by hypothesizing that there is a mathematical object, O, perhaps a solution to a final, unified set of equations, which is a mirror of the history of the universe, U. This means that for every property P of U, there is a theorem T, about O, and vice versa. We can call this *the mathematical universe hypothesis* (Tegmark, 2014). Making sense of this is harder than it looks, and indeed has not so far been achieved. It is, to say the least, unclear to us that the full explanation of many of the processes and structures we observe in the world can be made solely by means of mathematical deduction from the laws of physics.

There are several distinct challenges to the success of the mathematical universe hypothesis. Functional explanations, which we will discuss below, seem necessary to explain many facts about biological and ecological systems (Cortês *et al.*, 2021a,b,c). Many of us argue that there is no algorithm that predicts all functions of a physical system. One biological system that resists our efforts to reduce to physics is the brain: we certainly know enough to be confident we don't know how brains bring forth minds. On top of which it remains plausible that quantum physics is genuinely indefinite. Nor should we ever forget the problem of the initial conditions, which seem both improbable and inexplicable (Penrose, 1979).

The hypothesis of a mathematical universe may seem an innocent bit of physicalism, but it has astounding consequences (Mangabiera-Unger and Smolin, 2014; Smolin, 2013). It implies that every causal influence, observed to play out in time, can be mirrored by a theorem, which is a timeless truth. A complete mathematical correspondence in all cases reduces laws to theorems concerning the properties of an imagined mathematical object that represents the history of the universe. By doing so, it reduces causation to timeless truths. If everything that will happen in future time can be shown necessary, by a time-less logical argument, then the activity of time is reduced to a computation. So the crime went much deeper than Galileo's expulsion of sensation and qualities from the world, for the chain of implications leads to a conception of a static world, because, where logic timelessly reigns, bright colours and sweet music may be the first illusions, but the last illusion certainly must be time itself.

We can make the argument from the other side, where it takes the form of a direct refutation of the claim that there could exist a mathematical object, O, which is a mirror of nature, in the sense we described above. The claim is refuted by showing the existence of properties of our world that no mathematical object could explain or describe, let alone mirror. Here is one: in the real world it is always some present moment, which is one of a continual becoming of present moments (Mangabiera-Unger and Smolin, 2014; Smolin, 2013).

Mathematical logic is genuinely timeless — there is no system in mathematics which allows the possibility of theorems whose truth value depended on whether the subject was irreducible to our past, present, or future. The key word is irreducible, i.e. the truth value cannot be translated without loss of meaning to replace the tensed words past, present, and future with relations to relative clock readings. Or, to put it in McTaggart's sense: while temporal relations in nature are A-series, mathematics can only express B-series relations (McTaggart, 1927).

The second kinds of fact about the world that cannot be expressed or explained by mathematical truths are those about qualia, such as what it's like to see blue; or what is the unforgettable sound of a B minor seventh chord played on a Gibson SG made before 1960.

3. Quantum Mechanics with an Active Time

We don't often emphasize the centrality of the tensed distinction between past, present, and future in the measurement theory of quantum mechanics, but it is there — as indeed Heisenberg, Schrödinger, Dyson, and others discussed (Heisenberg, 1949; Dyson, 2004).

Physics as developed so far is about the past. All the measurements we analyse were taken in the past. They are compared with the predictions of theories made in the past. Some physicists like to talk about being B-theorists but the actual practice of science is purely A-series. It is easy to see that the A-series structure of past, present, and future is baked into the organization of science. The testing of theories by experiment relies on it, as the following argument shows. Let us look at a simple example. A scientific description of a subsystem, say the solar system, begins by taking records of a series of runs of the system. Each run is defined by initial conditions, with respect to some approximately fixed frame, and we follow and record the resulting trajectory, always relative to the frame of reference, until the experiment ends. This means that the actual context for comparing the results of an experiment with a theory is that the experiment necessarily happened in the past of the analysis. This is important because the prediction of the theory will have been some kind of mathematical object — say a trajectory in some space of motions with respect to external idealized elements. What it is compared to is not the system itself, but records of past relative positions. These records are also mathematical objects. What we do when we test the theory is to compare two mathematical objects to each other: the one being records of the past, which is compared to the second, which is a mathematical model of the past. When we compare two mathematical objects, time is not involved because you can check your work at any future time. This is trivially true if what are being compared are timeless records from the past with timeless logical deduction. But it is fallacious when the derivation is compared to actual nature unfolding, becoming in time (Mangabeira-Unger and Smolin, 2014).

3.1. Active time versus passive time

We can call the B-series measures of time that figure in records of past motions *weak time*, or passive time. It plays no generative role, it is just used to label and then order records of the past. In Smolin and Verde (2021) we define an active conception of time as the process that resolves indefinite circumstances to definite. We call the version of time we champion *resolving time*. It is what Bergson called *creative time* (Bergson, 2002). The world is literally remade over and over again, event by event, by the work of active time. By contrast, the 'time' that theorists talk about being emergent is not directly related to the resolving time we were just discussing. That passive, weak time is not an aspect of the universe at all. It is a mathematical coordinate that we assign to our records of experiments, which by definition is a mathematical object that was created when the runs of the experiment were made in the past.

We are not dealing here with a conflict between two different definitions of a 'time coordinate'. We are dealing with two very different conceptions of what a universe is. A time-created universe is constructed by the repeated actions of a function that creates novel events, out of the material of present events — which then become past events, or more simply part of the past, which exists only in records that give testimony to what they did in their present moment. According to Smolin and Verde (2021) an event is an action which resolves an indefiniteness or ambiguity in the world, which, step by step, creates our world. What then exists, moment by moment, is a network of these actions of resolution, whose connections are nothing less and nothing more than causation. That is why we see ourselves to be part of a definite world — because all that exists are these movements of resolution. An event is a process, each of which does its job of resolution, then vanishes as it creates a few of the next generation. Then, its work being done, it is gone. A time-created world is not a four-dimensional manifold, it is a continually recreated, roughly three-dimensional network of processes.

4. The Problem of Mental Causation

From a physicalist point of view, it is very puzzling that there is consciousness at all. Since physicalists assume that the complete explanation of the behaviour of an animal will be found in a full reduction to the underlying laws satisfied by the elementary particles that constitute it, any additional properties would be inessential.[4] But the knowledge argument (Jackson, 1982) demonstrates that what it's like to see blue is exactly such a property.

This of course depends on the laws of physics being deterministic so that a complete specification of the initial state yields a unique final state. Since the laws of physics acting on the initial state already completely determine the future state, there seems to be nothing left for qualia to do, no role for them to play in causing behaviour. Qualia are rendered 'epiphenomenal', properties that have no causal impact on the physical universe. Even if they exist in a weak sense of a Russell-inspired panpsychism (Russell, 1927), in which they represent other aspects of matter, not involved in the dynamical laws — perhaps essential or internal aspects of being — so long as they play no role in dynamics, it can't matter one whit whether they are there or not.

This would then imply that qualia are completely irrelevant for everything an animal may or may not do. The outcomes will be what will be — qualia or none. But then, why are there qualia? If being a zombie would be every bit as good for our evolution — living and thriving — why does nature go to such lengths, just to give us a show?

At this point we have a clear choice. We can continue to insist on the divisions Galileo and Newton made, and study only those aspects and subsystems of the universe that lend themselves to mathematical description, within which we hypothesize causal closure. Or we can bring some of those other parts and aspects of the world back into the universe, and search for causal closure within this larger set. It is neither anti-reductionist nor anti-scientific to take this second route.

We have already learned in biology, ecology, medicine, etc. how reductionist and functional styles of explanation work together (Cortês *et al.*, 2021a,b,c) to give what David Deutsch calls the 'best explanation' (Deutsch, 2014). Once we have a design for a world in which qualia and conscious events may have causal effectiveness in the

---

[4] We are grateful to Sean Carroll for a remark on this issue that changed our minds.

physical universe, we will work like biologists and pose functional questions about how the different parts of our world work together.

Our central hypothesis is as follows: *there is a mixed functionalist and reductive explanation for why humans and other animals experience qualia (or just experience).*

In simple physical systems, many successful explanations are completely reductive, in the sense of relying only on the fundamental laws and initial conditions. But there are cases where the arguments based on reduction alone fail. In many of these cases the right explanation has a functional component. For example, let us ask why a particular molecule, haemoglobin, is present in the biosphere in large quantities. Part of the answer is that the molecule is stable under the laws of physics, and those laws are seen to act in multiple settings during which haemoglobin is made.

It is pretty well understood how functional properties play an essential role in explaining the behaviour and constitution of living creatures. There are many questions, for example, about the very sparse distribution of actual proteins in the space of physically allowed proteins that are explained partly by satisfaction of the laws of physics, but also by what functions the proteins do for the living creatures that create them. But an essential part of the answer is that haemoglobin is a protein which performs a crucial function for a large number of creatures — namely, spreading oxygen through the blood. Because of this, animals that produced haemoglobin had greater fitness and were selected for. But as those creatures thrived, so did the proteins that contributed to their fitness.

4.1. Does consciousness serve a function?

Among all the aspects of an animal or human being that contribute to its fitness, few make a greater contribution than consciousness. There is much evidence that the focus of a person's consciousness can be trained and that a trained attention underlies the skills of an athlete, a musician, or a hunter. It seems possible, if not likely, that consciousness or awareness had and has a lot to do with the thriving of our species.

It is then very natural to suppose that, if the existence of consciousness is to be explicable, it must perform some function that increases the fitness of the creature that is endowed with it. But this requires that consciousness can intervene in the network of causes in the physical universe. But when we try to develop this idea, we run immediately into a very strong argument that the physicalists have to their credit, based on the causal completeness — or 'causal closure' — of the of the standard Newtonian paradigm (see Carroll, this issue).

We are at an impasse. To proceed we must give up the founding idea of the physical sciences, which is the universal governance of nature by a handful of physical principles, expressed mathematically by a causally closed set of equations. However, we do not want to give up universality altogether, just enough to allow new physical phenomena associated with qualia to play a role in the complete circle of causes. Our theory is framed in an events ontology, and hence we treat some events differently.

We accept the arguments (Chalmers, 1996; Jackson, 1982) that qualia cannot be reductively explained in physical terms; but we also accept the argument of Carroll (this issue) and others that qualia cannot be anything other than epiphenomenal within the currently understood physical laws and empirically known phenomena. There remains only the hope that innovations in physics may be more hospitable to consciousness: this seems to be the only road if we are to have an embedding of qualia into physics.

We introduce a new regime of physics that is related to qualia in a way that allows qualia to be consequential. We will speak of this as Mode II physics, as opposed to the well understood Mode I. This is far from a new strategy; it has been followed by most of the attempts to understand the role of awareness within quantum physics. Most of the approaches to quantum foundations do split the laws into two parts, the first being described by unitary Schrödinger evolution in a fixed Hilbert space, which we identify with Mode I.

In most formulations, quantum mechanics is more than this. Collapse of the wave function, whether spontaneous or based on a law of some kind, is strictly Mode II. Indeed, in any approach to a completion of quantum mechanics, from Bohr to Bohm, there is a part of the dynamics besides the unitary time evolution. This is where the Mode II physics is to be found. In all these cases, Mode I deals with evolution in Hilbert space, which yields generally indefinite states, given any basis relevant to the macroscopic world. In quantum theory, Mode II deals with everything else, including how definite states are produced. Mode II is also where Born's rule for probabilities, or whatever replaces it, is found.

Hence, our proposal is that: *Mode II is where we will find the physical correlates of consciousness*. Some of the reasons to suppose this include:

- Mode II is where indefinite states are made definite. But consciousness is always definite.
- All past proposals to connect quantum mechanics to consciousness, from Wigner (Wigner and Margenau, 1967) to Penrose (1979), and Chalmers and McQueen (2014), invoke a version of Mode II as the site of that connection.

The next question to be posed is whether consciousness and qualia are correlated with all Mode II events. Or are such correspondences rare, so that to be in correspondence with qualia requires further special circumstances.

4.2. Evolving laws and unprecedented events

We will not comment on the large literature that debates the plausibility of a general panpsychism. We propose that the general implausibility of attributing awareness to every last rock and molecule is avoided when qualia are associated with very special events, or clusters of events.

Part of our overall motivation is to situate the hard problem of consciousness within a theory of a time-made universe. This makes the overall argument for the reality of time at the core of our new conception of qualia. A key part of those general arguments (Pierce, 1891; Smolin, 1992; 1997; Alexander and Cunningham, 2021) is to establish that the laws must evolve. In particular, as some of us have argued previously (Cortês and Smolin, 2014; 2015), it is hard to square an active time that constructs the universe's history by events, each of which resolves something indefinite, from within the Newtonian paradigm based on unchanging laws. The conclusion of this line of argument is a claim that there are certain events whose causal future cannot be deduced from a complete knowledge of its causal past.

Actually, we can say something quite a bit stronger. *There can be no completely precedented events.* Leibniz's Principle of the Identity of the Indiscernible mandates that there can be no two events in the history of the universe with identical causal pasts and causal futures. Suppose we have two events whose causal pasts are identical. Then Leibniz's principle mandates that their causal futures must be different. Such a world cannot be governed by a deterministic theory, as that would require that, if the pasts are identical, so must be the futures. Even if the futures are different in each case, the statistical distribution of outcomes may be fixed over time.

We must then make a distinction between events which generate a constant statistical distribution of outcomes, whose causal future is at least on a statistical level a consequence of their causal past, and those which are not governed by any evolution law, deterministic or stochastic. We will call the first kind 'precedented' or 'habitual' events; the latter 'unprecedented' or 'free' events.

This distinction emerges naturally in several theories in which laws are allowed to evolve. One type of theory where the distinction emerges naturally is those governed by the Principle of Precedent.

4.3. The Principle of Precedent

The Principle of Precedent (Smolin, 2005) is an idea about the origin of laws, or rather how the notion of dynamical law could be replaced by a simpler hypothesis. It has an especially clear presentation in an operational formulation of quantum mechanics, such as Hardy (2001). Each quantum process is broken up into three stages: (i) a preparation, by which the experimenter picks out an initial state at an initial time, (ii) a unitary evolution generated over an elapsed time T by a Hermitian hamiltonian, $U(T)$, and (iii) a measurement made at a final time, where the system ends up in one of a number of output states. Given a set of possible input and output states, the experimenter measures the probabilities for each of the input states to become one of the output states:

p(output, input)

They arrive at this table of numbers by doing the experiment many times, with different choices of the input states. The theorists see their challenge as having a theory that can predict these probabilities.

A very important property of these probabilities is that they do not depend on the starting or ending time of the experiments. These probabilities for the different possible outcomes depend only on the elapsed time, T, and not on the initial time, so that the probabilities measured in the next year will converge with those measured over the last 100 years. Given this, we could posit a precedence law:

Law of precedence: Given a preparation for a physical system, choose the output state randomly from the set of past precedents.

The habitual states are those that have a large number of precedents. The free states are those without precedents. In the case of a preparation that has many precedents, the principle just stated does reproduce an evolution that matches that of quantum theory with a fixed Hamiltonian. How does the universe choose the outcomes of preparations which have no or few precedents? We propose that *the novel states or events are the physical correlates of conscious events.* At these moments, the universe has some degree of freedom to choose what happens next. It is these moments of freedom which make up conscious experience.

Those unprecedented moments are presumably common near the universe's origin, and spread throughout the universe, meaning the early universe resembled the universe as the panpsychist conceives of it. As the universe ages, it takes a higher degree of complexity for a state to be unprecedented. But we can wonder whether complex biomolecules might serve as a reservoir of novel states. Might the biosphere and the brain have evolved to make use of the special properties of novel states, including the freedom present at those moments to choose a small part of the future. It is not difficult to see that this access to novel states might give an animal a selective advantage.[5]

Note that large molecules are made up of smaller subsystems, such as atoms. The component atoms will not be novel. What we want to suggest is that, if there are entangled or coherent states which are made of many atoms which are sufficiently large and complex to be without precedent, these may serve as free states. The freedom in choosing the unitary evolution operator acting on such states will not impinge on the microscopic local dynamics governing each routine component, it will have to act non-locally, on the whole molecule, and be sufficiently weak so as to not have been discovered. Such a term might, for example, favour one folding of a protein over others.

5. Can the Definiteness of Qualia Arise in an Indefinite Physics?

In the time-created world we propose, qualia are properties of the events, or their causal relations. The creation of an event is a process of resolving an ambiguous or indefinite situation (Smolin and Verde, 2021; Kauffman, 2021). We claim qualia are associated with the outputs of these processes, which entails that qualia are always definite.

---

[5] The fact that the brain might have evolved to exploit currently unknown physical laws, with the accompanying selective advantage, was first pointed out by Marina Cortês (Cortês, in preparation).

A structural realist, who is a realist about qualia, would attempt to bridge the gaps between sensation and motion opened by Galileo's error by looking for structural parallels between the organization of qualia and the organization of their physical and neurological counterparts. We have to identify those aspects of the structure of qualia which are explained by organization imposed by the brain. Some examples are as follows:

- The moments of awareness seem to define a thick present. There is also a duration of each experienced moment in time of about 0.5 of a second. Experiments show that the order of two sounds heard within that interval may not be faithfully reproduced. There is also a delay effect.
- We experience qualia, never singly, but bundled together with others in a way that identifies and explains. 'Oh, that is an injured red bird, you can see it in the way it holds its wing.' These identifications and explanations seem to require the resources of a brain to organize.
- The brain constantly tries to give sense to sensory inputs, resulting in a coherent meaning. Neuroscientists tell us that there is a struggle within the brain to filter from all the senses what should be paid attention to, i.e. which should be part of each moment's bundle of conscious perceptions. But, almost all the time, the brain resolves the signal in order to avoid giving us ambiguity.
- Similar mechanisms appear to resolve contradictions between our present experience and the memory of past experiences. This resonance between the experience and the memory of the past experiences gives sense to our now. This is also why we see patterns.

There is much to learn about the structures imposed on perceptions which we may use to peer into those neurological systems. But we believe that it may be possible to look past these and search for evidence of those aspects of conscious experience that might be due to the fundamental nature of the world. The latter aspects we may call the irreducible structural aspects of experience.

Here is a short and preliminary list of those:

- Qualia are always definite. We never experience at different times a state, a contrary state, and then their superposition.

- Normally we have at most one conscious experience at a time per person.
- Qualia and conscious moments seem to express the structure of an event in an active time construction of the universe. They turn on and enter the stream of consciousness as soon as they are noticed; they sharpen and resolve something and then are gone. While they are present they are a heterodox mix of modalities and sensations.
- There are no pure qualia, in isolation. Each conscious experience seems to be a complex perception consisting of an array of colours, sounds, sensations of touch, and smell, all bound together. They seem at times to be organized around directions going outwards from us, as if there is a phenomenal sphere surrounding each of us on which the colours and other sensations are projected. We will refer to the bundled qualia and thoughts as *moments of awareness*.
- We are conscious of novelties, while unconscious of habitual patterns.
- Our experience seems to provide a background, on which a focus brings some connected qualia in high definition. While you are reading this sentence, your focus is on the text, and you have only a dim awareness of the birds singing outside your window. But at their mention your attention is captured by the birdsong, then an old memory of a friend who worked on birdsongs, and then...

This is clearly just a start — no, actually a rehearsal for a start. There is much to be done to develop the programme we have just outlined to discover or construct physics-hospitable qualia.

6. Causal Closure and Qualia

We close by returning to the themes of our first paper, now extended to incorporate the role of awareness and qualia in the construction of a time-created universe:

1. This is an ontology of present events. Nothing exists or persists, things only happen.
2. The universe is indefinite and underdetermined. What we mean by becoming or 'to happen' is for *something indefinite to become definite*. This is what we call an event.

3. The quantities that become definite at an event are called the view of the event. The views are real.
4. The causal future of some events is determined by their causal past; we say these are precedented. Others, not determined by their past, are unprecedented.
5. Unprecedented events must choose their next steps. We experience this creativity as awareness.
6. The universe often surprises itself. Qualia are expressions of the universe to surprise. Pleasures are expressions of acceptance of the surprise.
7. The quantities that become definite at any event are its endowments, which are passed to each from preceding events and become definite on reception. These endowments include energy and momentum. Each event is brought to happen by the passage of the endowments, from their immediate predecessors to them. These instances of 'passing on' define the causal relations, which are definite.
8. The direction from indefinite to definite gives the universe an arrow of time.
9. The indefinite is also called the future. This is because, being indefinite, it can at any time become definite — in an event which is definite and real. If it does, it may influence the present moments to come.
10 The quantum state is nothing but an expression of what we can best forecast or bet about the future, taking fully into account both what is indefinite and definite at this present moment.
11. The world recreates itself in every moment, as indefinites flash into momentary definites, after which they are nothing. Everything we see around us exists or did just exist, but was gone in the blink of an eye.
12. Consciousness is connected with — in fact, created by — the resolution of indefinite states. This ties qualia tightly to quantum theory — especially when that is looked at with the perspective of a world created by an active time. This implies a heightened sensitivity to novelties. The ability to detect novelty is not a peripheral or optional feature of the mind/brain — it is its main function. Qualia, we conjecture, are signals of the recognition of novel situations. We and other creatures have evolved the ability to do so through evolution — as a creature that can resolve ambiguities quickly will, all things being equal, survive better.

By integrating qualia into the history of fundamental physics, we may have resolved the problem of causal closure. Qualia, as part of the history of fundamental physics, play a role in the causal evolution of the universe.

Acknowledgments

We are especially grateful to Philip Goff for his invitation to contribute to this special issue and for encouragement and wise editing. We are grateful to Stephon Alexander, Julian Barbour, Arnaldo Benini, Herbert Bernstein, Saint Clair Cemin, Stuart Kauffman, Jaron Lanier, Joao Magueijo, Roberto Mangabeira-Unger, Simon Saunders, Carlo Rovelli, Steve Weinstein, for critical questions and encouragement, on this or its companion paper. We would like to make a special thanks to Ethan Coen for reminding us that it's all about gravity. This research was supported in part by Perimeter Institute for Theoretical Physics. Research at Perimeter Institute is supported by the Government of Canada through Industry Canada and by the Province of Ontario through the Ministry of Research and Innovation. This research was also partly supported by grants from NSERC, FQXi, and the John Templeton Foundation.

References

Alexander, S. & Cunningham, W.J. (2021) The autodidactic universe, *arXiv*, 2104.03902.
Bergson, H. (2002) *Bergson: Key Writings*, Pearson, K.A. & Mullarkey, J. (eds.), London: Continuum.
Carroll, S. (this issue) Consciousness and the laws of physics, *Journal of Consciousness Studies*, **28** (9–10).
Chalmers, D.J. (1996) *The Conscious Mind: In Search of a Fundamental Theory*, New York: Oxford University Press.
Chalmers, D.J. & McQueen, K.J. (2014) Consciousness and the collapse of the wave function, in Gao, S. (ed.) *Consciousness and Quantum Mechanics*, Oxford: Oxford University Press.
Cortês, M. (in preparation), The brain knows more physics than we do.
Cortês, M. & Smolin, L. (2014) Energetic causal sets, *arXiv*, 1308.2206. doi: 10.1103/PhysRevD.90.044035.
Cortês, M. & Smolin, L. (2015) The universe as a process of unique events, *arXiv*, 1307.6167. doi: 10.1103/PhysRevD.90.084007.
Cortês, M., Liddle, A., Kauffman, S. & Smolin, L. (2021a,b,c) Biocosmology 1, 2 & 3, in preparation.
Deutsch, D. (2014) *The Beginning of Infinity*, London: Allen Lane, Viking Press.
Dyson, F. (2004) Thought experiments dedicated to John Archibald Wheeler, in *Science and Ultimate Reality*, New York: Cambridge University Press.

Goff, P. (2019) *Galileo's Error: Foundations for a New Science of Consciousness*, New York: Vintage Books.

Hardy, L. (2001) Quantum theory from five reasonable axioms, *arXiv*, quant-ph/0101012.

Heisenberg, W. (1949) *The Physical Principles of the Quantum Theory*, New York: Dover.

Jackson, F. (1982) Epiphenomenal qualia, *Philosophical Quarterly*, **32** (127), pp. 127–136. doi: 10.2307/2960077.

Kauffman, S. (2021) On gravity as a recorded quantum arrow of time, *ArXiv*, 2108.13905.

Mangabiera-Unger, R. & Smolin, L. (2014) *The Singular Universe and the Reality of Time*, Cambridge: Cambridge University Press.

McTaggart, J.M.E. (1927) *The Nature of Existence, Vol. II*, Cambridge: Cambridge University Press.

Peirce, C.S. (1891) The architecture of theories, *The Monist*, **1** (2), pp. 161–176.

Penrose, R. (1979) Singularities and time-asymmetry, in Hawking, S. (ed.) *General Relativity: An Einstein Centenary Survey*, pp. 581–638, Cambridge: Cambridge University Press.

Russell, B. (1927) *The Analysis of Matter*, London: Kegan Paul.

Smolin, L. (1992) Did the universe evolve?, *Classical and Quantum Gravity*, **9**, pp. 173–191.

Smolin, L. (1997) *The Life of the Cosmos*, New York: Oxford University Press.

Smolin, L. (2005) Precedence and freedom in quantum physics, *arXiv*, 1205.3707.

Smolin, L. (2013) *Time Reborn*, New York: Houghton Mifflin Harcourt.

Smolin, L. & Verde, C. (2021) The quantum mechanics of the present, *arXiv*, 2104.09945.

Tegmark, M. (2014) *Our Mathematical Universe*, New York: Knopff.

Wigner, E. & Margenau, H. (1967) Remarks on the mind body question, in symmetries and reflections, scientific essays, *American Journal of Physics*, **35** (12), pp. 1169–1170. doi: 10.1119/1.1973829.

Anil K. Seth[1,2]

The Real Problem(s) with Panpsychism

Abstract: *Panpsychism is the notion that consciousness is fundamental and ubiquitous. Defenders of panpsychism appeal, at least in part, to the apparent implausibility of materialist accounts of consciousness. However, materialist accounts are more resourceful than often assumed by proponents of panpsychism, while panpsychism has insurmountable problems of its own. The real problem with panpsychism is not that it seems crazy. It is that it explains nothing and does not generate testable predictions.*

1. Introduction

Consciousness — subjective experience — has long puzzled philosophers and scientists. How can the 'what-it-is-likeness' of conscious experiences (qualia) be understood in relation to physical, material systems such as brains (and bodies)? The now standard statement of this question is in terms of David Chalmers' 'hard problem'. As he puts it, 'Why should physical processing give rise to a rich inner life at all? It seems objectively unreasonable that it should, and yet it does' (Chalmers, 1995, p. 201).

The materialist response to this challenge is that consciousness will, in the end, turn out to be a (perhaps emergent) property of particular arrangements of, or dynamical interactions among, physical elements,

Correspondence:
Email: a.k.seth@sussex.ac.uk

---

[1] Sackler Centre for Consciousness Science, School of Engineering and Informatics, University of Sussex, Brighton, BN1 9QJ, UK.
[2] Program for Brain, Mind, and Consciousness, Canadian Institute for Advanced Research (CIFAR), Toronto, Ontario, M5G 1M1, Canada.

in ways that are yet to be discovered, and which may well require dramatically new concepts and methods. And the suspicion that this response will forever be inadequate plays a dominant role in motivating alternative metaphysical perspectives for considering consciousness, of which panpsychism is a prominent example. (The relevant history is more complicated, but this will do for now.)

Panpsychism is the notion that consciousness is fundamental and ubiquitous in the natural world. This notion is sometimes caricatured, or misunderstood, as proposing that objects such as spoons, stones, and atoms have conscious experiences that are in some way relatable to the sorts of conscious experiences had by human beings.[3] But sophisticated panpsychists do not say this at all. In his book *Galileo's Error*, Philip Goff (2019) notes that panpsychism does not propose that every 'thing' is conscious. Your socks may not be conscious, though — for Goff and other sophisticated panpsychists — they may be ultimately composed of things that *are* conscious. Equally, the kind of consciousness envisaged as being fundamental and ubiquitous is not consciousness 'like ours'. As Goff says, 'If electrons have experience, then it is of some unimaginably simple form' (p. 113; page numbers from here on refer to Goff, 2019). As we will see, the qualifier 'unimaginable' poses a challenge for the utility of panpsychism.

The immediate appeal of panpsychism appears to be that, by building in consciousness from the 'ground up', there is no longer any hard problem to solve. But, for Goff and others, there is a deeper motivation. Noting that scientific theories post-Galileo concern the *disposition* of entities — be they electrons or socks — to behave in particular ways, there is apparently nothing at all to be said about the *intrinsic nature* of anything. At the same time, conscious experiences seem to be all about intrinsic natures, and — on at least some views — resistant to characterization in terms of dispositions. Goff traces the final step to the physicist Arthur Eddington: the proposal that consciousness *is* the intrinsic nature of matter, filling the unseemly hole at the centre of the scientific story (p. 132). This is an elegant move, but being elegant does not mean being right, or being useful.[4]

---

[3] See https://neurobanter.com/2018/02/01/conscious-spoons-really-pushing-back-against-panpsychism/ for a caricature of this caricature.

[4] A third, less frequently aired motivation is the 'argument from causation', which justifies panpsychism on the grounds that the phenomenology of will, motivation, and agency somehow reveals that all causation is mental, and therefore panpsychism must be true (see Mørch, 2019, for a discussion). This seems a particularly weak argument.

Panpsychism — though it remains unpopular among many consciousness researchers[5] and tends to attract overblown media coverage[6] — is enjoying a day in the sun. Besides the influential philosophical expositions by Goff and others (see, for example, Seager, 2020), there are prominent neuroscientific theories of consciousness which indirectly imply some form of panpsychism. Here I am referring to Giulio Tononi's 'integrated information theory of consciousness' (IIT), according to which consciousness is identical to local maxima of irreducible integrated information (Tononi *et al.*, 2016; Tononi and Koch, 2015). Because such maxima are found in many places, but not everywhere, the panpsychism implied by IIT is relatively restrained: consciousness is fundamental, but not ubiquitous. But it is panpsychism nonetheless.[7]

In this paper, I will first examine the motivation for (at least some) panpsychist views, arguing that materialism is more resourceful than panpsychists tend to claim. I will then outline some weaknesses of panpsychism, focusing on explanatory power and empirical testability. I will finish by recognizing that both consciousness and matter are more complicated than they might at first appear.

2. The Resourcefulness of Materialism

What should a successful materialist account of consciousness accomplish? At first glance, one might be tempted to set the bar at a full solution to the hard problem of consciousness. On this view, the path

---

For starters, the phenomenology of experiences of will, motivation, and agency no more justifies belief in the existence of actual 'mental causation' than the phenomenology of 'seeing red' justifies belief in external objectively existing 'redness'. See Seth (2021, chapter 11).

[5] In a recent survey of 232 attendees of annual meetings of the Association for the Scientific Study of Consciousness — a meeting which attracts neuroscientists, psychologists, and philosophers — a large majority agreed with the proposition that a complete biophysical explanation of consciousness is either definitely or probably possible, and an overwhelming majority thought that rocks and thermostats lack consciousness; see https://psyarxiv.com/8mbsk/.

[6] For example: https://www.dailymail.co.uk/news/article-9820481/Scientists-debating-bizarre-theory-consciousness-including-inanimate-objects.html.

[7] The non-ubiquity implied by IIT is quite specific. Feedforward neural networks, for example, are not conscious, regardless of how complex their input–output mappings might be. And when a macroscopic entity such as a human brain is conscious, in virtue of instantiating irreducible maxima of integrated information, the parts that make it up — neurons, for example — lack consciousness. See Tononi and Koch (2015).

towards a successful account might consist of increasingly detailed attempts to directly address this problem.

An alternative view is that a materialist account of consciousness is best pursued, not by addressing the hard problem 'head on', but instead by building increasingly sturdy explanatory bridges between properties of physical (e.g. neural) mechanisms and properties of conscious experience. The latter are both functional (i.e. those which are readily explained in terms of mechanisms that produce effects) and, critically, phenomenological. (Note that I am not claiming that these properties are independent — it is likely they are not independent.) Accounting for the phenomenological properties of consciousness means, at least in part, explaining why a particular experience (or class of experiences) is the way it is, and not some other way: why the experience of redness is not like the experience of blueness, or jealousy, or toothache.

I have called this (with tongue a little bit in cheek) the 'real problem' of consciousness: how to explain, predict, and control the various properties of consciousness in terms of physical processes in the brain and body (Hohwy and Seth, 2020; Seth, 2016; 2021). The real problem is distinct from the hard problem because it is not, at least not in the first instance, about explaining why and how consciousness is part of our universe. (Already you can see that panpsychism does not address itself to the real problem.) And it is distinct from Chalmers' panoply of so-called 'easy problems' because it does not sweep the experiential, subjective aspects of consciousness away under the carpet.

The real problem approach is by no means new — precursors and complementary perspectives include neurophenomenology (Varela, Thompson and Rosch, 1993), and Chalmers' own 'mapping problem' (Chalmers, 1996). Another useful way to think of this approach is as generalizing the popular and empirically productive notion of 'neural correlates' of consciousness to that of 'explanatory' or 'systematic' correlates of consciousness (Hohwy and Seth, 2020; Seth, 2009).

Worries about the adequacy of materialism do not apply to the real problem, as they might potentially apply to the hard problem.[8] There are many ways to leverage the resources of materialism to develop

---

[8] Such worries are often misplaced even when applied to the hard problem. The infamous zombie argument against materialism, for example, is intrinsically weak because it is a conceivability argument.

and test explanatory bridges between mechanism and phenomenology. Two examples from my own work include the use of predictive models to account for the characteristic phenomenology of different perceptual experiences (Hohwy and Seth, 2020; Seth, 2014), and the development of quantitative measures of emergence to account for the ways in which unified conscious experiences seem to be more than the sum of their parts (Barnett and Seth, 2021; Rosas *et al.*, 2020). Rather than reviewing the details of these and other examples (Seth, 2021), I will instead highlight three more general points.

First, explanation, prediction, and control are the criteria by which many scientific enterprises are assessed. To be successful, it is not always essential — at least not at first — for a science of X to account for the presence of X in the universe in the first place. Physics has not (yet) accounted for the presence of the universe, the properties of which it does such a brilliant job of explaining.

Second, the explanations provided by a science of X need not be intuitively satisfying. Quantum mechanics, for example, is not intuitively satisfying, which is putting it mildly since there is no consensus on what it even means (we'll come back to this). The case in which X equals 'consciousness' should not imply a different standard for intuitive satisfaction. Indeed, the very fact that we instantiate phenomenological properties ourselves might mean that a scientifically satisfactory account of consciousness will *never* generate the intuitive feeling that 'ah yes, this is right, it has to be this way'. These first two points mean that the supposed inadequacy of materialism that motivates panpsychism in part results from setting the explanatory bar too high.

Third, an intuitively satisfying and hard-problem-complete account of consciousness is not ruled out by the real problem approach. It is just not guaranteed by it. It could well be, and I suspect that it will be, that addressing the real problem will *dissolve* rather than outright *solve* the hard problem. I have motivated this prospect (like others, including especially Patricia Churchland, 1996) by appealing to the analogy with the scientific understanding of life.

Briefly, the vitalist notion that life could not be explained in terms of biophysical mechanisms was neither directly solved (by finding the elusive 'spark of life') nor eradicated (by discovering that life does not exist). It was *dissolved* when biologists stopped treating life as one big scary mystery, and instead started accounting for (i.e. explaining, predicting, and controlling) the properties of living systems (reproduction, homeostasis, and so on) in terms of physical and chemical processes. We still don't understand everything about life, but what seemed at one time beyond the reach of materialism no longer does. By analogy, the fact that consciousness seems hard-problem mysterious now, with the tools and concepts we have now, does not mean it will always seem hard-problem mysterious — and the best way forward is to build the sturdiest explanatory bridges that we can, and see how far we get.

This brings us back to Goff, who notes my use of this analogy in his book, and is unmoved. Goff argues that the problem with the analogy is that sometimes science progresses otherwise. He notes that Darwin's theory of evolution by natural selection solved an apparent mystery — the nature and origin of biological species — through a dramatic insight, rather than through incremental dissolution (pp. 7–8). This is a curious argument. I am not claiming that dramatic insights do not matter or never happen. To the contrary, it is likely that dramatic insights into how explanatory bridges can be built will be needed if the real problem approach is to succeed. Moreover, Darwin's insight did not come out of nowhere. Besides being independently discovered by Alfred Russell Wallace, a strong case can be made that the conditions needed for Darwin (and Wallace) to have their insights depended on a long history of gradual alterations in how people thought about incremental change and the depth of historical time — from Charles Lyell's geology, to the breakdown of previously fixed social classes (Levins and Lewontin, 1987).

A more common objection is that life and consciousness are simply too different for the analogy to hold water. The usual argument here is that living systems — however mysterious they may once have seemed — are objectively describable, whereas consciousness is by definition a subjective phenomenon. Because consciousness is intrinsically subjective, the argument goes, mechanistic explanations are inadequate in a more problematic sense than was the case for life.

We can concede that this difference is important, without conceding the utility of the analogy, for two reasons. First, the analogy is useful primarily in an historical sense, through underlining the imprudence of pronouncing the insufficiency of mechanistic, materialist accounts of a putatively mysterious phenomenon before such accounts have been taken as far as they can go. Second, the difference should not be overstated. The subjective mode of existence of conscious experiences mostly means the relevant data are difficult to collect and compare. This is challenging (Phillips, 2018), but it does not exclude the possibility of a successful science of consciousness. Methodological developments in (for example) applied or computational phenomenology hold great promise in this regard (Suzuki *et al.*, 2017), as does the practice of phenomenology more generally. As John Searle noted long ago, it is possible to be epistemologically objective about something that is ontologically subjective (Searle, 1998); see also (Dennett, 1991).

Goff raises one other objection to materialism which is worth covering off here. He argues that 'physical science tries to give a purely *quantitative* characterisation of reality, a description involving only mathematical terms' (p. 68). Noting that conscious experiences involve *qualitative* concepts such as 'yellow' or 'sour', Goff concludes that '[m]aterialists who claim both that reality can be exhaustively described in the quantitative language of physical science and that there is quality-rich consciousness contradict themselves' (p. 68).

There is no contradiction. A successful science of consciousness does not require *exhaustive* description (it is not entirely clear what 'exhaustive' means here — one suspects it might mean something close to instantiation, or perhaps to '*a priori* scrutability')[9] — it requires *sufficient* description for the purposes of explanation, prediction, and control. In addition, quantitative elements can be readily introduced into phenomenological methods — as can already be seen in examples such as quantitative colour spaces (Bird *et al.*, 2014). Finally, many successful scientific explanations operate with qualitative rather than, or as well as, quantitative concepts. Darwin's theory — highlighted by Goff as a paragon of materialist science — provides one striking example of such an explanation. (Note that the meaning of 'qualitative' is different in evolution than in consciousness; this is a distinction that Goff glosses over.)

Altogether, there is much more to materialism than the proposal of 'brute identities' between the physical and the phenomenal that Goff finds 'very unsatisfying' (p. 108). The real problem approach promises to repair Galileo's error — an error elegantly diagnosed by Goff as that of removing sensory qualities from the remit of science — without leaning on brute identities, and, equally, without

---

[9] *A priori* scrutability, as articulated by Chalmers (2012), is a way of making precise the notion that a phenomenon is in principle fully deducible from a description. Understood this way, an exhaustive description would be an end point of either fully solving, or fully dissolving, the hard problem.

succumbing to the temptations of easy big 'solutions' like panpsychism. Put simply, the best strategy is to identify and account for properties of consciousness, and repeat. Following this strategy may ultimately deliver identities between the physical and the phenomenal, but — given their ability to explain, predict, and control — these identities would no longer be 'brute'.

3. The Impotence of Panpsychism

What if we took panpsychism seriously, and diverted all our intellectual resources to pursuing a panpsychist solution to the problem of consciousness?

According to Goff, this would mainly involve devising solutions to the so-called 'combination problem'. Broadly, this is the problem of how to get from little conscious things (such as electrons) to big conscious things (such as philosophers). Goff sketches a variety of philosophical approaches to closing this 'panpsychist gap', which may or may not make progress (p. 147). But the combination problem is only a problem if you assume panpsychism in the first place. While other (perhaps materialist) approaches may face related problems — such as explaining the unity of consciousness (Bayne, 2010) — these are substantially different from the combination problem. The panpsychist gap is therefore different from the widely recognized 'explanatory gap' in consciousness research (Levine, 1983), which can be understood as a relevant challenge from a variety of theoretical and metaphysical starting points. In other words, the main problem that the panpsychist methodology addresses is a problem of its own making.

This observation leads to the related point that panpsychism offers no useful insight into consensus explanatory targets for consciousness science. Asserting that consciousness is fundamental and ubiquitous does nothing to shed light on why an experience of blueness is the way it is, and not some other way. Nor does it explain anything about the possible functions of consciousness, nor why consciousness is lost in states such as dreamless sleep, general anaesthesia, and coma.[10] In acknowledging that the putative consciousness of a single electron

---

[10] Goff wonders whether such states might induce a disruption of subsequent memory, rather than an interruption of consciousness itself (p. 141). However, while *some* reported absences of consciousness may reflect amnesia, to defend panpsychism on the basis of there being no explanatorily interesting difference in consciousness between coma and wakeful awareness is a high price to pay.

will be unimaginably different from the consciousness of a human being (p. 113), we find the implicit concession that adopting panpsychism has no implications for explaining anything about subjective properties of consciousness as they pertain to conscious organisms.

Solving the combination problem does not address this criticism, since the combination problem concerns how already-conscious things are combined, or dissociated, into other conscious things. At most, it would offer a highly indirect way of explaining the basis of characteristic properties of consciousness, real problem style, but to which the core panpsychist assumption adds nothing besides other, uniquely panpsychist, problems.

Worst of all for panpsychism is that it is not itself testable, and that it does not lead to testable predictions. The challenges are overwhelming, going far beyond the difficulties of relating subjective reports to objective (e.g. neural) data. How can one begin to assess whether an electron is conscious? How does assuming panpsychism lead to any new testable predictions about consciousness whatsoever, whether in electrons, octopuses, or philosophers? The absence of responses to questions like these in Goff's otherwise comprehensive manifesto is striking.

One possible response to the testability challenge is that theories which indirectly imply some form of panpsychism may generate testable predictions. Here, again, I am thinking of IIT. However, in the case of IIT, the panpsychist interpretation makes experimental predictions *less* testable, rather than more testable. Specifically, empirical measurements of the core quantity of Φ (phi) are generally infeasible on the 'strong' (panpsychist) version of IIT, according to which one needs to measure all the states a system could potentially be in (its maximum entropy distribution). However, there are 'weaker' readings of IIT, better aligned with the real problem approach, in which integrated information is treated as an explanatory correlate linking properties of neural mechanisms to pervasive phenomenological features of global conscious states. Measures of Φ based on the weaker reading only require knowledge of the actual occurent dynamics of the system (its empirical distribution), and so are empirically tractable (Barrett and Seth, 2011; Mediano, Seth and Barrett, 2019).[11]

---

[11] One intriguing prediction flowing from IIT is that neurons may contribute to conscious experiences if they are *inactive*, but not if they are explicitly *inactivated* (Tononi *et al.*,

Ultimately, it is precisely the broader appeal of panpsychism, which Goff traces to Arthur Eddington, that guarantees its empirical impotence. Associating consciousness with the intrinsic — rather than the dispositional — nature of things guarantees that the property of 'being conscious' cannot itself make an experimentally measurable difference to anything, since to do so would require a dispositional aspect. (The flip side of this conclusion is also worrying. As Sean Carroll (this issue) points out, if the property of being conscious were to make a difference to things at a fundamental level, this would mean modifying the known laws of physics, which is not to be undertaken lightly.)

There is something ironic in Arthur Eddington's central role in all this, since Eddington's acclaim was largely due to his formulating testable predictions deriving from Einstein's general theory of relativity, and it was the dramatic experimental validation of these predictions that led to Einstein's theory playing a dominant role ever since (Kennefick, 2019). The contrast with panpsychism is clear: the success of modern physics is entirely due to its weird ideas — be they from relativity or quantum mechanics — being experimentally testable.

4. What is Matter, Anyway

Returning to materialism, I've argued against underestimating its powers from two related perspectives: shifting the explanatory target from the hard problem to the real problem, and noting the rich resources available for building explanatory bridges from neural mechanisms to properties of phenomenology. But materialism is fundamentally a claim about matter in general, rather than neurons in particular, and it is worth bearing in mind that the fundamental nature of matter is far from being understood.

Quantum mechanics — our current best theory of matter, and one that has stood up to every experimental interrogation that has been thrown at it — admits several interpretations, each bizarre, all controversial, and every one of which undermines the naïve materialistic picture of a universe ultimately consisting of particles of various kinds interacting via forces. It could be that a richer picture of matter itself

---

2016). This prediction is difficult, though perhaps not impossible, to test. However, should the experiments turn out to be feasible and to favour IIT, this by no means rules out non-panpsychist accounts of the data.

might further deepen the resources of materialism. For example, the 'relational' interpretation of quantum mechanics argues that the fundamental nature of matter is in the form of interactions (Rovelli, 1996). It is conceivable, though by no means guaranteed, that adopting such a perspective may in the long run change one's views about the possibilities of materialist explanations — not only of consciousness, but of many other phenomena too (see Carlo Rovelli's enchanting *Helgoland*, 2021, for something along these lines; see also Rovelli, this issue).

I stress that I am not falling into the lazy and false syllogism that consciousness and quantum mechanics are both mysterious, so therefore they must be related. Nor am I condoning theories that propose a specific basis for consciousness in terms of quantum mechanical processes (Hameroff and Penrose, 2014), nor those that propose that consciousness brings the classical world into existence by collapsing the wave function (Chalmers and McQueen, in press; Wigner, 1961). There is plenty of work to be done in building explanatory bridges between mechanism and phenomenology using the uncontroversial tools of classical physics, neuroscience, and psychology, and this project may well be sufficient to dissolve the hard problem. My argument is simply that respecting the complexities of both consciousness *and* matter provides yet more reason to believe that resorting to panpsychism reflects a failure of imagination rather than an insight into necessity.

5. Summary

Goff offers a four-point manifesto for a post-Galilean science of consciousness (p. 174). I'm completely on board with the first three: realism about consciousness, empiricism, and anti-dualism. The fourth — a panpsychist methodology — is where I get off the bus. Panpsychism explains nothing, cannot be tested in itself, does not lead to testable predictions, and may actively discourage the generation of such predictions. It is a seductive easy-out to the hard problem, and there is no need for it.

A real problem approach, rooted in materialism, but going far beyond establishing mere correlations or prematurely proposing brute identities, has more than enough gumption to explain, predict, and control properties of consciousness in terms of their underlying mechanisms. It remains to be seen whether following this approach will fully dissolve the hard problem, or whether residues of mystery will persist, but the journey on this particular bus is going to be beautiful and enlightening, and there are plenty of tickets still for sale.

Acknowledgments

I am grateful to Philip Goff both for writing his accessible and illuminating book, and for inviting me to write this article. Many thanks to Tim Bayne for comments on a draft of this paper. I am also grateful to the Dr Mortimer and Theresa Sackler Foundation and to the Canadian Institute for Advanced Research (CIFAR) Program on Brain, Mind, and Consciousness.

References

Barnett, L.C. & Seth, A.K. (2021) Dynamical independence: discovering emergent macroscopic processes in complex dynamical systems, *ArXiv*, [Online], https://arxiv.org/abs/2106.06511..

Barrett, A.B. & Seth, A.K. (2011) Practical measures of integrated information for time-series data, *PLoS Computational Biology*, **7** (1), e1001052. doi: 10.1371/journal.pcbi.1001052.

Bayne, T. (2010) *The Unity of Consciousness*, Oxford: Oxford University Press.

Bird, C.M., Berens, S.C., Horner, A.J. & Franklin, A. (2014) Categorical encoding of color in the brain, *Proceedings of the National Academy of Sciences USA*, **111** (12), pp. 4590–4595. doi: 10.1073/pnas.1315275111.

Carroll, S. (this issue) Consciousness and the laws of physics, *Journal of Consciousness Studies*, **28** (9–10).

Chalmers, D.J. (1995) Facing up to the problem of consciousness, *Journal of Consciousness Studies*, **2** (3), pp. 200–219.

Chalmers, D.J. (1996) *The Conscious Mind: In Search of a Fundamental Theory*, New York: Oxford University Press.

Chalmers, D.J. (2012) *Constructing the World*, Oxford: Oxford University Press.

Chalmers, D.J. & McQueen, K. (in press) Consciousness and the collapse of the wave function, in Gao, S. (ed.) *Consciousness and Quantum Mechanics*, Oxford: Oxford University Press.

Churchland, P.S. (1996) The hornswoggle problem, *Journal of Consciousness Studies*, **3** (5–6), pp. 402–408.

Dennett, D.C. (1991) *Consciousness Explained*, Boston, MA: Little, Brown.

Goff, P. (2019) *Galileo's Error: Foundations for a New Science of Consciousness*, London: Rider.

Hameroff, S. & Penrose, R. (2014) Consciousness in the universe: A review of the 'Orch OR' theory, *Physics of Life Reviews*, **11** (1), pp. 39–78. doi: 10.1016/j.plrev.2013.08.002.

Hohwy, J. & Seth, A.K. (2020) Predictive processing as a systematic basis for identifying the neural correlates of consciousness, *Philosophy and the Mind Sciences*, **1** (2), art. 3.

Kennefick, D. (2019) *No Shadow of a Doubt: The 1919 Eclipse that Confirmed Einstein's Theory of Relativity*, Princeton, NJ: Princeton University Press.

Levine, J. (1983) Materialism and qualia: The explanatory gap, *Pacific Philosophical Quarterly*, **64**, pp. 354–361.

Levins, R. & Lewontin, R.C. (1987) *The Dialectical Biologist*, Cambridge, MA: Harvard University Press.

Mediano, P.A.M., Seth, A.K. & Barrett, A. B. (2019) Measuring integrated information: Comparison of candidate measures in theory and simulation, *Entropy*, **21** (1), art. 17. doi: 10.3390/e21010017.

Mørch, H.H. (2019) The argument for panpsychism from experience of causation, in Seager, W. (ed.) *The Routledge Handbook of Panpsychism*, London: Routledge.

Phillips, I. (2018) The methodological puzzle of phenomenal consciousness, *Philosophical Transactions of the Royal Society of London B: Biological Sciences*, **373** (1755). doi: 10.1098/rstb.2017.0347.

Rosas, F., Mediano, P.A.M., Jensen, H.J., Seth, A.K., Carhart-Harris, R. & Bor, D. (2020) Reconciling emergences: An information-theoretic approach to identify causal emergence in multivariate data, *PLoS Computational Biology*, **16** (12), e1008289. doi: 10.1371/journal.pcbi.1008289.

Rovelli, C. (1996) Relational quantum mechanics, *International Journal of Theoretical Physics*, **35**, pp. 1637–1678.

Rovelli, C. (2021) *Helgoland*, London: Allen Lane.

Rovelli, C. (this issue) Relations and panpsychism, *Journal of Consciousness Studies*, **28** (9–10).

Seager, W. (ed.) (2020) *The Routledge Handbook of Panpsychism*, London: Routledge.

Searle, J. (1998) *Mind, Language, and Society: Doing Philosophy in the Real World*, New York: Basic Books.

Seth, A.K. (2009) Explanatory correlates of consciousness: Theoretical and computational challenges, *Cognitive Computation*, **1** (1), pp. 50–63.

Seth, A.K. (2014) A predictive processing theory of sensorimotor contingencies: Explaining the puzzle of perceptual presence and its absence in synesthesia, *Cognitive Neuroscience*, **5** (2), pp. 97–118. doi: 10.1080/17588928.2013.877880

Seth, A.K. (2016) The real problem, *Aeon*, [Online], https://aeon.co/essays/the-hard-problem-of-consciousness-is-a-distraction-from-the-real-one.

Seth, A.K. (2021) *Being You: A New Science of Consciousness*, London: Faber & Faber.

Suzuki, K., Roseboom, W., Schwartzman, D.J. & Seth, A.K. (2017) A deep-dream virtual reality platform for studying altered perceptual phenomenology, *Scientific Reports*, **7** (1), 15982. doi: 10.1038/s41598-017-16316-2.

Tononi, G. & Koch, C. (2015) Consciousness: Here, there and everywhere?, *Philosophical Transactions of the Royal Society of London B: Biological Sciences*, **370** (1668). doi: 10.1098/rstb.2014.0167.

Tononi, G., Boly, M., Massimini, M. & Koch, C. (2016) Integrated information theory: From consciousness to its physical substrate, *Nature Reviews Neuroscience*, **17** (7), pp. 450–461. doi: 10.1038/nrn.2016.44.

Varela, F.J., Thompson, E. & Rosch, E. (1993) *The Embodied Mind: Cognitive Science and Human Experience*, Cambridge, MA: MIT Press.

Wigner, E. (1961) Remarks on the mind–body problem, in Good, I.J. (ed.) *The Scientist Speculates*, Portsmouth, NH: Heinemann.

Christof Koch[1]

Reflections of a Natural Scientist on Panpsychism

Abstract: *Panpsychism shares many intuitions with integrated information theory (IIT), in particular that consciousness is an intrinsic fundamental property of reality, is graded, and can be found in small amounts in simple physical systems. Unlike panpsychism, however, IIT clearly articulates which systems are conscious and which ones are not (resolving panpsychism's combination problem) and why consciousness can be adaptive. The systemic weakness of panpsychism, or any other -ism, is that they fail to offer a protracted conceptual, let alone empirical, research programme that yields novel insights or proposes new experiments. Without those, progress on the mind–body problem will not occur.*

1. Introduction

As a romantic reductionist, I've always had a secret crush on the singular beauty of panpsychism (Koch, 2012). Yet, when enthusing about panpsychism's elegance to scientists, clinicians, and technologists, I encounter for the most part blank stares of incomprehension; graduate students, finely attuned to the unspoken beliefs of their elders, roll their eyes and talk of conscious spoons and chairs. Few living scholars take panpsychism seriously. Much to my surprise, though, this is changing. By how much is readily apparent when reading Philip Goff's *Galileo's Error* (2019; see also the bestseller by the writer Harris, 2019).

Correspondence:
Email: koch.christof@gmail.com

---

[1] Allen Institute for Brain Science, Seattle, WA, USA.

Until about the end of the twentieth century, the prevailing orthodoxy in Anglo-American philosophy departments was materialism, and its modern variant, physicalism. The eagle of physicalism, having vanquished its enemy, dualism, screamed supremacy. In 1978, Daniel Dennett crows:

> Since it is widely granted these days that dualism is not a serious view to contend with, but rather a cliff over which to push one's opponents, a capsule 'refutation' of dualism, to alert if not convince the uninitiated, is perhaps in order. (1978, p. 252)

In Goff's compact book written for an educated lay audience, Goff forcefully argues for why physicalism and dualism are both mistaken about the nature of consciousness and for the validity of a third view that was not on the radar of the twentieth-century metaphysical world order — panpsychism, the ancient belief that the mental is both fundamental and ubiquitous in the world, including in non-living entities.

2. Galileo's Mistake?

First, what was Galileo's error that the title refers to? About 400 years ago, in 1632, the Italian physicist Galileo Galilei wrote a short polemic, *Il Saggiatore* or *The Assayer*, dedicated to his friend and benefactor who had just been elected Pope Urban VIII in Rome. Galileo addressed a critic, a mathematician and astronomer, who argued from prior religious conviction and dogma rather from observation. Galileo elevates mathematics as the true language describing nature in a rightfully famous quote:

> Philosophy is written in this grand book, the universe, which stands continually open to our gaze. But the book cannot be understood unless one first learns to comprehend the language and read the letters in which it is composed. It is written in the language of mathematics, and its characters are triangles, circles, and other geometric figures without which it is humanly impossible to understand a single word of it; without these, one wanders about in a dark labyrinth.

Galileo goes on to distinguish the material properties of something — its size, shape, location, motion, and so on — from its sensory qualities such as being warm and cold, white or red, bitter or sweet, noisy or silent, sweet or foul odour. The former are inherent to the object, while the latter are in the mind of the observer. The chemical properties of sweetened hazelnut cocoa spread are distinct from the yummy taste of Nutella that no one should do without. Following John Locke, these two classes of attributes, physical and mental, are known as primary and secondary qualities. Experimental science can study primary qualities without having to worry about secondary ones. This then is the successful path of science that today bestrides the globe — focus on the objective world, leaving subjective feelings aside. But, according to Goff, this is pernicious, for turning away from investigating the mental leaves science unable to explain how the mental comes about:

> So long as we follow Galileo in thinking (A) that natural science is essentially quantitative and (B) that qualitative cannot be explained in terms of the quantitative, then consciousness, as an essentially qualitative phenomenon, will be forever locked out of the area of scientific understanding. (Goff, 2019)

This is the error that Galileo committed, creating four centuries of confusion.

At this point, I need to raise an objection. Throughout *Galileo's Error*, and following Locke, Goff distinguishes between the quantitative aspects of primary qualities and the qualitative aspects of secondary, sensory qualities. Fine — but don't blame this distinction on Galileo. There is not a single statement in *The Assayer* in which Galileo argues that mathematics cannot be applied to secondary qualities. Galileo simply does not address this point. Furthermore, the *corpo sensitivo* that Galileo refers to in the Italian original is part of nature (*pace* Aristotle's sensitive soul) and mathematics is needed to read nature's open book (*per* Galileo). Therefore, this absence of evidence cannot be taken as evidence of absence.

Galileo never asserts that mathematics cannot explain sensory qualities, such as tastes, odours, and colours. Rather, Galileo adopts a *pragmatic stance* of exploring the physical universe, because he had pioneered two powerful new tools, the telescope and mathematical physics, and was eager to quantify phenomena in the sky using them and get credit for that (as every working scientist can appreciate). To me, it seems that Goff is retrofitting his ideas of the mind onto Galileo.

2. Dualism, Physicalism, and their Failures

The next two chapters of *Galileo's Error* introduce the two poles around which modernity's mind–body debates are organized — dualism and physicalism. The most famous version of dualism comes to us from Rene Descartes, postulating the existence of two sorts of substances in the universe — physical stuff (*res extensa*) and mental stuff (*res cogitans*). Dualism, like any -ism, comes in many shades, such as property-, substance-, and naturalistic-dualism, for which Goff provides brief sketches. All suffer from the *interaction* problem; that is, how does the mental stuff interact with the physical stuff making up brain tissue in a way that would still satisfy physical law?

Next, Goff demonstrates through a series of philosophical arguments that the proposition that physical sciences as currently conceived can explain consciousness is not correct. There is nothing in the foundational equations of quantum mechanics nor of special or general relativity that would account for the existence of conscious percepts, memories, thoughts, feelings, and so on. That neuroelectrical activity of a particular kind in the cortex is closely linked to particular feelings is not in dispute. But why this should be so (both in the sense of efficient causes and final causes) is the open question.

The book succinctly introduces the reader to the standard canon of thought experiments, starting with the mill argument of Gottfried Wilhelm Leibniz from 1702, and expanding into a roll call of philosophers of the second half of the last century — Thomas Nagel's what-is-it-like-to-be-a-bat argument, Frank Jackson's knowledge argument about the hypothetical neuroscientist Mary who knows everything there is to know about colour but grows up in an entirely black-and-white environment, the conceivability argument a.k.a the totemic zombie thought experiment developed most famously by David Chalmers, and the Chinese room argument of John Searle.

Physicalism is simply inadequate to deal with the reality of lived experience, from the banal taste of cold French fries to the most exalted encounter with the numinous. One extreme reaction to the inability of physicalism to explain the mental is simply to deny that the mental exists (a strategy familiar to psychotherapists). Goff labels this illusionism, a truly remarkable form of abnegation. Goff quickly dispenses with this bizarre view, approvingly citing his mentor, Galen Strawson, who calls illusionism 'the silliest claim ever made'. Indeed, Strawson opines:

> If there is any sense in which these philosophers are rejecting the ordinary view of the nature of things like pain... their view seems to be one of the most amazing manifestations of human irrationality on record. It is much less irrational to postulate the existence of a divine being whom we cannot perceive than to deny the truth of the common sense view of experience. (1994, p. 53)

3. Enter Panpsychism

Having demolished, on the one hand, the belief that physics as currently conceived can explain the mental and, on the other, the belief that a soul-like substance can account for consciousness, Goff turns to the view that the mental and the physical are closely allied, two sides of the same coin. Goff came to panpsychism by way of a Pauline conversion in a crowded bar — rather than on the road to Damascus — when he was overcome by the reality of his conscious experience, a reality that could neither be denied nor elided as an illusion. Given the inadequacy of standard views of the mind–body problem, he turned — first covertly and then publicly — into a panpsychist.

He is not alone. Historically, many of the brightest minds in the West took the position that matter and soul are one substance, starting with the pre-Socratic philosophers, Thales and Anaxagoras, as well as Plato. The Renaissance cosmologist Giordano Bruno espoused such ideas, as did Leibniz, Arthur Schopenhauer, and the palaeontologist and Jesuit Teilhard de Chardin (Skrbina, 2017).

Particularly striking are the many scientists with panpsychist views, such as the founders of psychology and psychophysics — Gustav Fechner, Wilhelm Wundt, and William James — and the astronomers and mathematicians Arthur Eddington, Alfred North Whitehead, and Bertrand Russell. *Galileo's Error* marks the return of one form of panpsychism to academic philosophy in which physics and consciousness are dual aspects of the same reality. Physics has nothing to say about the intrinsic nature of matter but describes how bits and pieces of matter relate to each other, say via Coulomb's law, while consciousness describes the intrinsic nature of matter. Alan Watts (1999) expresses it poetically as:

> For every inside there is an outside, and for every outside there is an inside; though they are different, they go together.

The brain can be studied as a physical object; it can be placed inside a magnetic scanner and probed. The experience of the brain of a volunteer lying inside the scanner, the machine-gun-like loud knocking of the magnetic coils and the narrow field of view, is the way this object, the brain, feels from the inside. Physics tells us how physical systems behave from the outside, while experience is the interior view of these systems.

Panpsychism is unitary. There is only one substance, not two. This elegantly eliminates the need to explain how the mental emerges out of the physical and vice versa.

As Goff makes clear, variants of panpsychism (neutral monism is currently the most popular one) have returned to the academe that, like a monopolistic landlord (*pace* the Dennett quote at the beginning), tried to evict the mental entirely from university philosophy departments as well as from the universe at large.

Goff rightfully highlights the Achilles heel of panpsychism, the combination problem. What determines the boundaries of consciousness? I, the author, am conscious, as are you, the reader. Yet there isn't a single, integrated author–reader mind. Why not? By what principle are the monadic boundaries decided? Two lovers whose bodies are intermingled are still two distinct conscious entities, each with their own feelings, as Tristan and Isolde famously bemoan in Wagner's eponymous opera. There is no Tristan–Isolde Über-mind. Experiences do not aggregate into larger, superordinate experiences.

Essentially the only clinical or psychological observations discussed by Goff in *Galileo's Error* are the split-brain patients studied by Roger Sperry in the 1960s (more than half a century ago, an eternity in a rapidly moving field!). Sperry discovered that severing the cerebrum along its midline, cutting the 200 million fibres connecting the two cortical hemispheres, leaves two conscious minds in its wake, with two distinct streams of conscious experience, one of which can speak while the other one can sing or answer simple yes/no questions. Incongruently, Goff refers to these two distinct minds as an example of disunified consciousness rather than simply two distinct conscious minds.

It is in the context of panpsychism that Goff briefly mentions the integrated information theory (IIT) of consciousness as an empirically testable scientific theory of consciousness. IIT shares many insights with panpsychism (Tononi and Koch, 2015).

The technical part of the book ends with a manifesto for a post-Galilean science of consciousness that embodies four commitments to account for reality *tout court* — to accept the fundamental nature of conscious experiences and of physics on an equal footing, to avoid dualism, and to account for higher-order consciousness of people and non-human animals in terms of more basic forms of consciousness.

In a coda, Goff speculates on a number of topics of personal interest, including climate change, mystical experiences (which are, by their nature, difficult to write about for others), the existence or otherwise of moral truth, and the meaning of life. In a heart-warming conclusion, he expresses the hope that a widespread acceptance of the view that many things under the sun are 'ensouled' offers a way to re-enchant the world.

4. Integrated Information Theory as a Testable Theory of Consciousness

I am sympathetic to Goff's positive appraisal of panpsychism, having arrived at a somewhat related position, but starting from a very different vantage point: I am a neuroscientist with a background in physics in pursuit of an empirically testable theory of consciousness.

IIT postulates five phenomenological axioms of consciousness (intrinsic existence, composition, information, integration, and exclusion), and derives five associated postulates from which arises a principled account of both the quantity and the quality of any one experience of a system in a particular state. The quantity of consciousness is specified by the system's *integrated information* (phi or Φ), a pure, non-negative number. If it is zero, the system does not feel like anything and does not, strictly speaking, exist for itself (in IIT, consciousness is closely tied to intrinsic causal power; that is, intrinsic existence). The larger the Φ, the more the system exists for itself, the more it is conscious. The theory provides an explicit calculus to evaluate the integrated information of any one system.

IIT shares many intuitions with panpsychism — consciousness is an intrinsic, fundamental property of reality; it is graded and may well be widespread among animals, not just mammals. That does not imply that a bee feels obese or plans for the weekend, but that it too may feel content when returning pollen-laden in the sun to its hive and that, when it dies, it ceases to experience anything. According to IIT, the mental may be found throughout the tree of life and, possibly, even in non-evolved physical systems, violating the 'no brain, never mind' intuition most of us have.

IIT addresses several major shortcomings of panpsychism — it explains why consciousness is adaptive, it explains the different qualitative aspects of consciousness (why a 'kind of blue' feels different from a stinky Limburger cheese), and it head-on addresses the combination problem so ably outlined by Goff — per IIT's exclusion postulate, only systems with a maximum of Φ have intrinsic existence, are conscious. Thus, to return to our fateful lovers — Tristan's brain has its own maximum of Φ, as does Isolde's brain. To the soundscape of some of the most rapturous music ever composed, Tristan yearns to be Isolde, exclaiming 'Tristan you, I Isolde, no longer Tristan', evoking Isolde to echo, 'You Isolde, Tristan I, no longer Isolde'. As they respond to each other, the joint integrated information of their interacting brains is non-zero but is dwarfed by the massive amount of integrated information within each of their brains. As only maximum of Φ have intrinsic experience, are conscious, there is no Tristan–Isolde Über-mind. However, IIT does leave open the possibility of the merging of two brains into one mind, provided an artificial brain-bridging technology directly links millions of neurons in one cortex to those of the other (Koch and House, 2020), a sort of artificial corpus callosum. It is more challenging to put this into an ecstatic duet.

The exclusion postulate dictates whether or not an aggregate of entities — ants in a colony, cells making up a tree, bees in a hive, starlings in a murmurating flock, an octopus with its eight semi-autonomous arms, and so on — exist as a unitary conscious entity or not.

IIT offers a startling counter-example to Goff's claim that qualitative aspects of conscious experience cannot be captured by quantitative considerations. Indeed, Haun and Tononi (2019) published a detailed, mathematical account of how the phenomenology of two-dimensional space, say an empty canvas, can be fully accounted for in terms of intrinsic causal powers of the associated physical substrate, here a very simple, grid-like neural network. Integrated information theory may well be wrong (and can be disproven by relevant experiments, such as those in the ongoing adversarial collaboration between IIT and global neuronal workspace theory; Melloni *et al.*, 2021), but provides proof-of-principle for how quantitative primary qualities (here intrinsic causal power of simple model neurons that can be numerically computed; it doesn't get more quantitative than that) correspond to secondary qualities — the experience of looking at a blank wall.

5. The Eternal Return in Philosophy of Mind

This brings me to a systemic weakness in *Galileo's Error*. It fails to provide for a protracted conceptual, let alone empirical, research programme to validate panpsychism, to provide proof that it is the best possible explanation of all available facts. With panpsychism 'in mind', where does the field go next? What puzzling clinical or laboratory anomaly yields to an explanation? Besides claiming that everything has both intrinsic and extrinsic aspects, panpsychism is barren and has nothing constructive to say about the relationship between the two.

For instance, the heart is an organ of deep complexity, a highly evolved biochemical and bioelectric system, just like the brain. Yet only the latter is associated with consciousness; indeed, the recipient of a transplanted heart retains their own consciousness rather than inheriting it from the heart donor. Why? Many regions of the central nervous system, such as the spinal cord or the cerebellum, perform complex 'computations' with elaborate maps, and input and output representations (to adopt the common idiom of computational neuroscience), yet do not contribute to consciousness. Why? In neurotypical humans, it is the cerebral cortex, and in particular its posterior regions, that contribute to consciousness (Koch *et al.*, 2016). Why? The phenomenology of extended space is completely different from the flow of time, the agony of visceral pain, or the impenetrable beauty of ultramarine blue (Haun and Tononi, 2019). Why? Clearly, it must relate to the organization — dare I say, the *form* — of the underlying physical substrate. Yet *Galileo's Error* eschews any discussion of such questions.

The last 2,500 years of Western philosophy of mind have seen the rise and fall (and rise and fall) of many schools of thought concerning the mental and the physical. Highly polished arguments and counter-arguments are exchanged in a never-ending cycle that results in drawn-out sophisticated disagreements but no resolution. Thus, despite Dennett's casual one-paragraph dismissal of dualism, it is making its way back into the academe (Bogardus, 2013; Swinburne, 2013; Lavazza and Robinson, 2014; Guta, 2018; Loose, Menuge and Mooreland, 2018), while Aristotelian *hylomorphism*, a sophisticated variant of non-reductive physicalism, is enjoying a renaissance in biology (e.g. Marmodoro, 2014; Owen, 2018). Purely logical, mathematical, or linguistical arguments are, in the absence of empirical verification, not compelling enough to break this impasse.

Indeed, it is not even clear that there is a generally accepted notion of progress within philosophy of mind (Bourget and Chalmers, 2014; Chalmers, 2015). While historical confusions in the subject have been cleared up, antecedent assumptions for particular premises are better understood, connections between concepts and ideas have been clarified, there is no communal convergence on the perennial Big Questions, such as 'What is the relationship between the mind and the body?', 'Do we have free will?', never mind the more mechanistic and phenomenal questions I listed earlier.

For a field of intellectual enquiry that has posed these questions since antiquity, this is disappointing. The science of consciousness must break out of these endless epicycles of arguments by formulating a sustainable programme of hypothesis formulation and experimental validation or falsification in the context of a quantitative theory of consciousness. In this manner, philosophers, natural scientists, and other scholars can finally come to understand the mind–body conundrum and embrace both the mental and the physical as two, apparently distinct, aspects of a single underlying reality.

References

Bogardus, T. (2013) Undefeated dualism, *Philosophical Studies*, **165.2**, pp. 445–466.

Bourget, D. & Chalmers, D.J. (2014) What do philosophers believe?, *Philosophical Studies*, **170**, pp. 465–500.

Chalmers, D.J. (2015) Why isn't there more progress in philosophy?, *Philosophy*, **90**, pp. 3–31.

Dennett, D. (1978) Current issues in the philosophy of mind, *American Philosophical Quarterly*, **15**, pp. 249–261.

Guta, M.P. (ed.) (2018) *Consciousness and the Ontology of Properties*, London: Routledge.

Goff, P. (2019) *Galileo's Error: Foundations for a New Science of Consciousness*, New York: Pantheon.

Harris, A. (2019) *Conscious: A Brief Guide to the Fundamental Mystery of the Mind*, New York: Harper Collins.

Haun, A. & Tononi, G. (2019) Why does space feel the way it does? Towards a principled account of spatial experience, *Entropy*, **21** (12), art. 1160. doi: 10.3390/e21121160.

Koch, C. (2012) *Consciousness: Confessions of a Romantic Reductionist*, Cambridge, MA: MIT Press.

Koch, C., Massimini, M., Boly, M. & Tononi, G. (2016) The neural correlates of consciousness: Progress and problems, *Nature Reviews Neuroscience*, **17**, pp. 307–321.

Koch, C. & House, P. (2020) Brain bridging: A more perfect union, *Nature*, [Online], https://www.nature.com/articles/d4_586-020-02469-0.

Lavazza, A. & Robinson, H. (eds.) (2014) *Contemporary Dualism: A Defense*, London: Routledge.

Loose, J.J., Menuge, A.J.L. & Moreland, J.P. (eds.) (2018) *The Blackwell Companion to Substance Dualism*, Oxford: Wiley Blackwell.

Marmodoro, A. (2014) *Aristotle on Perceiving Objects*, New York: Oxford University Press.

Melloni, L., Mudrik, L., Pitts, M. & Koch, C. (2021) Making the hard problem of consciousness easier, *Science*, **372** (6545), pp. 911–912.

Owen, M. (2018) Aristotelian causation and neural correlates of consciousness, *Topoi*, [Online], https://doi.org/10.1007/s11245-018-9606-9.

Skrbina, D.F. (2017) *Panpsychism in the West*, revised ed., Cambridge, MA: MIT Press.
Strawson, G. (1994) *Mental Reality*, Cambridge, MA: MIT Press.
Swinburne, R. (2013) *Mind, Brain, and Free Will*, Oxford: Oxford University Press.
Tononi, G. and Koch, C. (2015) Consciousness: Here, there and everywhere?, *Philosophical Transactions of the Royal Society B*, **370**, 20140167.
Watts, A. (1999) *Man, Nature, and the Nature of Man*, London: Macmillan.

Jonathan Delafield-Butt[1]

Autism and Panpsychism

Putting Process in Mind

Abstract: Panpsychism is a metaphysical framework around which science can understand the nature of subjective experience. It affords a scientific view of mind and body as a coherent mind–body unity, with agentive purpose. Fundamental to minds is motor control, a core aspect that combines sensory experience, its evaluation in choice of agent action, and extension into the public expression of intentional movement. This primary mind–body process appears disturbed in autistic individuals. Empirical analysis of the spatio-temporal properties of intentional movement in autism shows a disruption to the efficient prospective integration and control of movement, a core aspect of mind. This paper examines the capacity of a panpsychist metaphysic to explain mind as fundamentally constituted by units of mind–body sensorimotor agency, which can be understood as the basic building blocks of embodied experience. The implications of a post-Cartesian metaphysic in scientific understanding of minds allows for deeper consideration of the role of movement in subjective experience, and its disturbance in autism as a disturbance to the organization of conscious sensorimotor experience and agency. It's impact on modes of cognition and neural substrates is discussed.

1. Autism and Panpsychism

Panpsychism and autism may seem unexpected bedfellows, but autism is fundamentally a condition of disturbance of subjective experience

Correspondence:
Email: jonathan.delafield-butt@strath.ac.uk

[1] University of Strathclyde, Glasgow, Scotland, UK.

and panpsychism is a metaphysical framework around which science can better analyse and understand the structure of subjective experience, of mind. Panpsychism affords a view of the so-called 'mind–body duality' as instead a combined 'mind–body singularity', a monad. By doing so, panpsychism may afford a deeper view of the structure of the mind–body relationship than what the conventional vision of a mind *and* a body — a split Cartesian mind–body duality — can.

Panpsychism has been neglected over the twentieth century as its main metaphysical competitor, materialism, made exceptional gains in powering scientific understanding, and because the idea of 'mind everywhere' seems on the surface of it to be simply untrue. But powerful arguments suggest panpsychism must be true (Basile, 2010; Strawson, 2006; 2017). More recently in the twenty-first century, the limits of materialism to explain conscious experience and to serve as a useful explanatory framework for the psychological sciences, especially in mental health, are drawing in. Panpsychism is now enjoying a resurgence as a favoured solution to the 'hard problem' of consciousness (Chalmers, 1995; Goff, 2019), which is ultimately at the root of our scientific explanations of mind. Panpsychism is also finding new explanatory use in biology as mounting evidence of a fundamental agency operative in many simple, non-human, and even non-neural organisms comes to the foreground (Calvo *et al.*, 2019; Delafield-Butt, 2007; 2008; Delafield-Butt *et al.*, 2012; Trewavas *et al.*, 2020).

For our purposes here, one of panpsychism's most useful features is identification and characterization of *fundamental mind–body singularities* known as 'monads' or 'actual occasions', following two of the great panpsychist philosophers from the eighteenth and twentieth centuries, Leibniz (1716) and Whitehead (1929), respectively. Panpsychism is not a flat theory of an omnipresent, simple, and ubiquitous mind that is everywhere. Rather, panpsychism is fundamentally a process metaphysic deeply structured by an ontology of units.

1.1. The ontological unit, an actual occasion

These fundamental 'mind–matter' units in Leibniz's and Whitehead's ontology are thought to be invariant no matter their scale or complexity. Whitehead (1929) spent some considerable time describing them, borrowing from the contemporary biology and physics of his day. He named these units 'actual occasions' or 'organisms', each one alive with sentience, its own subjectivity, agency, and purpose. He reasoned each one 'lives' only for a finite period before it 'dies', the effects of its life appropriated into many other occasions, or organisms, touched during its life. These 'occasions' are the 'fundamental drops of experience', they are 'what there is' in nature — mind–body events.

The aim of this paper is to bring this philosophical framework to bear on autistic experience and scientific understanding of autism. It follows the assertion that panpsychism can afford new thinking in examination of experience useful for empirical scientific advance (Delafield-Butt, 2007; 2008). Science can benefit if it can take 'consciousness... as an *epistemological starting point* on par with the epistemological starting points we get from observation and experiment' (Goff, 2019, p. 173). Here, I take as my starting point the metaphysical assertion of panpsychism as a process ontology, bringing it together with first-person accounts of autistic experience, with empirical data from autism research and motor control psychology, and with theoretical gains in affective neuroscience, embodied cognition, and autism research. In this way, I will map a fundamental mind–matter ontology (metaphysics) onto empirical data from autistic subjective experience (first person) and experimental data on its physical manifestation (third person).

1.2. Movement in mind

Fundamental to this repositioning of perspective is motor control, and the psychological nature of animated action. Animal action is always prospectively organized, organized with an eye to the future (Lee, 2000; von Hofsten, 1993). Each motor impulse generates forces of inertia and momentum that must be known and compensated for ahead of time. Each action is coherent and unified to achieve its efficient purpose, its goal (Bernstein, 1967; Trevarthen, 1984). Additional goals, ideas, aspirations, and ambitions may be woven into that singular moment, that singular goal, but in the moment there is only one direction, one purpose, one goal embodied in the actions of the agent, with mind and body operating as a cohesive and singular entity.

We now understand autism to be a disruption to the efficient integration and control of movement subserving this singular, embodied purpose. Empirical evidence collected over two decades of sensitive motion-capture studies shows a subtle, but significant, autistic disturbance to intentional body movement (Anzulewicz, Sobota and Delafield-Butt, 2016; Casartelli *et al.*, 2020; Cook, Blakemore and Press, 2013; Glazebrook, Elliott and Lyons, 2006; Mari *et al.*, 2003; Schmitz *et al.*, 2003; Torres *et al.*, 2013). The evidence suggests this motor disturbance underpins the disorder altogether, and that this motor aspect of autism ought to be considered a core clinical feature of the disorder (Fournier *et al.*, 2010; Trevarthen and Delafield-Butt, 2013a). Purposive movement is the most fundamental aspect of mind lived and experienced in a body (Sheets-Johnstone, 2011). And it is clear its efficient action is disrupted in autism.

Movement normally initiates towards a goal, proceeds towards that goal with small, subtle corrective adjustments until the goal is reached, and successfully acquired. In autism, these small corrective manoeuvres are larger, both increasing and decreasing accelerative forces with exaggerated under- and over-compensations. Because purposeful movement occurs very quickly, usually under one second, these small corrections are not noticeable by the human eye, but they are evident using high-precision motion capture with subsecond kinematic analysis. And they tell us that, in autism spectrum disorders (ASD), it is as if the trajectory of the movement is always a surprise to the system, that it is forced to constantly monitor and correct its trajectory (Torres *et al.*, 2013). This means that, at a fundamental level, sensing and engaging the environment in small, purposeful acts is disturbed (Brincker and Torres, 2013; Whyatt and Craig, 2013).

This 'autism motor signature' (Anzulewicz, Sobota and Delafield-Butt, 2016; Parma and De Marchena, 2016) is characterized by a velocity and acceleration–deceleration profile that is regularly adjusted with an increase in velocity at the end of a movement when normally it would reduce to a near-zero at its goal. From a traditional cognitive perspective, this simply implies a functional error in motor execution with disturbance to the sensory processing monitoring it. In this standard paradigm, conscious experience is not thought to be entwined with simple motor control, but as something above it that instructs a motor command to deliver an automated, non-conscious execution of a motor program. In contrast, panpsychism aligns with an embodied and ecological psychology that recognizes action as fundamental to psychological construction, and the composition of consciousness (Reed, 1996; Trevarthen and Delafield-Butt, 2017).

If we follow this idea that 'fundamental drops' of mind–matter experience are the ontological units on which experience is built, we can see how simple action units can begin to compose the more complex conscious experiences of normal adult life. These units ought to become an important, central unit of analysis for understanding the psychophysical composition of any particular organism of study, too, offering a deeper perspective on experience as embodied and enactive (Gangopadhyay and Kiverstein, 2009; Stewart, Gapenne and Di Paolo, 2011; Varela, Thompson and Rosch, 1991). It can take theory of the role of the body in mind further than even the very welcome gains of embodiment in cognitive science and in enactivism (De Jaegher and Di Paolo, 2007), especially for autism research (De Jaegher, 2013; Grohmann, 2017; Roberts, Krueger and Glackin, 2019). Panpsychism deepens the shift in emphasis advanced by the extended, enactive, embodied, and emotional view of mind (Clark, 2008; Newen, Bruin and Gallagher, 2018), into a richer embodiment of lived experience, structured by a nested ontology of moments, or occasions, of experience.

2. The Ontological Unit of Purposeful Agent Action

Panpsychism teaches us that the apparent continuous stream of experience is actually composed of discreet elements of experience that have a particular regular and lawful form of process within them. These mind–body elements incorporate the dual aspect of our experience, with its private mental side and public physical side, into a singular element, or occasion. This is the ontological unit that Whitehead (1929) defined at length in his technical opus, *Process and Reality*.

2.1. Goal-directedness in mind

Whitehead's 'actual occasion' is defined as the critical element in his 'philosophy of organism' (*ibid.*). It explains goal-directed processes as ontological universals, prehending and acting to achieve an intended future, a subjective aim. Whitehead's ontological unit senses, integrates, and acts. It is very similar, if not identical, to what psychology identifies as a goal-directed action (Delafield-Butt, 2014). The juxtaposition of the ontological unit and goal-directed action defines more precisely how embodied subjective experience can be composed by complexes of discreet actions of the body in everyday life.

Each goal-directed action obtains data in the *specious* present for a *prospective* purpose. These units integrate and reintegrate new data into a single act to achieve a goal; guided by the subject's aim, the intention of the agent determines the course of the integration and the act. These units are individually discreet and serve as building blocks on which more complex experience can be built. In practice these mind–matter building blocks are interwoven to form 'fluid', coherent wholes. They can be serially organized and nested within more complex, far-reaching projects to serve ambitions for a distal future. For example, reaching to grasp a cup of coffee may flow fluidly into a next step, a next occasion, of drawing it up to the mouth for drinking. Such serially organized action can become integrated and chained together to serve greater, more spatio-temporally distal goals, such as energizing oneself in the moment to do well in an exam in ten minutes, that will altogether serve an ambition for a degree in the future (Delafield-Butt and Gangopadhyay, 2013; Pezzulo and Castelfranchi, 2009). Each atomic piece nests together to present actions and their experience as continuous and flowing (Whitehead, 1929, p. 317).

2.2. Actions units are mind–matter units

Each unit is a 'fundamental drop of experience' (*ibid.*, p. 28), both matter and mind together. Each action unit involves a mental pole, which is the perceptual and intentional part, together with a physical pole, which is the acting out of the subject into the material of action. Experience as we know it, according to Whitehead, is a society of these actual occasions, one succeeding the next and interrelating with others. By understanding this unit as an 'atom' of experience, we place ourselves in a better position to understand how this part contributes to the integrated experience of the whole person.

For a more detailed description of the ontological unit in relation to sensorimotor control, see Delafield-Butt (2014).

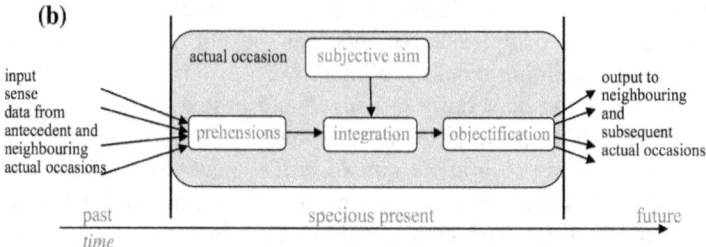

Figure 1. Schematic of the ontological unit. Sense-data is prehended and integrated until satisfaction is reached, guided by the subjective aim. The final act of the unit is an objectification of itself into neighbouring and subsequent actual occasions. The comparable components between the ontological unit and a unit of action are, respectively, (i) prehending and sensing, (ii) integrating and also integrating, or comparing, (iii) subjective aim and goal, and (iv) objectification and goal acquisition. In both cases, successful acquisition of the aim of the unit yields satisfaction, the aim being reached, and the unit perishes, its effects now a part of the foundation for future events. Adapted from Delafield-Butt (2014).

3. Autistic Disturbance to Ontological Units of Agent Action

According to the system set out above, every unit of experience requires a goal, a subjective aim. But in autism, that goal can be unclear, with an uncertain path from the present moment to the goal. This makes the present moment also unclear, and uncertain. Adjacent events, such as sounds from passing cars or questions from social partners, are perceived, but may not readily be subsumed into the present moment to make integrated sense of it. Autistic disruption to the ontological unit can be characterized by disruption to the subsecond-by-subsecond prospective integration of the senses and adaptive adjustment of motor impulses to serve the immediate subjective aim, or purpose of the act in movement.

Conscious perceptions and their integration in time to serve the purpose of the present moment may be thwarted. The senses of the outside world through vision, hearing, smell, or touch may be heightened, decontextualized, and surprising (Bogdashina, 2003; Feldman *et al.*, 2018; Jones *et al.*, 2009; Robertson and Simmons, 2015). Or, in other cases they may be attenuated and imperceptible (Ben-Sasson *et al.*, 2009). The same is true of perception of the body's vital states (Trevisan, Parker and McPartland, 2021), and probably also of the proprioceptive system that monitors the body-in-movement through direct sense of the skeletomusculature (Torres *et al.*, 2013; Torres and Denisova, 2016). In each of these cases, perception of events appears to lack context, they are not integrated into the occasion at hand, and so can be felt as surprising or hostile, or missed altogether. Surprises overstimulate a sensitive and volatile arousal system, leading to emotional reactions (Delafield-Butt, Dunbar and Trevarthen, 2021). This decontextualized, unintegrated autistic perception and capacity for adaptive response can become overwhelming, destabilizing the integrative and purposeful fluidity of the moment as it seeks satisfaction in adaptive, integrative action.

One result of this disturbance to the present moment may be a compensatory proximal spatio-temporal focus, i.e. a focus that is closer to the present moment, allowing that moment to be understood and controlled, both in mind and in body. This proximal focus can be seen in the stereotypes common in some children with autism, such as hand flapping. The regular, repetitive nature of it produces a degree of coherence within contiguous actions, a flap forward and flap back. By keeping each unit, each flap, contiguous with the other, the action runs continuously seamlessly in a known, predictable manner. This can give a sense of coherence, certainty, and calm.

Such proximal focus has advantages and disadvantages. On the negative side, it can mean restricted interests early in life that thwart learning about the varieties of experiences possible in the world, and this can compromise development and learning. It can also disrupt intra-personal coherence of basic states of being as moving-with-feeling in self-awareness, affecting internal subjective coherence of consciousness, and challenging one's capacity to self-regulate arousal and interest, especially in communication with others (*ibid.*). And on the positive side, it's temporal discontinuity with a typical linear sense of time can present new modes of thought that allow special intelligences to develop.

4. Modes of Thought in Autism

Temporal extension is what child psychologist Margaret Donaldson (1992) called the 'line mode' of thought. It is made in active participation through events in time that themselves are structured with an aim, or purpose (Delafield-Butt and Trevarthen, 2013; Trevarthen and Delafield-Butt, 2013b). And although apparently continuous and fluid, each of life's daily tasks itself presents as a discreet unit, and is composed of units (Delafield-Butt, 2018). For example, in my case getting out of bed requires knowing where the coffee is and the state of readiness of the stove-top machine, then the action of moving out of bed, opening the bedroom door, walking through the corridor (quietly, so as not to disturb the children), navigating the cats (also anticipating the kitchen routine), then preparing the machine and coffee for making coffee, and food for the cats. This whole effort is organized to achieve a goal — a cup of coffee — which itself is the first step in starting my day, which itself will be composed of similar projects: breakfast, readying the children for school, working on a paper, teaching a class, attending meetings, each meeting of which is a step along a protracted, multi-person social project with a goal of its own, and to which I contribute my conscious intelligence, energy, and skills. All of this is extensive and organized in 'future' time. Its actions organized to serve the future goals of the subject in action in the present moment.

In autism, this line mode of thought and the plans that it enables is disturbed (Delafield-Butt and Gangopadhyay, 2013; Trevarthen and Delafield-Butt, 2013a). But it is not only thought that is disturbed, it is the basic element of consciousness on which thought rests that is disturbed. The morning ritual of coffee proceeds linearly from bed to kitchen to coffee-making, and its final, beautiful result. As a neurotypical person, the coffee is in mind from the beginning, and each step along the way falls in naturally to bring me from my present moment without coffee, informed and structured by implicit knowledge of the steps required to get there, how they can be carried out, and approximately how each step will feel, to bring me to the goal efficiently. This future-oriented organization of knowledge that structures conscious experience of a living person in a body and that occupies and navigates space-time is not organized with the same cohesive, prospective, and linear manner in autism. Rather, all things may come at once.

As a result, memories of experiences may not be contextualized into a line extensive in time, but as discreet events decontextualized and available as objects. The self-identifying Asperger's entrepreneur Elon Musk describes his mental 'vector space' with ready access to a store of memories at any one moment (Musk, 2021). This is not dissimilar to the experience of Naoki Higashida (2007), a non-verbal autistic boy who describes his lack of experience of time:

> Inside my head there really isn't such a big difference between what I was told just now, and what I heard a long, long time ago. So, I do understand things, but my way of remembering them works differently

> from everyone else's. I imagine a normal person's memory is arranged continuously like a line. My memory, however, is more like a pool of dots. (pp. 23–4)

Naoki's pool of dots, although hard to contextualize into the present moment with its particular goal in mind, nevertheless can allow for powerful abstract knowledge without the restrictions of lived linearity.

The same disruption found at the level of subsecond prospective action may be present up and down the levels of organization of agent action. Fragmentation of normal linear sequences into discreet features is observable down the scale with observation of perceptual incoherence identified under the weak central coherence hypothesis (Happé and Frith, 2006). Weak central coherence may be one manifestation of a more fundamental, core issue of ontological integration.

The same ontological characteristics appear to radiate up levels of action organization in autism, keeping the same form. For example, a large project spanning many tens of minutes, such as walking to the swimming pool for a swim, may present as a project with many uncertainties. Each step in the process — preparing one's clothes for the pool, leaving the house, walking down the street, making each change of direction appropriately while walking to arrive at the pool, entering the styles, changing, and entering the water — can be presented in autism at the start of the project as a cacophony of individual elements without sequence, leaving the individual uncertain of how each step can be put together linearly in time. This uncertainty creates anxieties, elevates arousal, and generates resistance to start the project altogether. The parallel with a single act of movement, reaching-to-grasp, for example, is clear. Each project — reach-to-grasp or walk-to-pool — presents a linear sequence of coordinated actions that must be known and performed at a suitable moment in time — over many tens of milliseconds in the case of reaching-to-grasp, and over many tens of minutes in the case of walking to the pool. Thus, disruption to the ontological structure appears the same, irrespective of its spatio-temporal scale.

In this way, conscious experience itself is different in autism from neurotypical individuals. And although autism is not considered a clinical disorder of consciousness (American Psychiatric Association, 2013; World Health Organization, 2019), conscious content becomes differently organized and conscious experience altered.

5. Conclusions

Understanding the mind–body relation from the position of panpsychism lends a different structure with which to frame, analyse, and understand embodied experience. That frame allows empirical analysis of mind–body data as subsecond kinematics of goal-directed action to be interpreted *vis-à-vis* subjective experience. Without this framework, the structure of embodied experience and the role of sensorimotor goal-directedness as fundamental to it can be missed. Instead, one may be left with a flat, mechanical framework that views mind as a machine, presenting specialized modules for awareness, planning, and execution of motor programs. The deterministic model may make good sense of the same empirical data, but it also leads discovery into its own predetermined areas of interest in cortically-mediated modules, each of which is without conscious experience on its own.

Panpsychism, on the other hand, casts a different light on the matter altogether, one where mind, purpose, and consciousness are present at all levels of the living organism. Defining these levels, how they structure experience, and how they may be disrupted in cases of psychopathology such as autism is the work of good science to follow, with exactly the same empirical and experimental basis as what has come before, but with new direction and a deeper sense of the role of agency in psychophysiological organization and health.

This revision of our metaphysical assumptions may ultimately help to identify the neuropsychological systems that compose mind–body units of conscious experience, with a deeper appreciation of brainstem integrations of neuromotor control and affective evaluations as a source of primary conscious experience (Damasio, 2010; Panksepp and Biven, 2012; Solms, 2021; Solms and Panksepp, 2012; Vandekerckhove and Panksepp, 2009; 2011). We are beginning to understand that brainstem-mediated primary consciousness, responsible for the sensorimotor integrations in skilled action, and its affective evaluation, is disrupted in autism from early life (Bosco *et al.*, 2019; Dadalko and Travers, 2018; Delafield-Butt, Dunbar and Trevarthen, 2021). This is an area ripe for investigation, informed by a deeper understanding of the metaphysics of embodiment, and how it may be organized in autism.

Acknowledgments

I am grateful to Penelope Dunbar for extensive discussion and insight into autism. And to Pauline Phemister, the late Wendy Wheeler, and the RSE Organisms and Their Choices group for discussion of panpsychism and agency in nature, and especially also to Ross Stein and Pierfrancesco Basile on Whitehead in earlier, lengthy discussion and whom I have never properly thanked. Finally, to Philip Goff who has reinvigorated panpsychism and its importance to science, not just to consciousness studies. There is more work to do.

References

American Psychiatric Association (2013) *Diagnostic and Statistical Manual of Mental Disorders (DSM-5)*, 5th ed., Washington, DC: American Psychiatric Association.

Anzulewicz, A., Sobota, K. & Delafield-Butt, J.T. (2016) Toward the autism motor signature: Gesture patterns during smart tablet gameplay identify children with autism, *Scientific Reports*, **6**.

Basile, P. (2010) It must be true — but how can it be? Some remarks on panpsychism and mental composition, in Basile, P., Kiverstein, J. & Phemister, P. (eds.) *The Metaphysics of Consciousness*, pp. 93–112, Cambridge: Cambridge University Press.

Ben-Sasson, A., Hen, L., Fluss, R., Cermak, S.A., Engel-Yeger, B. & Gal, E. (2009) A meta-analysis of sensory modulation symptoms in individuals with autism spectrum disorders, *Journal of Autism & Developmental Disorders*, **39** (1), pp. 1–11.

Bernstein, N.A. (1967) *The Co-ordination and Regulation of Movements*, Oxford: Pergamon Press.

Bogdashina, O. (2003) *Sensory Perceptual Issues in Autism and Aspergers Syndrome: Different Sensory Experiences, Different Perceptual Worlds*, London: Jessica Kingsley Publishers.

Bosco, P., Giuliano, A., Delafield-Butt, J., Muratori, F., Calderoni, S. & Retico, A. (2019) Brainstem enlargement in preschool children with autism: Results from an intermethod agreement study of segmentation algorithms, *Human Brain Mapping*, **40** (1), pp. 7–19.

Brincker, M. & Torres, E. (2013) Noise from the periphery in autism, *Frontiers in Integrative Neuroscience*, **7**, art. 34.

Calvo, P., Gagliano, M., Souza, G.M. & Trewavas, A. (2019) Plants are intelligent, here's how, *Annals of Botany*, **125** (1), pp. 11–28.

Casartelli, L., Cesareo, A., Biffi, E., Campione, G.C., Villa, L., Molteni, M. & Sinigaglia, C. (2020) Vitality form expression in autism, *Scientific Reports*, **10** (1), 17182.

Chalmers, D.J. (1995) Facing up to the problem of consciousness, *Journal of Consciousness Studies*, **2** (3), pp. 200–219.

Clark, A. (2008) *Supersizing the Mind: Embodiment, Action, and Cognitive Extension*, Oxford: Oxford University Press.

Cook, J.L., Blakemore, S.J. & Press, C. (2013) Atypical basic movement kinematics in autism spectrum conditions, *Brain*, **136** (Pt 9), pp. 2816–2824.

Dadalko, O.I. & Travers, B.G. (2018) Evidence for brainstem contributions to autism spectrum disorders, *Frontiers in Integrative Neuroscience*, **12**, art. 47.

Damasio, A. (2010) *Self Comes to Mind: Constructing the Conscious Brain*, New York: Pantheon Books.

De Jaegher, H. (2013) Embodiment and sense-making in autism, *Frontiers in Integrative Neuroscience*, **7**. doi: 10.3389/fnint.2013.00015.

De Jaegher, H. & Di Paolo, E. (2007) Participatory sense-making, *Phenomenology and the Cognitive Sciences*, **6**, pp. 485–507.

Delafield-Butt, J.T. (2007) Towards a process ontology of organism: Explaining behaviour in a cell, in Kelly, T. & Dibben, M. (eds.) *Applied Process Thought: Frontiers of Theory and Research*, Paris: Ontos Verlag.

Delafield-Butt, J.T. (2008) Biology, in Weber, M., Seibt, J. & Rescher, N. (eds.) *Handbook of Whiteheadian Process Thought*, Paris: Ontos Verlag.

Delafield-Butt, J.T. (2014) Process and action: Whitehead's ontological units and perceptuomotor control units, in Koutroufinis, S. (ed.) *Life and Process*, pp. 133–156, Berlin: De Gruyter Ontos.

Delafield-Butt, J. (2018) The emotional and embodied nature of human understanding: Sharing narratives of meaning, in Trevarthen, C., Delafield-Butt, J. & Dunlop, A.-W. (eds.) *The Child's Curriculum: Working with the Natural Voices of Young Children*, Oxford: Oxford University Press.

Delafield-Butt, J.T., Pepping, G.-J., McCaig, C.D. & Lee, D.N. (2012) Prospective guidance in a free-swimming cell, *Biological Cybernetics*, **106**, pp. 283–293.

Delafield-Butt, J.T. & Gangopadhyay, N. (2013) Sensorimotor intentionality: The origins of intentionality in prospective agent action, *Developmental Review*, **33** (4), pp. 399–425.

Delafield-Butt, J.T. & Trevarthen, C. (2013) Theories of the development of human communication, in Cobley, P. & Schultz, P. (eds.) *Theories and Models of Communication*, pp. 199–222, Berlin: De Gruyter Mouton.

Delafield-Butt, J.T. & Trevarthen, C. (2017) On the brainstem origin of autism: Disruption to movements of the primary self, in Torres, E. & Whyatt, C. (eds.) *Autism: The Movement Sensing Perspective*, Boca Raton, FL: Taylor & Francis CRC Press.

Delafield-Butt, J.T., Dunbar, P. & Trevarthen, C. (2021) Disruption to embodiment in autism, and its repair, in Papaneophytou, N. & Das, U. (eds.) *Emerging Programs for Autism Spectrum Disorder*, Amsterdam: Elsevier Academic Press.

Donaldson, M. (1992) *Human Minds: An Exploration*, London: Allen Lane.

Feldman, J.I., Dunham, K., Cassidy, M., Wallace, M.T., Liu, Y. & Woynaroski, T.G. (2018) Audiovisual multisensory integration in individuals with autism spectrum disorder: A systematic review and meta-analysis, *Neuroscience & Biobehavioral Reviews*, **95**, pp. 220–234.

Fournier, K.A., Hass, C.J., Naik, S.K., Lodha, N. & Cauraugh, J.H. (2010) Motor coordination in autism spectrum disorders: A synthesis and meta-analysis, *Journal of Autism & Developmental Disorders*, **40** (10), pp. 1227–1240.

Gangopadhyay, N. & Kiverstein, J. (2009) Enactivism and the unity of perception and action, *Topoi*, **28** (1), pp. 63–73.

Glazebrook, C.M., Elliott, D. & Lyons, J. (2006) A kinematic analysis of how young adults with and without autism plan and control goal-directed movements, *Motor Control*, **10** (3), pp. 244–264.

Goff, P. (2019) *Galileo's Error: Foundations for a New Science of Consciousness*, New York: Pantheon Books.

Grohmann, T.D.A. (2017) A phenomenological account of sensorimotor difficulties in autism: Intentionality, movement, and proprioception, *Psychopathology*, **50** (6), pp. 408–415.

Happé, F. & Frith, U. (2006) The weak coherence account: Detail-focused cognitive style in autism spectrum disorders, *Journal of Autism and Developmental Disorders*, **36** (1), pp. 5–25.

Higashida, N. (2007) *The Reason I Jump*, London: Hodder & Stoughton.

Jones, C.R.G., Happé, F., Baird, G., Simonoff, E., Marsden, A.J.S., Tregay, J., Phillips, R.J., Goswami, U., Thomson, J.M. & Charman, T. (2009) Auditory discrimination and auditory sensory behaviours in autism spectrum disorders, *Neuropsychologia*, **47**, pp. 2850–2858.

Lee, D.N. (2000) Prospective guidance of movement, *International Journal of Psychology*, **35** (3–4), p. 186.

Leibniz, G. (1716) *La Monadologie*.

Mari, M., Castiello, U., Marks, D., Marraffa, C. & Prior, M. (2003) The reach-to-grasp movement in children with autism spectrum disorder, *Philosophical Transactions of the Royal Society B: Biological Sciences*, **358**, pp. 393–403.

Musk, E. (2021) Interview, *Youtube*, [Online], https://www.youtube.com/watch?v=vXuWbPRgflg.

Newen, A., Bruin, L.D. & Gallagher, S. (eds.) (2018) *The Oxford Handbook of 4E Cognition*, Oxford: Oxford University Press.

Panksepp, J. & Biven, L. (2012) *The Archaeology of Mind: Neuroevolutionary Origins of Human Emotions*, New York: Norton.

Parma, V. & de Marchena, A.B. (2016) Motor signatures in autism spectrum disorder: The importance of variability, *Journal of Neurophysiology*, **115** (3), pp. 1081–1084.

Pezzulo, G. & Castelfranchi, C. (2009) Thinking as the control of imagination: A conceptual framework for goal-directed systems, *Psychological Research*, **73**, pp. 559–577.

Reed, E.S. (1996) *Encountering the World: Toward an Ecological Psychology*, Oxford: Oxford University Press.

Roberts, T., Krueger, J. & Glackin, S. (2019) Psychiatry beyond the brain: Externalism, mental health, and autistic spectrum disorder, *Philosophy, Psychiatry & Psychology*, **26**, E-51.

Robertson, A. & Simmons, D. (2015) The sensory experiences of adults with autism spectrum disorder: A qualitative analysis, *Perception*, **44**, pp. 569–586.

Schmitz, C., Martineau, J., Barthélémy, C. & Assaiante, C. (2003) Motor control and children with autism: Deficit of anticipatory function?, *Neuroscience Letters*, **348**, pp. 17–20.

Sheets-Johnstone, M. (2011) *The Primacy of Movement*, 2nd ed., New York: John Benjamins.

Solms, M. (2021) *The Hidden Spring: A Journey to the Source of Consciousness*, London: Profile Books.

Solms, M. & Panksepp, J. (2012) The 'id' knows more than the 'ego' admits: Neuropsychoanalytic and primal consciousness perspectives on the interface between affective and cognitive neuroscience, *Brain Science*, **2**, pp. 147–174.

Stewart, J., Gapenne, O. & Di Paolo, E. (eds.) (2011) *Enaction: Toward a New Paradigm for Cognitive Science*, Cambridge, MA: MIT Press.

Strawson, G. (2006) Realistic monism: Why physicalism entails panpsychism, in Freeman, A. (ed.) *Consciousness and Its Place in Nature*, pp. 3–31, Exeter: Imprint Academic.

Strawson, G. (2017) Mind and being: The primacy of panpsychism, in Brüntrup, G. & Jaskolla, L. (eds.) *Panpsychism: Contemporary Perspectives*, Oxford: Oxford University Press.

Torres, E.B., Brincker, M., Isenhower, R.W., Yanovich, P., Stigler, K.A., Nurnberger, J.I., Metaxas, D.N. & José, J.V. (2013) Autism: The micromovement perspective, *Frontiers in Integrative Neuroscience*, **7**, art. 32.

Torres, E.B. & Denisova, K. (2016) Motor noise is rich signal in autism research and pharmacological treatments, *Scientific Reports*, **6**, 37422.

Trevarthen, C. (1984) How control of movement develops, in Whiting, H.T.A. (ed.) *Human Motor Actions: Bernstein Reassessed*, pp. 223–261, Amsterdam: Elsevier (North Holland).

Trevarthen, C. & Delafield-Butt, J.T. (2013a) Autism as a developmental disorder in intentional movement and affective engagement, *Frontiers in Integrative Neuroscience*, **7**, art. 49.

Trevarthen, C. & Delafield-Butt, J.T. (2013b) Biology of shared meaning and language development: Regulating the life of narratives, in Legerstee, M., Haley, D. & Bornstein, M. (eds.) *The Infant Mind: Origins of the Social Brain*, pp. 167–199, New York: Guildford Press.

Trevarthen, C. & Delafield-Butt, J.T. (2017) Development of consciousness, in Hopkins, B., Geangu, E. & Linkenauger, S. (eds.) *Cambridge Encyclopedia of Child Development*, pp. 821–835, Cambridge: Cambridge University Press.

Trevisan, D.A., Parker, T. & McPartland, J.C. (2021) First-hand accounts of interoceptive difficulties in autistic adults, *Journal of Autism & Developmental Disorders*, preprint. doi: 10.1007/s10803-020-04811-x.

Trewavas, A., Baluška, F., Mancuso, S. & Calvo, P. (2020) Consciousness facilitates plant behavior, *Trends in Plant Science*, **25** (3), pp. 216–217.

Vandekerckhove, M. & Panksepp, J. (2009) The flow of anoetic to noetic and autonoetic consciousness: A vision of unknowing (anoetic) and knowing (noetic) consciousness in the remembrance of things past and imagined futures, *Consciousness and Cognition*, **18**, pp. 1018–1028.

Vandekerckhove, M. & Panksepp, J. (2011) A neurocognitive theory of higher mental emergence: From anoetic affective experiences to noetic knowledge and autonoetic awareness, *Neuroscience & Biobehavioral Reviews*, **35** (9), pp. 2017–2025.

Varela, F.J., Thompson, E.T. & Rosch, E. (1991) *The Embodied Mind*, Cambridge, MA: MIT Press.

von Hofsten, C. (1993) Prospective control — a basic aspect of action development, *Human Development*, **36**, pp. 253–270.

Whitehead, A.N. (1929) *Process and Reality*, New York: Macmillan.

Whyatt, C. & Craig, C. (2013) Sensory-motor problems in autism, *Frontiers in Integrative Neuroscience*, **7**. doi: 10.3389/fnint.2013.00051.

World Health Organization (2019) *International Statistical Classification of Diseases and Related Health Problems (ICD-11)*, 11th ed., Geneva: World Health Organization.

Robert Prentner[1]

Dr Goff, Tear Down This Wall!

The Interface Theory of Perception and the Science of Consciousness

Abstract: Here I outline an alternative to panpsychism, premised on the interface theory of perception, that too subscribes to a 'post-Galilean' research programme. However, interface theorists disagree along several lines. (1) They note that Galileo's distinction should be replaced by a truly non-dual account, referring to a difference of degree only. (2) They highly appreciate the role of mathematics, in particular when it comes to actually engaging scientifically with consciousness. Some notable features of the interface theory are its scepticism towards our epistemic capacities and its rejection of the existence of a public, physical reality. In addition, some interface theorists further employ a thin concept of 'conscious agency' to ground their theory. The interface theory leaves open many of the problems of consciousness science (e.g. what is a 'self'?) as questions for further (scientific, mathematical) research.

1. Introduction

Being a non-dualist is hard. Really hard. On the surface, it appears that simply rejecting the distinction into the categories of 'mind' and 'matter' suffices for being a non-dualist. But this is not the whole

Correspondence:
Email: robert.prentner@amcs.science

---

[1] Center for the Future Mind, Florida Atlantic University, Boca Raton, FL, USA.

story. Most often, what happens is that dualisms of some sort are smuggled into one's theory through the backdoor. Daniel Dennett famously declared people who reformulated Cartesian assumptions within a materialistic theory (e.g. that 'consciousness' is to be found in a single unified centre of physical processing) to be 'Cartesian materialists' (Dennett, 1991). Philip Goff, on the other hand, calls for a 'post-Galilean science of consciousness' that rests on the assumptions of 'realism about consciousness', 'empiricism', and 'non-dualism' (p. 174).[2] I applaud Goff for this bold declaration.

Yet, the devil is in the details. For example, his clear-cut distinction into categorical ('intrinsic') and dispositional ('structural') might seem to be just a 'dualism in disguise'.[3] A lot of this is motivated by an 'error' made by Galileo Galilei, who distinguished between the mathematically describable properties of public physical objects and the subjective properties of consciousness. This might seem an act of ignorance on the part of Galileo. However, one of the main reasons that science has been so successful is precisely because it excluded conscious experience right from the start.

But what to do if the goal is now to construct a science of consciousness? I appreciate the point made by Arthur Eddington and others that physical knowledge is ultimately only about the relation of pointer readings. This kind of knowledge is incomplete and leaves out any knowledge about the intrinsic nature of things. Goff's suggestion: 'Plug the hole with consciousness' (p. 132).

However, if one wants to stick to a truly non-dual solution, then adopting something similar to the following might be helpful (if not necessary). I would like to regard both types of statements, the ones about public physical[4] objects and the ones about the subjective properties of consciousness, as differing by degree and not category.

---

[2] When not otherwise indicated, page numbers pertain to Goff (2019). I also use 'panpsychism' to refer to the specific position that is (I believe) held by Goff. It has been pointed out (e.g. by Skrbina, 2005) that panpsychism should be understood more broadly, in terms of a framework that encompasses a variety of views that all agree on mind being ubiquitous and fundamental. Given this more liberal reading, the view presented here qualifies as an instance of a 'pro-panpsychism'.

[3] Though to be fair, Goff's self-understanding is explicitly non-dual (pp. 135f.). But when he says that 'physical properties *are themselves* forms of consciousness' (emphasis PG), isn't this rather 'idealism in disguise'?

[4] Note that I intend 'physical' to mean: void of any form of experience; 'fully external', so to say (Prentner, 2018). There exist different uses of the word 'physical' (*cf.* Strawson, 2006; Stoljar, 2017) about which I do not speak here.

Another, related, issue pertains to the role of mathematics for the science of consciousness — and I explicitly encourage the use of mathematics to study consciousness, rather than pointing to its limitations. True enough, mathematical knowledge might be incomplete in the sense that it only ever captures mere structure. However, I claim that it is precisely mathematics that lets us understand how some (apparently) 'objective' structures could have emerged in the first place.[5]

2. The Interface Theory of Perception

The account discussed here is premised on the idea that we have limited insight into the 'true nature of reality', eventually tearing down the categorical wall that separates 'mind' from 'matter'. I believe that some of the crucial questions that we encounter along the way could be answered with the help of mathematical concepts such as 'structure preserving maps', 'information geometry', or 'higher categories'. It is out of the scope of this brief discussion to work out an answer to any of these questions. Still, I hope to invoke a sense of urgency to deal with them but also the conviction that this can be done in principle.

One defining property of consciousness expresses an epistemic limitation: facts about consciousness are accessible (at least in part) only from what is called a 'first-person perspective' (Chalmers, 2004). No amount of scientific, objective knowledge seems to make it intelligible why experience appears to have certain properties (e.g. qualia) that we attribute to them based on this first-person perspective. Goff (pp. 69ff.) makes us aware of this when discussing Frank Jackson's knowledge argument (Jackson, 1982; 1986). Whether or not one agrees with any ontological conclusion drawn from this, it poses some interesting epistemic challenges.

The first-person perspective has been subject to a lot of controversy, with some people shedding doubt on the coherence of the very concept of unmediated forms of knowledge (or 'givens') derivable from it

---

[5] Without having a firm, mathematical handle on what 'structure' is supposed to mean, the questions of its emergence are meaningless. The idea is that 'objective' properties arise as 'transformation invariants' of experiential processes; for related ideas see Robert Nozick's (2001) *Invariances*, Alfred N. Whitehead's 'method of extensive abstraction' (Whitehead, 2015), or the 'objective idealisms' discussed in Atmanspacher (2020).

(e.g. Sellars, 1963; Dennett, 1991; Metzinger, 2003; Prentner, 2019). Recently, there has been much interest in the question how a creature (or, its brain) comes up with a model of itself (Metzinger, 2003). Some people have even speculated that this could explain how consciousness is nothing but the brain's self-attribution of a private inner life (Graziano and Kastner, 2011; Frankish, 2017). By contrast, less attention has been directed to the complementary question of how a creature comes up with a model of its world, thereby attributing a public, physical existence to it. The catch is that such a model would not need to resemble the 'objective' state of the world. It would not even need to explicitly encode a *specific* worldly state.[6] Basic evolutionary thinking merely tells us that all a creature must do is to act successfully in its world, and its internal (cognitive) architecture must be such that this is possible (Mark, Marion and Hoffman, 2010; Hoffman, Singh and Prakash, 2015a; Guez *et al.*, 2019; Prakash *et al.*, 2020; in press) — metaphorically speaking: it needs to have an interface that lets it deal with its world. Truthful representation takes the backseat.

We are nothing but such creatures. There are two consequences of this if one is willing to think it through radically. First, it extends to empirical knowledge as such, *dispensing with the availability of a physical 'ground truth'*: space-time is the species-specific data format of our interface (Hoffman, Singh and Prakash, 2015a), objects are error-corrected representations of fitness consequences (Fields *et al.*, 2017), and even quantum mechanics ultimately plays out on this interface (Prakash, 2019). Second, there is *no in-principle distinction between 'public' and 'private'* forms of knowledge, since both refer to procedures relative to our interface.

What is left might be a mere collection of processes that relate perceptions to actions. This additional claim seems quite natural, since it is premised on the assumption that we all have experience and (try to) do something about it — arguably, nothing could be more basic than that. Since this basic assumption has an active connotation where perceptions have (often unintended) consequences, one might refer to

---

[6] It turns out that experiences need not be in a one-to-one correspondence to (assumed) physical states but could correspond to probabilistic combinations thereof; *cf.* the example in Prakash *et al.* (in press).

these processes as 'agents'[7] (Hoffman and Prakash, 2014, Fields *et al.*, 2017). The main task is then to show how one could recover the appearance of a public, physical world (inhabited by non-physical selves) from interactions between such agents.[8]

3. What is 'Really Real'?

Imagine you are immersed in a virtual reality version of *Grand Theft Auto* (GTA). You are interacting with computer-generated content such as steering wheels, cars, policemen, and pedestrians. There is an 'external reality', namely the computer that governs the objects that you see, but its description does not involve steering wheels, cars, policemen, or pedestrians. Trying to understand the behaviour of the computer (its program) using these categories would be foolish. Lines of computer code do not resemble steering wheels. You cannot even rely on causality. Intervening with the steering wheel might have the observable consequence of driving into a wall (all other things being equal), but steering wheels do not cause crashes in GTA, the program does. Perhaps you still think that using steering wheels and other types of interventions lets you figure out what program the computer is running. Good luck.

Moreover, GTA employs algorithms (a so-called 'physics-engine') that simulate the realistic behaviour of objects, even those you do not interact with. For example, objects that land in water will create ripples and wave patterns in response. Whether you are successful at the game has nothing to do with your grasp of the nature of these underlying mechanisms but only with your ability to manipulate virtual contents in a way that lets you score points.

Is simulated physics unreal? What is 'really real'? And how could we find out? Rather than pondering the true nature of this external reality (is it physical? Is it a mere simulation by a highly-evolved alien species? Has it something to do with consciousness?), it might initially be a better strategy to ask about the principles and

---

[7] Note that this usage of the word 'agent' refers to a very thin notion, unlike a stronger notion of 'agency' that carries, say, connotations of embodiment or environmental embedding (Prentner and Fields, 2019), or spatio-temporal realization.

[8] Note that this is not *subjective* idealism, if one believes in the objective reality of these processes; also note that the idea is not a variant of sense-datum theory, even though it might appear as such when interface theorists speak of 'perceptual icons'. There are only (perceptual) experiences, and not 'experience-objects' that we are aware of (Hoffman, Singh and Prakash, 2015b).

mechanisms underlying its appearance as being *stable* (steering wheels generally do not turn into gear shifters), *consistent* (killing pedestrians leads to being chased by policemen), and *law-like* (objects that fall into water create ripples) — properties we typically attribute to the world around us. Philosophically speaking, we have shifted from a transcendent question about the nature of reality to a transcendental one about the conditions of possibility of its experience (Kant, 1998).

Of course, computer games and virtual reality simulations are not the real world, and the metaphor above breaks down eventually. In the following, I therefore wish to translate the question about its appearance into the language of the interface theory of perception. This will intentionally be done non-mathematically.[9]

The interface theory assumes a mapping P from an external world (the computer running the program) to the contents of our perceptions that comprise steering wheels and pedestrians. Likewise, when we see pedestrians, we can decide to turn the steering wheel to the left to avoid hitting them. Abstractly, this could be represented by a mapping D that connects perceptions to actions. Finally, certain actions affect (A) the external world of the computer and feed back into our perceptions at some later time (P′). For example, we might see that the steering wheel indeed moved to the left and we consequently avoided hitting the pedestrian. Note here that we are never in perfect control of the situation. While we can choose actions, those actions might or might not be successful. This is life, unfortunately. We also, strictly speaking, do not perceive these actions, but only their consequences at future times. Seeing the steering wheel turning left is a perceptual response; 'making a left turn' refers to a consequence, not the action as such. We *initially did not intend* to make a left turn, but we *learned* to execute actions that typically result in perceptions as of left turns. These are some of the basic ingredients of the theory — in a nutshell: an agent's perception is the representation of its external world, based on the consequences of past actions. A 'decision-process' mediates between perceptions and actions and could incorporate, in more sophisticated settings, things like memory, goals, or predictions.

An interesting question is now: what properties do these structures need to have in order to guarantee future perceptions to appear as if

---

[9] The interested reader is referred to Hoffman, Singh and Prakash (2015a) for mathematical details.

they were consistent with a representation of an objective world? Possible answers would specify the types of mappings and representations that make good interfaces. They would be specifications about mathematical entities, for example:

- **Stability**. The mapping from this world to our experience is relatively stable, although the mapping need not be structure-preserving in any substantial sense.[10] Otherwise we were not able to do anything useful. In particular, a system would not be able to decide on any given course of action (if steering wheels might suddenly turn into gear shifters, then there is no point in deciding to turn the wheel to the left, *cf.* Durham, 2020).
- **Consistency**. This is about action-consequences, and how they feed back into future perceptions. Future perceptions should be consistent with the actions an agent previously took (*modulo* the uncertainty whether they succeeded or not). But not all our actions have an effect, and not all effects we perceive are due to our actions. Thus, we also need to ensure:
- **Law-like behaviour**. Things that cannot be subjectively influenced should enter in a 'nice' (predictable) way into our experience. This does not mean that, just because we could predict something, there exists some objective (i.e. agent-independent) law that governs its behaviour. There can be 'laws without laws', based on predictability (Müller, 2020).

More refined models will likely capture some structural aspects of our perceptual interface, and will extend them using symmetry considerations (e.g. we perceive space as locally Euclidean; using translation invariance, we end up with a Euclidean model of space).

4. How Do Agents Agree?

So far, the discussion mainly centred around the appearance of regularities (stability, consistency, law-likeness), discussed against the background of a single agent that perceives, decides, and acts.

An entirely different question pertains to the coordination of the actions of a collection of agents. If we go back to the GTA metaphor, we now look at multiplayer games. (This is unlike the situation where, say, policemen resembled mere icons on our interfaces.) What makes

---

[10] Of course, gradual changes (e.g. over 'phylogenetic' periods of time) are possible.

coordination possible? Answering this question by postulating public physical objects, while intuitive, is explanatorily lazy. This would not answer the question, for example, in terms of a mechanism that guarantees (perhaps surprisingly) inter-agent agreement. Instead, it would refer to an underlying agent-independent reality that is somehow mirrored in the agents' experiences and serves as a common point of reference.

A motivating example is given by synaesthesia. Synaesthetes can have vastly different experiences compared to 'normal' people when encountering the same stimuli. For example, whereas I hear a middle C, a synaesthete might see, in addition, the colour blue. In the language of the interface theory, the mapping from the external world goes to very different perceptions.

An interesting question is: how different can such mappings be and still allow meaningful interaction between agents? Does this, for example, require a homomorphism between their two interfaces? Can we articulate a more refined or weaker constraint in information-theoretic terms (e.g. based on mutual information)? Could we use the category-theoretical tool of a 2-morphism — that is, a morphism that exists between morphisms (e.g. Dobson and Prentner 2021) — letting go of the assumption that there is a single same world underlying the agents' experiences?

Finally, we would not only want to compare the experiences of two (or more) interacting agents, but also ask about the status of the universe that the two observers inhabit. One famous argument in the philosophy of science proceeds via scrutinizing our knowledge (Putnam, 1975). In particular, it suggests that the success of science is best explained by the idea that scientific knowledge in fact mirrors (to some extent) an agent-independent reality:

> The main reason for believing scientific theories (at least the best-verified ones) is that... it would be a miracle if science said nothing true — or at least approximately true — about the world. The experimental confirmations of the best-established scientific theories, taken together, are evidence that we really have acquired an objective (albeit approximate and incomplete) knowledge of the natural world. (Sokal and Bricmont, 1998, p. 57)

A possible counter to this argument for 'scientific realism' revolves around the notion of 'success' and its relativity (Feyerabend, 1975). Another counter would grant science its success but reject the claim that one therefore needs to accept that our theories somehow depict an agent-independent world — or at least that this would be the best explanation for science's success. Again, the assumption of unguarded (naïve) realism strikes me as explanatorily lazy.

Of course, the philosophical literature on this is huge, and I have to limit myself to emphasizing certain key points. Typically, our best scientific theories are being evaluated against the results of measurements that single out the 'best-verified ones' (using the terminology from above). As such, the most pressing questions will pertain to measurements and how measurements between different agents are related to each other. As a caveat, note that one might wish to limit oneself to the situation where some (perhaps evolutionary formulated) criteria of success are assumed,[11] rather than spelling out an account of what amounts to 'acting successfully' — one simply stipulates that there are success criteria and that agents get feedback on whether or not they are met. Any successful interaction between agents would be reinforced, so as to produce stable, seemingly objective structures. While it is highly unlikely, following the interface theory, that such structures would correspond to anything 'out there', a certain subset could give the appearance of a stable backdrop for successful 'measurements' (i.e. repeatable, quantifiable, comparable, and consistent interactions).

This raises many open and potentially fruitful questions for the science of consciousness: how do agents decide they are both looking 'at the same thing'? How do they agree on any particular 'measurement unit' like a yard-stick? What makes the measurements replicable when conducted by different observers at different times? On the face of it, this suggests that 'the thing being measured' is the same for all observers, and that 'the measurement unit' and 'the measurement procedure' are the same for all observers at all times. But is this really the case? How can it be justified? Yet it seems that the concepts involved can be meaningfully translated into mathematical language. For example, replicability might be taken as the possibility to observe a repeatedly occurring event, defined within the context of a stable measurable space.[12]

---

[11] This renders this stance somewhat close to pragmatism, taking the notion of a 'successful interaction' as primitive.

[12] In interface language: experience exhibits the same partition over time, and the same subset of experience is 'lighting' up repeatedly.

5. Consciousness

I have speculated on necessary structural properties of procedures, understood as a sequence of perceptions and actions, relative to interfaces, which could give rise to regularities within and between agents — regularities that might ground measurements and perhaps even the appearance of an agent-independent, public world. However, I have said nothing about 'qualia' or 'what-it-is-like' to be such an agent. Since these properties take a central role in Goff's metaphysics, I would like to finally address them.

For example, it has been proposed that the basic ingredients of the universe are 'conscious agents', for which going through sequences of perception and action comes with a distinct raw feeling — it is like something to be such an agent (Hoffman and Prakash, 2014; Fields et al., 2018). Yet, one might wish to exactly state what properties typically attributed to consciousness could be recovered from such a description (Kleiner, 2020). If one is convinced that any property of consciousness (save a basic 'what-it-is-likeness' which is stipulated to begin with) could be recovered from a dynamical system of such agents, this gives rise to the following recipe:

1. Identify the particular property and formalize it.
2. Show how it could be recovered from the dynamics (of a network) of such agents.

That such a reconstruction is indeed possible has been conjectured in the form of a 'conscious agent thesis' (Hoffman and Prakash, 2014). It is the truth of this conjecture that would turn the minimal concept of a 'conscious agent' into a full-blown description of (typically human) consciousness. This leaves open many questions as to what those properties are (e.g. 'selfhood'), what the explananda precisely look like (e.g. how 'selfhood' could be formalized), and what their explanations would be (e.g. relatively stable configurations of entangled agents that share a common history). I do not see how progress on such problems could be made without mathematics.

Finally, connecting back to the original distinction of Galileo, let me briefly sketch some alternative positions. (1) Subjective and objective properties are somehow completely different. This is the option that Galileo preferred, but seemingly also Goff when he distinguishes categorically between 'intrinsic natures' and structural properties. Both Galileio's materialism and Goff's panpsychism seem to presuppose a shared, objective world. (2) There is only individual subjective experience and no shared world at all. This is the position that is traditionally associated with solipsism. (3) A position that would sit somewhere in between, where there are only the distinct interfaces of individual agents as well as something that is shared. This something is arbitrarily different from what is displayed on any of the interfaces, but it would not exist in the absence of agents.

Why should one adopt the latter position? First, it is explicitly non-dual. Whether or not this is really warranted, panpsychists often receive backlash from people who see in panpsychism a way of 'injecting' phenomenal properties into physical objects that leads to statements about conscious tables and chairs. Interface theorists arguably do not fall prey to this charge. There is no question of whether matter is 'intrinsically' conscious, since material objects are icons on interfaces. An icon is not conscious, and it does not exist in the absence of agents. Second, and more importantly, it stays close to the conventional scientific (mathematical) method: identify a theoretical primitive, give a mathematical description of it (to the extent possible), and derive stuff from there.[13] One of the largest open problems for panpsychism is to explain not just '*any old conscious experience*' but '*our* conscious experience' (Goff, 2009). The interface theory promises to tackle this problem.[14]

Acknowledgments

I thank Shanna Dobson, Chris Fields, Philip Goff, Don Hoffman, Chetan Prakash, and Jan Westerhoff for comments on an earlier draft. I am particularly indebted to Manish Singh, who helped greatly with conceiving of the overall structure of this piece.

References

Atmanspacher, H. (2020) The Pauli–Jung conjecture and its relatives: A formally augmented outline, *Open Philosophy*, **3** (1), pp. 527–549.

---

[13] There is a trade-off. Whereas panpsychism might be thought to be closer to the (established) scientific worldview, e.g. the standard model with its bottom-up ontology, the interface theory might be thought to align more closely with science's method of mathematization and the resolution of apparent conflicts (say, between mind and matter) by postulating a more fundamental level of description.

[14] Relatedly, it has been argued that the combination problem does not in fact exist when the interface theory is applied not just to the environment, but also to the observing agent (*cf.* Fields, this issue).

Chalmers, D.J. (2004) How can we construct a science of consciousness?, in Gazzaniga, M. (ed.) *The Cognitive Neurosciences III*, pp. 1111–1120, Cambridge, MA: MIT Press.

Dennett, D.C. (1991) *Consciousness Explained*, Boston, MA: Little, Brown & Company.

Dobson, S. & Prentner, R. (2021) Perfectoid diamonds and n-awareness: A meta-model of subjective experience, *ArXiv*, [Online], https://arxiv.org/abs/2102.07620.

Durham, I. (2020) A formal model for adaptive free choice in complex systems, *Entropy*, **22** (5), art. 568.

Feyerabend, P.K. (1975) *Against Method*, New York: New Left Books.

Fields, C. (this issue) What is a theory of consciousness for?, *Journal of Consciousness Studies*, **28** (9–10).

Fields, C., Hoffman, D.D., Prakash, C. & Prentner, R. (2017) Eigenforms, interfaces and holographic encoding, *Constructivist Foundations*, **12** (3), pp. 265–291.

Fields, C., Hoffman, D.D., Prakash, C. & Singh, M. (2018) Conscious agent networks: Formal analysis and application to cognition, *Cognitive Systems Research*, **47**, pp. 186–213.

Frankish, K. (ed.) (2017) *Illusionism as a Theory of Consciousness*, Exeter: Imprint Academic.

Goff, P. (2009) Why panpsychism doesn't help us explain consciousness, *Dialectica*, **63** (3), pp. 289–311.

Goff, P. (2019) *Galileo's Error: Foundations for a New Science of Consciousness*, New York: Vintage Books.

Graziano, M.S.A. & Kastner, S. (2011) Human consciousness and its relationship to social neuroscience: A novel hypothesis, *Cognitive Neuroscience*, **2** (2), pp. 98–113.

Guez, A., Mirza, M., Gregor, K., Kabra, R., Racanière, S., Weber, T., Raposo, D., Santoro, A., Orseau, L., Eccles, T., Wayne, G., Silver, D. & Lillicrap, T. (2019) An investigation of model-free planning, *ArXiv*, [Online], https://arxiv.org/abs/1901.03559.

Hoffman, D.D. & Prakash, C. (2014) Objects of consciousness, *Frontiers in Psychology*, **5**, art. 577.

Hoffman, D.D., Singh, M. & Prakash, C. (2015a) The interface theory of perception, *Psychonomic Bulletin & Review*, **22** (6), pp. 1480–1506.

Hoffman, D.D., Singh, M. & Prakash, C. (2015b) Probing the interface theory of perception: Reply to commentaries, *Psychonomic Bulletin & Review*, **22** (6), pp. 1551–1576.

Jackson, F. (1982) Epiphenomenal qualia, *The Philosophical Quarterly*, **32** (127), pp. 127–136.

Jackson, F. (1986) What Mary didn't know, *The Journal of Philosophy*, **83** (5), pp. 291–295.

Kant, I. (1998) *Critique of Pure Reason*, Guyer, P. & Wood, A.W. (eds.), Cambridge: Cambridge University Press.

Kleiner, J. (2020) Mathematical models of consciousness, *Entropy*, **22** (6), pp. 609–653.

Mark, J.T., Marion, B.B. & Hoffman, D.D. (2010) Natural selection and veridical perceptions, *Journal of Theoretical Biology*, **266** (4), pp. 504–515.

Metzinger, T. (2003) *Being No One: The Self-Model Theory of Subjectivity*, Cambridge, MA: MIT Press.

Müller, M.P. (2020) Law without law: From observer states to physics via algorithmic information theory, *Quantum*, **4**, art. 301.

Nozick, R. (2001) *Invariances*, Cambridge, MA: Harvard University Press.

Prakash, C. (2019) On invention of structure in the world: Interfaces and conscious agents, *Foundations of Science*, **134** (3), pp. 105–116.

Prakash, C., Fields, C., Hoffman, D.D., Prentner, R. & Singh, M. (2020) Fact, fiction, and fitness, *Entropy*, **22** (5), art. 514.

Prakash, C., Stepehens, K.D., Hoffman, D.D., Singh, M. & Fields, C. (in press) Fitness beats truth in the evolution of perception, *Acta Biotheoretica*, [Online], https://doi.org/10. 1007/s10441-020-09400-0.

Prentner, R. (2018) The natural philosophy of experiencing, *Philosophies*, **3** (4), pp. 35–14.

Prentner, R. (2019) Consciousness and topologically structured phenomenal spaces, *Consciousness and Cognition*, **12** (1), pp. 93–118.

Prentner, R. & Fields, C. (2019) Using AI methods to evaluate a minimal model for perception, *Open Philosophy*, **2**, pp. 503–524.

Putnam, H. (1975) *Mathematics, Matter, and Method*, Cambridge: Cambridge University Press.

Sellars, W. (1963) Philosophy and the scientific image of man, in *Science, Perception and Reality*, pp. 1–14, London: Routledge & Kegan Paul.

Skrbina, D. (2005) *Panpsychism in the West*, Cambridge, MA: MIT Press.

Sokal, A. & Bricmont, J. (1998) *Fashionable Nonsense: Postmodern Intellectuals' Abuse of Science*, New York: Picador.

Stoljar, D. (2017) Physicalism, in Zalta, E.N. (ed.) *The Stanford Encyclopedia of Philosophy, Winter 2017 Edition*, [Online], https://plato.stanford.edu/archives/win2017/entries/physicalism/.

Strawson, G. (2006) Realistic monism: Why physicalism entails panpsychism, in Freeman, A. (ed.) *Consciousness and its Place in Nature: Does Physicalism Entail Panpsychism?*, pp. 3–31, Exeter: Imprint Academic.

Whitehead, A.N. (2015) *The Concept of Nature: The Tarner Lectures Delivered in Trinity College November 1919*, Hampe, M. (preface), Cambridge: Cambridge University Press.

Chris Fields[1]

What is a Theory of Consciousness for?

Abstract: *Galileo's Error (Goff, 2019) leaves important questions unasked and hence unanswered. I focus on two of these: the question of what a theory of consciousness is supposed to accomplish, and the question of what the materialism–dualism–panpsychism debate is actually about.*

1. Introduction

Philip Goff's book, *Galileo's Error*, advances a cogent and, as the text progresses, increasingly passionate argument for panpsychism. Panpsychism, not dualism or materialism, is Goff's proposed 'foundation for a new science of consciousness'. I am quite happy with panpsychism and will not contest Goff's arguments supporting it. What, however, is the 'new science of consciousness' for which panpsychism is meant to provide a foundation? What is its theoretical structure? Most importantly, what questions does it answer? What is it for?

Goff saddles Galileo with creating the 'problem of consciousness' (p. 3).[2] The problem of consciousness is that 'we now seek a scientific explanation… of the conscious mind' (p. 21) but neither have one nor have a good idea how to develop one. In particular, according to Goff and many others (see e.g. Dietrich and Hardcastle, 2005), it is not clear that neuroscience can explain consciousness. But what does it

Correspondence:
Email: fieldsres@gmail.com

---

[1] Independent researcher, France.
[2] All otherwise unspecified page references are to Goff (2019).

mean to 'explain consciousness'? What exactly is an explanatory theory of consciousness supposed to provide?

I offer two arguments in this paper. The first is that the explanatory target of most scientists working on consciousness differs in significant ways from the explanatory target of the 'science of consciousness' that Goff seems to envisage. Goff, like many other philosophers, appears to want an explanatory science of qualia as such, a science that not only explains why phenomenal experiences occur, but also why particular experiences have both the particular implementations and the phenomenal characters they have, e.g. why, at least for neurotypical humans, red things look red. He also quite explicitly wants a solution to the combination problem. It is not clear, however, whether such a science is possible, or what it would accomplish if it is. I employ an alternative radical panpsychism, the Minimal Physicalism (MP) of Fields, Glazebrook and Levin (2021), as a contrast to Goff's approach in exploring these issues. While Goff's panpsychism is fundamentally ontological and is based on materiality, MP is fundamentally functional and is based on a characterization of physical interaction as information exchange (Fields, Glazebrook and Marcianò, 2021). Because MP employs a mathematical formalism, that of quantum information theory, it is Galilean in Goff's sense. Because it represents the contents of consciousness as subject to quantitative physical constraints, it has substantial predictive power. It predicts nothing, however, about qualia as such, and, as discussed below, rejects most versions of the combination problem as ill-posed.

If the explananda for a science of consciousness are still fundamentally in question, what is the current 'consciousness war' between dualism, materialism, and panpsychism even about? My second argument is that it is in fact a proxy war: a minor campaign in a broader cultural conflict about human specialness. Dualism, materialism, and panpsychism as Goff presents them are each uneasy alliances that fragment along deeper fault lines when examined. The deep conflict about specialness commands fiercer loyalties and more clearly reveals the underlying motivations of the contestants. Understanding these motivations helps reveal what a theory of consciousness of the sort Goff proposes might be *for*.

2. What Does it Mean to 'Explain Consciousness'?

2.1. Is a theory of qualia even possible?

Explaining consciousness is regularly presented as a 'grand challenge' for twenty-first-century neuroscience (Hougan and Altevogt, 2008; see also Altevogt, Hanson and Leshner, 2008; Seth, 2010). What excites neuroscientists, however, is often regarded as technical and mundane (as 'easy problems' in the terminology of Chalmers, 1996) by philosophers (see Signorelli, Szczotka and Prenter, 2021, for a more nuanced analysis). As LeDoux, Michel and Lau (2020) point out, neuroscience has now spent well over fifty years trying to understand the difference between what subjects can report awareness of, as introspectively accessible contents of consciousness, and what they can demonstrate awareness of independently of such introspective access. Such studies dodge the ontological question of what awareness is in the first place. Even theories that identify consciousness with some independently-defined construct do not explain *why* the proposed identification holds: integrated information theory (IIT; Oizumi, Albantakis and Tononi, 2014), for example, does not explain *why* locally-maximal integrated information, Φ, is sufficient for qualia (*cf.* Cerrulo, 2015). The question that most interests philosophers, the 'hard problem' of why any experiences occur at all, is not just not answered; it is never even posed by the neuroscience of consciousness. Neuroscience takes it for granted that organisms — at least those with brains — are not philosophical zombies (e.g. Lamme, 2018). Indeed Klein and Barron (2020) argue that the 'hard problem' is not a problem for neuroscience or any other science to solve, but rather a set of folk intuitions to be overcome.

Why this disagreement, often tacit, about fundamental explananda? Why do many if not most neuroscientists dismiss the hard problem as uninteresting or irrelevant? Is it really Galileo's fault that neuroscientists, and indeed scientists working on more phylogenetically basal, non-neural organisms, e.g. bacteria (Levin, 2020; Lyon, 2020), are more concerned with what their subjects are aware of and how their awareness informs their behaviour than with what awareness *is* in some fundamental ontological sense? Proposals that awareness is, or is made possible by, quantum entanglement, or by its removal via the 'collapse of the wave function' (Wigner, 1961; Georgiev, 2020), suggest, at any rate, that employing a quantitative mathematical formalism does not inhibit theorizing about the ontology of consciousness. Everyday observations about qualia motivate both the fundamental postulates and the mathematical formalism of IIT. The question is what such theorizing is meant to accomplish. Surgeons and physicians treating the comatose need to know whether their patients are having experiences, introspectively reportable or otherwise. Ethicists face similar questions about animals, plants, AI systems, even Nature as a whole (Dietrich *et al.*, 2021). It is this kind of question that IIT, as well as other theories based in neural or, more broadly, cellular or even abstract information-processing activity try to answer. Goff dismisses IIT as an 'easy problem' theory (p. 35). What is the question, then, for a hard problem theory?

Goff's extended discussion of zombies suggests that whether some system X is capable of experience — capable of any kind of experience at all — is the question of interest. His discussion of black-and-white Mary suggests that the question is not whether X is having an experience but rather what kind of experience X is having. Or, perhaps, it is the question of what distinguishes black-and-white experiences from colour experiences, or of what makes colour experiences *colour* experiences, and not, say, aural or tactile experiences. A bit later we have: 'The job of science [of consciousness], then is... to give an account of the place of feelings in a general theory of reality' (p. 109). Goff's real goal is finally made explicit on pp. 130–38: it is to reposition consciousness as the *intrinsic nature* of matter, something foundational, beyond explanation, that just exists. Consciousness in this case is neither derived nor emergent, but rather fundamental, a primitive brute fact about the structure of the world.

As mentioned earlier, I agree with Goff on this point: consciousness is neither derived nor emergent, but fundamental. Indeed, in MP, consciousness is a fundamental characteristic not of matter, but of the information exchange implemented by physical interactions; hence it characterizes any system that interacts with any other system, material or otherwise. The non-zombiehood of organisms or other systems of interest is, in this setting, not an explanandum for neuroscience or any other science, but rather a foundational assumption. Where does this leave the hard problem, the problem of explaining '*why* are certain kinds of brain activity correlated with consciousness?' (p. 35, emphasis in original). Or, in the language of IIT, why is $\Phi > 0$ correlated with consciousness? Surely the plain answer is that these are no longer questions: for a panpsychist, *everything* is correlated with consciousness. Every material object in Goff's version, every physical interaction in MP. Why are we humans conscious? Because we exist.

Other than letting us stop worrying about the hard problem, however, what does the postulate that consciousness is fundamental buy us? Goff devotes considerable space to arguing that a successful, predictive science does not require intrinsic natures, so aside from satisfying a certain inchoate yearning for a settled ontology, what is the claim that consciousness is the intrinsic nature of humans, or of anything, good for? Goff criticizes Chalmers and McQueen for not providing '*enough* of a causal role for the conscious mind' (p. 47, emphasis in original). Making consciousness the intrinsic nature of physical objects, or, in MP, of physical interactions, appears to deprive it of any causal role whatsoever. Consciousness is a logical precondition for anything happening, indeed for anything existing, but it does not cause or prevent any particular happening. In Galilean language, consciousness is a characteristic, not a component, of a functional mechanism.

Consider black-and-white Mary. Assuming she is physical, she is conscious. So she is conscious of something. But what? What distinguishes her black-and-white experiences from her later colour experiences? Not consciousness *per se*, but something about its particular content at some time. Neuroscience describes the differences in implementation between black-and-white and colour experiences in exquisite detail, and explains how the two kinds of experiences can afford different behaviours. Computational models generalize such implementation-level differences; both MP and IIT, for example, provide formal specifications of what differences between inputs are detectable by a given system, and of how detection of one input versus another influences behaviour. What more do we want? An explanation of the qualia themselves, of the particular 'feel' of particular colour experiences, proponents of black-and-white Mary and similar thought experiments insist. Neuroscience notes the correlations between colour experiences and affective or motivational feelings, and evolutionary neuroscience explains why, for example, experiences of red correlate, in humans, with experiences of danger in some contexts and experiences of excitement in others, but neither has anything to say beyond this. Neither do computational models of such correlations. Neither does the claim that consciousness is fundamental.

Is there a 'why' question about the correlation between physical or biological states and qualia more fundamental than the ones answered by evolutionary and developmental neuroscience, and does this further question pose a new 'hard problem' that a science of consciousness should solve? Both evolutionary and developmental neuroscience are concerned with averages: is there an additional, real-time 'why' question about each individual quale? Are individual qualia — the redness of a particular apple, the taste of a particular chocolate — well-enough defined to even pose this question? Illusionists argue that qualia are ephemeral and lack empirically-accessible identities, and are therefore illusory (e.g. Frankish, 2016), but does ephemerality imply illusion? Physics is full of phenomena that are ephemeral but real: fluctuations in the quantum vacuum, for example. Why should qualia not have this same status? In MP, they do: qualia are ephemeral but real. The qualia of a Boltzmann Brain (Bousso and Freivogel, 2007) are just as real, for example, as Mary's or mine, and are ephemeral by definition. As every observer and every event of observation are unique, in principle, in MP, MP regards every quale as unique. The only science of qualia MP allows is a science of descriptive reports, third-party measurements, and averages: the kind of correlative science that neuroscience — indeed, biology in general — already offers.

Goff and I appear, at least, to deeply differ on this point: Goff appears to want a science of qualia *per se*, an explanation of why particular qualia have particular implementations and arise in particular circumstances, and to believe that such a science can be developed from a panpsychism anchored in materiality. There seem to be two motivations for this, one scientific and the other purely philosophical. The scientific motivation stems from an assumption that both experiences and experiencers are in some sense compositional, an assumption Goff shares with IIT (Oizumi, Albantakis and Tononi, 2014; see especially Figure 15). This assumption leads via well-trodden paths to the combination problem.

2.2. Does panpsychism need to solve the combination problem?

Panpsychism is regularly saddled with the combination problem; indeed, Goff characterizes it as 'the hard problem' for panpsychism (pp. 147ff.) and suggests that, absent a robust theory of combination, panpsychism is a 'lost cause' (p. 146). How do simple experiences combine to produce complex experiences (or in the 'emergentist' version, how do clusters or colocalizations of simple experiences enable the emergence of more complex experiences)? How do simple *experiencers* combine to produce, or enable the emergence of, complex experiencers? This problem is often regarded as intractable, but Goff is optimistic, particularly about the emergentist version.

Consider me, in surgery after a good wallop of anaesthetic. I am 'unconscious', but what does this now mean? That I no longer have an intrinsic nature? Or just that there has been some functional change, something neuroscience has a handle on (Kelz and Mashour, 2019)? 'I' may be anaesthetized, but plenty of my systems are still operating with full awareness, regulating heart rate and so forth. Turning the cortex off inhibits the kind of consciousness that philosophers and surgeons are most interested in, but another kind of consciousness, one at least as important to us as organisms, seems to remain. At least it does in MP, in which *every* bounded system has awareness. We know, however, essentially nothing about qualia for this kind of consciousness, just as we known nothing about qualia for fungi, plants, or bacteria. What is it like to be my basal ganglia? Such questions are rarely raised in either neuroscience or philosophy. What is it like to be a neuron in my basal ganglia? Cook (2008) provides a compelling account of what it's like to be a neuron; if he is right, it isn't pleasant.

Straightforward (i.e. reductionist) versions of the combination problem assume that the experiences of my basal ganglia are components of my experience, and experiences of neurons are components of my basal ganglia's experience. Solving this reductionist version would require saying how my experiences incorporate those of my basal ganglia. The exclusivity (or 'maximality' in Goff's usage, pp. 167–9) postulate of IIT is designed to prevent this kind of cross-level compositionality of experiences by denying that my basal ganglia have any experiences at all, i.e. by making them zombies, whenever I, as a containing system with larger Φ, am conscious. Emergentist versions of the combination problem may assume that proper components of an experiencer have no experiences on their own (as in IIT), but that they somehow enable high-level experiences. Low-level biological processes clearly enable high-level processes, conscious or not, in a well-understood physiological sense. Is there a further kind of enabling that is specific to consciousness?

Computation provides a useful analogy here. The theory of virtual machines (Smith and Nair, 2005) renders the 'emergence' of complex, system-scale computational behaviour from the simpler behaviours of logic gates or operating-system components well understood. A virtual machine stack is, however, a semantic hierarchy, not a causal hierarchy. Low-level operations do not 'combine' to form high-level operations in any straightforward, context-independent, or locally reproducible way. Computation poses an implementation problem, one solved by well-defined inter-level interfaces that enforce semantic constraints, but it poses no 'combination problem' over and above this. Why should this not be the case for conscious systems?

The combination problem as Goff formulates it does not arise in MP; indeed, the quantum-theoretic formalism renders it ill-posed (Fields, Glazebrook and Levin, 2021). Systems in MP execute hierarchical computations — indeed, hiearchical Bayesian inference — but do not have well-defined mereological decompositions. Experiences are compositional in MP under precisely specified conditions, but experiencers are not compositional. Agents in MP — and their environments — are virtual-machine hierarchies over underlying quantum computations. They are semantic constructs, not causal constructs.

If adopting a panpsychist position does not, as MP demonstrates, entail facing the combination problem, why is the combination problem so central to discussions of panpsychism? The answer, I believe, is ontology. Goff's motivations are fundamentally ontological; he describes the first four chapters of *Galileo's Error* as 'an essay in ontology' (p. 183). Here MP provides a somewhat extreme counter-example: it rejects not only the common-sense classical ontology of observer-independent material objects (*cf.* Hoffman, Singh and Prakash, 2015), but even the assumptions of observer-independent space, time, or information (*cf.* Wheeler, 1989). It replaces, as Galilean science does generally when pushed toward instrumentality, such ontological assumptions with assumptions about how systems or interactions can be formally described.

The materialist and dualist positions with which Goff contrasts his panpsychism are similarly ontological positions. The problem of qualia that they attempt to solve is an ontological problem, and the solutions they offer are ontological solutions. Goff never says so explicitly (though he comes close in Technical Appendix B), but one might speculate that the *real* error of Galilean science, in Goff's view, is its deep-seated disinterest in ontology. Aside from recognizing that consciousness is a real phenomenon (p. 174), however, what is the ontology of consciousness trying to accomplish? What is the problem of qualia actually about?

3. What is the 'Consciousness War' About?

The difficulty of pinning down exactly what a theory of consciousness is supposed to explain suggests that something not quite ordinary is going on. Sciences generally have straightforward explanatory goals

— why should a science of consciousness be different? Goff remarks that '[t]he brute identity theory [identifying consciousness with brain states] is very unsatisfying' (p. 108). How then is the idea that consciousness is the *intrinsic nature* of brain states satisfying? If not an ontological yearning, what is a theory identifying consciousness with the intrinsic nature of matter meant to satisfy?

Goff answers this question in Chapter 5: a panpsychist theory of consciousness is meant to satisfy an *existential* yearning, a yearning for meaning. It is meant to counter the 'cosmic alienation' (p. 216) of postmodern life. While many blame this cognitive and emotional dislocation on materialism, Goff correctly points to dualism as the culprit: it is modern dualism, particularly as expressed by Descartes, that isolated humans even from other animals as the only conscious entities on Earth. Is it fair, however, to blame Descartes' contemporary, Galileo, for this? Is it really the idea that science should employ the formal tools of mathematics that separated humans from their mammalian cousins, or from the idea of Gaia? Should we not look a bit deeper into history?

We can presume that at one time, say 15,000 years ago, everyone on Earth was an animist, and hence a panpsychist. What happened? The invention of human 'specialness' seems to have coincided, though at different times on different continents, with the near-simultaneous development of agriculture, urbanism, and overtly-hierarchical political power (Harari, 2014). Perhaps feeling special — as Marx suggested — made life as an expendable underling a bit more tolerable. Hence the abandonment of panpsychism in favour of human exceptionalism as a foundational cultural myth appears, at least in Eurasia, to have predated the European Enlightenment by several millennia. No one criticizes the exceptionalist myth or its cultural effects better than Gray (2002). It is unfortunate that Goff did not pursue the historical demise of animist panpsychism more deeply, as it perhaps offers some advice for our current predicament.

I would suggest that the pan-cultural replacement of animism by exceptionalism also offers some insight into the current battles about consciousness. Thoroughgoing materialists tend to align with thoroughgoing panpsychists in believing that humans are just part of the natural order, more cognitively capable and hence more interesting psychologically than rocks, but of no different ontological status. Such a view is highly deflationary; it implies that the science of human experience is the science of human experiential capabilities, an implication that MP joins evolutionary neuroscience in embracing.

Even dualists can support this deflationary view provided they believe, as animists do, that rocks have rock-specific souls. More sectarian panpsychists, however, limit the 'pan-' to some preferred class of entities, mammals for example, or more liberally, animals or even all organisms. Rocks and artefacts are out; they have the 'wrong organization' to be conscious. Perhaps as 'mere aggregates' they lack intrinsic natures; at any rate, they are zombies. Sectarian materialists are happy with this limitation, imagining that consciousness 'emerges' from zombiehood with the evolution of complex nervous systems, maybe only with human nervous systems. Or maybe, as Bell (1990) teases us, only with the evolution of PhDs. Contemporary dualists are, by and large, happy to go along with this: contemporary dualism is partial to zombies. Hence whether a theory admits zombies as logically possible is the acid test between exceptionalism and a truly universal, exception-free panpsychism, like MP, that views consciousness as a *logical* precondition for existence.

It is not clear from *Galileo's Error* where Goff's loyalties lie in this larger debate (see especially pp. 113–14). But if Goff does support a thoroughgoing, unrestricted panpsychism, why should mathematics pose a problem? Simple systems are just as conscious as complex ones, but understanding what their experiences are *like* may well benefit from — even require — a sophisticated mathematical formalism. We may need mathematics to solve the combination problem, and even if there is no such thing as combination, we may need mathematics — indeed approaches as different as IIT and MP demand mathematics — to understand how complex systems use high-bandwidth informational inputs to guide their behaviour. If, on the other hand, Goff supports a more limited, sectarian brand of panpsychism, e.g. one compliant with the maximality condition of IIT (pp. 167–9), a formal analysis may be required to identify those systems that neither contain nor are contained in systems with larger Φ, and hence to distinguish conscious systems from zombies. Either way, what Galileo taught us seems essential to a science of consciousness, not antithetical to it. True, Galileo dispensed with the 'what is it like' questions, but we may need his methods to answer them, however conditionally.

One is left, unfortunately, with the feeling that Galileo has been singled out as a bogeyman, as so many have singled out 'materialism' or 'Enlightenment culture'. Humans are humans, and humans are, fortunately or otherwise, descendants of a long and difficult mammalian, vertebrate, metazoan, and ultimately microbial lineage. It is no surprise that we look out for ourselves first and our kin or our neighbours second. It is no surprise that we think we're special. Science has been chipping away at this sense of specialness at least since Copernicus. Hopefully the developing science of consciousness can help.

References

Altevogt, B.M., Hanson, S.L. & Leshner, A.I. (2008) Molecules to minds: Grand challenges for the 21st century, *Neuron*, **60**, pp. 406–408.
Bell, J. (1990) Against 'measurement', *Physics World*, **3** (8), pp. 33–41.
Bousso, R. & Freivogel, B. (2007) A paradox in the global description of the multiverse, *Journal of High Energy Physics*, **06**, 018.
Cerullo, M.A. (2015) The problem with phi: A critique of integrated information theory, *PLoS Computational Biology*, **11**, e1004286.
Chalmers, D.J. (1996) *The Conscious Mind: In Search of a Fundamental Theory*, New York: Oxford University Press.
Cook, N.D. (2008) The neuron-level phenomena underlying cognition and consciousness: Synaptic activity and the action potential, *Neuroscience*, **153**, pp. 556–570.
Dietrich, E. & Hardcastle, V.G. (2005) *Sisyphus's Boulder: Consciousness and the Limits of the Knowable*, Amsterdam: Johns Benjamins.
Dietrich, E., Fields, C., Sullins, J. P., van Heuveln, B. & Zebrowski, R. (2021) *Great Philosophical Objections to Artificial Intelligence: The History and Legacy of the AI Wars*, London: Bloomsbury.
Fields, C., Glazebrook, J.F. & Levin, M. (2021) Minimal physicalism as a scale-free substrate for cognition and consciousness, *Neuroscience of Consiousness*, **2021**, niab013.
Fields, C., Glazebrook, J.F. & Marcianò, A. (2021) Reference frame induced symmetry breaking on holographic screens, *Symmetry*, **13**, art. 408.
Frankish, K. (2016) Illusionism as a theory of consciousness, *Journal of Consciousness Studies*, **23** (11–12), pp. 11–39. Reprinted in Frankish, K. (ed.) (2017) *Illusionism as a Theory of Consciousness*, Exeter: Imprint Academic.
Georgiev, D.D. (2020) Quantum information theoretic approach to the mind–brain problem, *Progress in Biophysics and Molecular Biology*, **158**, pp. 16–32.
Goff, P. (2019) *Galileo's Error: Foundations for a New Science of Consciousness*, New York: Vintage.
Gray, J.N. (2002) *Straw Dogs*, London: Granta.
Harari, Y.N. (2014) *Sapiens*, New York: Harper.
Hoffman, D.D., Singh, M. & Prakash, C. (2015) The interface theory of perception, *Psychonomic Bulletin & Review*, **22**, pp. 1480–1506.
Hougan, M. & Altevogt, B.M. (Rapporteurs) (2008) *From Molecules to Minds: Challenges for the 21st Century*, Washington, DC: National Academies Press.
Kelz, M.B. & Mashour, G.A. (2019) The biology of general anesthesia from paramecium to primate, *Current Biology*, **29**, pp. R1199–1210.
Klein, C. & Barron, A.B. (2020) How experimental neuroscientists can fix the hard problem of consciousness, *Neuroscience of Consciousness*, **2020**, niaa009.

Lamme, V.A.F. (2018) Challenges for theories of consciousness: Seeing or knowing, the missing ingredient and how to deal with panpsychism, *Philosophical Transactions of the Royal Society of London B*, **373**, 20170344.

LeDoux, J.E., Michel, M. & Lau, H. (2020) A little history goes a long way toward understanding why we study consciousness the way we do today, *Proceedings of the National Academy of Science USA*, **117**, pp. 6976–6984.

Levin, M. (2020) Life, death, and self: Fundamental questions of primitive cognition viewed through the lens of body plasticity and synthetic organisms, *Biochemical and Biophysical Research Communications*, [Online], https://doi.org/10.1016/j.bbrc.2020.10.077.

Lyon, P. (2020) Of what is 'minimal cognition' the half-baked version?, *Adaptive Behavior*, **28**, pp. 407–428.

Oizumi, M., Albantakis, L. & Tononi, G. (2014) From the phenomenology to the mechanisms of consciousness: Integrated Information Theory 3.0, *PLoS Computational Biology*, **10**, e1003588.

Seth, A. (2010) The grand challenge of consciousness, *Frontiers in Psychology*, **1**, art. 5.

Signorelli, C.M., Szczotka, J. & Prenter, R. (2021) Explanatory profiles of models of consciousness: Towards a systematic classification, *Neuroscience and Consciousness*, **2021**, niab021.

Smith, J.E. & Nair, R. (2005) The architecture of virtual machines, *IEEE Computer*, **38** (5), pp. 32–38.

Wheeler, J.A. (1989) Information, physics, quantum: The search for links, in Zurek, W.J. (ed.) *Complexity, Entropy, and the Physics of Information*, pp. 3–28, Boca Raton, FL: CRC Press.

Wigner, E.P. (1961) Remarks on the mind–body question, in Good, I.J. (ed.) *The Scientist Speculates*, pp. 284–302, London: Heinemann.

Luke Roelofs[1]

Is Panpsychism at Odds with Science?

Abstract: *Galileo's Error is a superlative work of public philosophy, particularly as a way of introducing modern academic panpsychism to a broader audience. In this commentary, I reflect on an issue that is prominent, though often with different background concerns, in both academic and popular discourse: what it means to be 'scientific' or 'unscientific'. Panpsychism is not itself a scientific hypothesis, but neither is it (as critics sometimes claim) in conflict with science. Indeed, Goff argues, and I agree, that panpsychism is an eminently scientific worldview, in the sense of a way of viewing reality that accords with and embraces what science reveals. But what exactly it means to 'accord with and embrace' science is disputed; this paper tries to untangle some of the threads.*

1. Introduction

Stop me if you've heard this one before. Science is great, but it has limits. We should accept its contributions but recognize those limits, the things it can't tell us, the mysteries it can't address, the questions it will never answer. For those things, we should look beyond science, to… and usually at this point, whoever is speaking will insert their favoured idea, product, or organization. Often the idea, product, or organization in question is somewhere on the spectrum from harmless nonsense to harmful nonsense. So it could be understandable for someone seeing Philip Goff give a similar-sounding spiel ('Galileo

Correspondence:
Email: luke.mf.roelofs@gmail.com

[1] New York University, NY, USA.

was great, but let me tell you about his Error...') to be immediately suspicious. Combining this with a self-confessedly weird pitch — electrons are conscious! — seems to warrant even more suspicion, and so does the vague affinity sometimes claimed (either by critics or defenders) between panpsychism and various other -isms often viewed with suspicion, like 'animism' or 'pantheism'. It could be enough to make one feel that the whole idea has 'the faintly sickening odour of something cooked up in the metaphysical laboratory' (Nagel, 1986, p. 49).[2]

And recently there's been no shortage of people ostentatiously holding their noses: panpsychism is 'a philosophically motivated pseudoscience', according to Barry Smith,[3] 'the consequence of knowing next to no science', according to Patricia Churchland,[4] a 'crazy hypothesis [with] not a shred of evidence supporting it', according to Jerry Coyne,[5] and 'one of several major steps backwards taken by philosophers over recent years', according to Dan Kaufman.[6] Anecdotally, I've been disappointed by people I know throwing panpsychism in with belief in literal deities, reliance on healing crystals, and opposition to vaccines: the low point was someone challenging me to explain what differentiates panpsychism from the reactionary conspiracy theory Q-anon. While Goff's excellent book is a remarkable achievement in bringing complex philosophical ideas to a wider audience, more exposure for panpsychism is likely to mean more of this kind of suspicion.

I think this suspicion is misplaced, but I do have a little sympathy: in a world where the internet seems to connect us with a million charlatans every day, we have to reject most things out of hand, and one could do worse than treating 'Science is great, but...' as a red flag. It's not too much of an exaggeration to say that unscientific beliefs nearly overthrew US democracy, hamstrung responses to a global pandemic, and are in the process of rendering the Earth uninhabitable for humans. A cultural shift towards refusing to give

---

[2] That this is said by someone in the course of offering an argument for panpsychism is, of course, somewhat ironic.
[3] https://twitter.com/smithbarryc/status/1320509697611497472.
[4] https://twitter.com/patchurchland/status/1320463304012173312.
[5] https://www.realclearscience.com/2020/01/06/panpsychism_makes_a_sneaky_return_288956.html.
[6] https://twitter.com/ElectricAgora/status/1320507706185306113.

unscientific-sounding ideas the benefit of the doubt would probably be a net good right about now.

One of the striking things about many of the attacks on panpsychism is how endlessly they recycle objections that Goff has already patiently dismantled, both in this book and, often, in direct conversations with people like Churchland. I don't think I would add much value by just repeating Goff's points, but perhaps there's value in spending a bit of time identifying and criticizing the things that *are* unacceptably unscientific, rather than just explaining why panpsychism isn't. So in this paper I want to focus on that — on analysing what sort of views a scientific outlook should rule out, and how panpsychism does and doesn't relate to them.

2. 'Being Scientific' as Accepting Current Science

We can distinguish a few ways for a view to try to be 'scientific'. The first and simplest is compatibility with science, specifically regarding the world's causal structure. Panpsychists of Goff's variety are very interested in certain things science can't tell us, but what it certainly *can* tell us is what causes what, and when, where, and how. A basic requirement for a worldview to be scientific is that it treat science as the final authority on this: if the best available science says that A can only cause B under X conditions, then A can only cause B under X conditions. That means, in particular, ruling out phenomena like telepathy, psychokinesis, ESP, spoonbending, precognition, past life regression, and communication with the dead, since all of these would require causal mechanisms sharply different from any of the gravitational, nuclear, and electromagnetic ones recognized by current science. Of course, there may turn out to be technological ways to produce phenomena like this — radiotransmitting neural implants that effect a sort of telepathy, for instance — but we'll have to devise them first, and in the meantime we should confidently reject parapsychology.

It should go without saying that panpsychists agree with all this (Goff certainly does). But I have sometimes encountered the idea, both from friends of panpsychism and from its enemies, that it might somehow make space for parapsychology in a way that materialism doesn't. After all, if the spoon in my hand is just consciousness — if, as on some versions of the view, it and I are just two permutations of the same cosmic consciousness — then shouldn't it make sense for *this* consciousness to be able to speak directly to *that* consciousness?

No. Unromantic as it may seem, the cosmic consciousness seems to be rigidly and inflexibly committed to certain laws of action, which dictate that we can bend the spoon only by applying the familiar sorts of forces, for instance by means of those permutations of the cosmic mind that we customarily call our 'hands'. After all, it's not as if we need panpsychism in order to unsettle our everyday view of spoons: we already know a spoon is just a complex of ripples in the universal quantum wave function, like us. But alas, grasping that truth will not help us bend it, except by means of those complex ripples in the universal quantum wave function that we customarily call our 'hands'.

3. 'Being Scientific' as Accepting Likely Future Science

Here is a second, slightly more aggressive, way in which a worldview might try to be 'scientific'. It might aim for compatibility, not just with presently established science, but with where science looks like it's going. That is, it might try to pre-emptively incorporate the most ambitious, complete way for present science to advance — as opposed to exploiting whatever 'gaps' and uncertainties may exist at present. There is obviously more room for controversy about what this demands: different philosophical assumptions will predict different future advances. Indeed, the central dispute between Goff and critics like Churchland is whether a particular gap — the explanatory gap between objective and subjective descriptions — should be expected to close with more scientific work. But it is crucial that panpsychists have well-understood arguments for the special status of this gap: they're not just observing that we don't have an answer yet, not just exploiting whatever gaps happen to exist.

As Goff notes (2019, p. 162), I defend a version of panpsychism committed to micro-reductionism: a complete account of the facts about the smallest components of nature, and their relations to one another, will determine all the facts about the larger components — the 'bottom level' fully determines all the higher levels.[7] This is

---

[7] This goes in reverse too: a full account of the largest thing, the cosmos itself, would determine every fact about its smaller portions. Micro-reductionism, thus understood, isn't opposed to holistic views of the universe as ultimately just one big thing, including the mysticism-influenced form of holism that Goff discusses over pp. 205–11. It just says that it's the same universe whether we think of it as one big thing or as trillions of little things: the forms of our description will change, but the reality described doesn't.

largely because, while I accept that micro-reductionism hasn't been (and perhaps can't be) decisively established (as Goff discusses over pp. 162–4), it seems to me the direction science is going, and to that extent a micro-reductionist philosophy seems to me more attractive than an emergentist one which posits a 'patchwork world' (*cf.* Cartwright, 1983; cited by Goff, 2019, p. 163; see also Moran, this issue).[8] On this score I think I aim for a more aggressive stance than Goff does: while he is agnostic about micro-reductionism, I strongly suspect that whatever gaps currently exist in our scientific reductions — of neuroscience to biochemstriy, of biochemistry to particle physics, etc. — reflect our own current limitations, not facts about reality. Hence where I see a way of being more vigorously and pre-emptively scientific, he sees 'dogmas which are simply accepted as part of the zeitgeist' (p. 163). (As a matter of intellectual psychology, I suspect it's not a coincidence that I'm a compatibilist about free will and determinism, and Goff is more sympathetic to incompatibilism.)

Another part of embracing not just the undeniable data of science but the world-picture they seem to be suggesting is embracing the absence of any really sharp boundaries between different kinds of objects: between a brain and a rock lies an indefinitely gradual continuum of intermediate forms, with no place on that continuum where anything fundamental appears or disappears. For panpsychists, this means there is a pressure towards the sort of 'universalist' panpsychism that Goff holds off from endorsing: as he says, 'Most panpsychists will deny that your socks are conscious, while asserting that they are ultimately composed of things that are conscious' (2019, p. 113). It's true that most panpsychists try to restrict which composite things are conscious, but I think this is ultimately a mistake, and I think Goff agrees; at least, he has argued for universalism forcefully in print (Goff, 2013; *cf.* Buchanan and Roelofs, 2019, pp. 3007–10).[9]

---

[8] It is worth noting here that quantum entanglement, far from being an inexplicable anomaly for the micro-reductionist perspective, is deducible from the laws operating the micro-level: from a full account of two particles and the way they interact, one can *predict* that they will become entangled, and how this entanglement will affect the behaviour of each.

[9] Partly because I think universalism is ultimately more defensible, and partly just for ease of exposition, I will sometimes put panpsychism as the idea that 'everything is conscious'. In strictness this should be 'everything is conscious or made out of conscious parts', but relative to the usual assumption that most things are utterly devoid of consciousness, 'conscious or made of conscious parts' is still a pretty striking claim.

4. An Interlude on Observational Equivalence

One upshot of (at least some forms of) panpsychism being an aggressively scientific view, in the two senses just discussed, is that panpsychism and materialism are robustly observationally equivalent: no current observations discriminate between them, nor will any foreseeable future observations. This claim of observational equivalence has provoked incredulity from some critics (Churchland calls it 'probably the stupidest thing I have ever read'),[10] while others regard it as correct, but fatally undermining panpsychism by rendering it unfalsifiable and therefore nonsensical (I've not yet seen a good response to the natural rejoinder that this standard implies the exact same verdict for materialism). I think the claim of observational equivalence is basically correct, and indeed central to properly understanding the dispute between panpsychism and its rivals, but in full strictness it requires two important qualifications.

First, it is not quite true that panpsychism and materialism predict all the same observations; they differ on one crucial observational prediction, though which one it is depends on the observer. From my perspective, they differ in their predictions about whether or not I am subjectively conscious: panpsychism predicts that I should be, while materialism gives no basis for predicting this. I say this because of the much-discussed explanatory gap: we cannot see any explanation of consciousness on a purely material basis, any reason why some amount of movements-and-forces-in-space should *feel* some way. And so if we did not know already, from the first person, that we are conscious, then nothing in the world-picture provided by materialism would give us reason to expect that we should be. At least, that is my take on what the respective theories predict: obviously there is no shortage of debate about this. But if panpsychists are right in their arguments — in particular, right that their theory explains human consciousness and materialism does not — then the upshot is that, so to speak, if I lived in a panpsychist world, I should expect to be conscious, and if I lived in a physicalist world, I should have no definite expectation on that score (or even expect not to be). Since I am conscious, that's a data-point in favour of panpsychism.

Of course, *you* can't observe whether a prediction that I should be conscious is borne out; from your perspective, the key observation is

---

[10] https://twitter.com/patchurchland/status/1320459305888280576.

whether *you* are subjectively conscious, and in general each of us can observe that we ourselves are conscious, as panpsychism predicts. This is an odd sort of prediction — there is no single observation that can be predicted and evaluated by the scientific establishment collectively, but only a mass of individual observations that each individual scientist has already made before they even look at a textbook. I don't think this observational difference between panpsychism and materialism is therefore unimportant, but it cannot be denied that it is atypical. If we confine ourselves to more standard, shareable, 'public' observations, panpsychism and materialism are observationally equivalent: they differ observationally only when we recognize this important but atypical set of observations.[11]

There is a second difference between panpsychism and materialism that relates to observation, but is not well captured in any summary of 'observable evidence'. Just now I said that 'each of us' can observe that we're conscious. Who does that cover? Who, that is, counts as an observer? This is a big question, that risks dropping out of view if we just ask 'what is observable?', and skim past the question 'to who?'. Theories which disagree about who is an 'observer' are not really 'observationally equivalent', even if they agree about what a given observer should expect to observe. In the words of MC Hammer, 'when you measure, include the measurer'.[12] Admittedly, neither panpsychism nor materialism directly provide a complete answer to this question. But panpsychism does suggest that, in so far as consciousness is one key requirement for being an observer, everything in the universe fulfils that requirement, and so is perhaps one step of the way towards observerhood.

5. 'Being Scientific' as Adding No New Elements to the Scientific Picture

A third sense in which we might want our philosophy to be scientific is that, as well as treating science as the authority on the world's

---

[11] Isn't it part of the scientific method to distrust private observations, to only trust what can be publicly shared? Perhaps in the sense that *I* should be wary of trusting *your* private observations (more precisely, your reports of them to me), but it makes no sense for *me* to distrust *my own* private observations until they have been publicly validated — after all, how would I learn that they had been publicly validated, except *via* some perception of my own?

[12] https://twitter.com/mchammer/status/1363908982289559553?lang=en.

causal structure, we shouldn't add any *new* structure, even if we stipulate that it has no direct causal impact. Of course, panpsychists like Goff are in some sense 'adding' something to the physical picture, namely consciousness (and lots of it!). But this isn't posited as something additional: it's posited as *what — without realizing it — we were talking about all along*. There are electrons, they move in space, they have negative charge and a little bit of mass: the panpsychist doesn't add anything to this picture, they just propose that the things already in this picture — objects like electrons, relational structures like space, properties like charge and mass — are in fact forms of consciousness. This isn't adding, it's interpreting. And it rests on recognizing that the physical picture itself is just under-specified: it tells us how this thing called an 'electron' behaves, and how these properties called 'charge' affect that behaviour, but never says (never could say) what any of this is in and of itself.

Goff talks about this idea at length, under the heading of 'simplicity' or 'Ockham's razor', and I think he does an excellent job of explaining why simplicity-based reasoning is an important, indispensable part of any scientific worldview, and why it tells in favour of panpsychism or materialism and against dualism. I think it's useful to belabour the point a little, because this concern not to add anything unnecessary tells not only against dualism, but also against virtually any sort of individual afterlife, and, I think, against belief in any number of gods — at least if they are understood as intelligent agents who knowingly pursue goals.[13] That is, part of the scientific spirit that drives panpsychism is a thorough-going opposition to anthropomorphism in our understanding of the universe and its non-human parts.

Panpsychism's opposition to anthropomorphism is, I suspect, sometimes a bit confusing. If we hear 'X is conscious', we naturally assume that a certain basic sort of intelligence goes along with this — desiring things, representing one's surroundings, choosing to do things that satisfy those desires in light of those representations, etc. If we hold onto that link between consciousness and intelligence, then attributing

---

[13] This sort of scientific spirit may still be compatible with a sort of impersonal pantheism, something like what Einstein expresses by saying 'I believe in Spinoza's God, who reveals himself in the harmony of all that exists, not in a God who concerns himself with the fate and the doings of mankind' (Isaacson, 2008, pp. 388–9, *cf. New York Times*, 1929). See also Leidenhag (this issue) for an attempt to argue that theist panpsychism does better than atheist panpsychism at accomodating the principle of simplicity.

consciousness more widely will imply attributing intelligence more widely, including to things whose behaviour and internal structure don't fit such a structure. And that would violate Ockham's razor: attributing intelligence to, say, a rock, or an electron, would mean positing a sort of complexity — the specific content of its representations, the specific things it desires, the mechanisms by which the one influences the other, and so on — that had no correspondence to anything scientifically determinable about its structure or behaviour.

Fortunately, attributing intelligence to rocks is not what panpsychism is about: it's about holding onto the link between intelligence and observable behaviour, while revising the link between consciousness and intelligence. Scientific observation and study are still the best and only ways to determine what sort of intelligence any natural system (a rock, a river, a plant, an insect, a frog) has, what sort of information it can absorb and process, what its cognitive capacities are. But at present, any account of a system's degree and forms of intelligence tends to provoke the further, awkward, question: is it intelligent *enough* to be conscious? Are its cognitive capacities 'advanced enough', do they cross the magic threshold for there to be some subjective experience accompanying them? Panpsychism's impact is here: it dismisses this question by throwing out the whole idea of a magic threshold, a boundary between the conscious and the non-conscious. It replaces it with the very different question: 'what *sorts* of consciousness accompany these cognitive or informational processes, whatever they are?' This is the same question we already have to ask about every system which we decide *is* conscious, and it's often a profoundly difficult question, especially as we get further and further from human-like minds. Panpsychism doesn't by itself answer this question: it just tells us that we can ask this question about every system we study, and thereby dissolves the binary question 'conscious or not?'.

6. An Interlude on 'Animism'

Some readers might take the last two paragraphs to constitute a rejection of something called 'animism', thought of as ascribing human-like minds to all sorts of natural things. If that's what you associate with the term 'animism', then panpsychism is opposed to animism. But I prefer not to use that term to express this point, because it is so entangled with a long history of contentious meanings: originally a term in the anthropology of religion, often used to caricature non-European beliefs, and given multiple contradictory definitions by different authors with different aims (Bird-David, 1999). To the extent that the term has currency now, it is often in the context of what is sometimes called 'the new animism' (see, for example, Rose, 1996; 2003; Harvey, 2006), and in this context it seems to express an orientation, a stance taken towards natural things, more than it expresses any truth-claim about their nature. That is, 'animism' in this context is about approaching forests, rivers, mountains, animal populations, and so on in a spirit of respect, with the goal of learning from them, finding out what they need, and providing it while taking what we need in turn. On this reading, being an animist is a matter of practice, not belief: the practice of caring about non-human things for their own sake. As a way of approaching things, this is not the sort of thing that can be true or false, nor conflict or agree with any given scientific theory or discovery. It contrasts not with doctrines, like panpsychism or materialism, but with rival approaches, like the exploitative instrumental approach that treats anything non-human as a resource to be owned and used.

But a practical approach can still benefit from theoretical underpinnings. As Freya Mathews, a panpsychist environmental philosopher, writes: '…animism does leave certain philosophical questions unaddressed: which things count as alive, in the animist sense?… In any case, what is it about animate things that entitles them to be treated with respect[?]' (2020, p. 133). To put it another way: the stance of approaching something with respect presupposes that it has some sort of needs, and that we have some way of determining what they are. Maybe these needs belong to it as a whole, or derive from interests of its parts or members, or inhere in its relationships to other things; maybe on some fundamental level these different options aren't as distinct as we tend to think. But what sort of needs a system is capable of having, and what its actual needs are, and how we can find out about them are all questions that demand philosophical attention.

I think Mathews and Goff are right that panpsychism has a lot to offer as a theoretical underpinning for the relational orientation that the new animists speak of: the belief that everything in nature is some form of consciousness pairs very nicely with the attempt to approach everything in nature with some form of respect.[14] But I also think the relationship between theory and practice here is a fairly loose relationship: a materialist cosmology could potentially pair just as neatly with animism, in so far as it tells us that human and non-human parts of nature are fundamentally the same kind of thing, just intricate arrangements of matter that, for a fleeting moment, becomes aware of itself (*cf.* Goff, 2019, pp. 189–90).[15] So too, perhaps, could a form of naturalistic dualism that saw the glimmer of an immaterial mind in bears and birds and beetles, or even in trees and rivers. The only view which is really deeply unsuited to this task is the sort of anthropocentric dualism which radically distinguishes human beings from all non-human life. Ironically, this sort of dualism has little following in academic philosophy but enormous reach in major world religions.

7. Towards a Monist United Front

The fact that panpsychism and materialism are both well-suited to provide a metaphysical backing to a respectful, 'animistic' orientation towards natural things, in a way that anthropocentric dualism is not, points towards an important difference between the context of academic philosophy and that of wider society. In academic philosophy of mind, materialism is the hegemonic view; everything else is defined by its departures from materialist orthodoxy. In this context, it's natural to group panpsychists, dualists, and other non-materialists together, united by their status as rebels against the system.

But that doesn't reflect the wider culture. It's hard to know what *is* the most widely accepted metaphysics of conscious, or even if most people have a coherent position on the question, but dualism — or at least, ideas about past lives, the afterlife, and so on, that seem to require dualism — is plausibly the most widespread actual belief. And, as Goff (2019, pp. 188–90) and Papineau (2020) suggest, many

---

[14] Mathews makes this claim in much of her work, see especially Mathews (2003; 2020), Goff (2019, pp. 188–95); *cf.* Skrbina (2005); Vetlesen (2019).

[15] For this reason, I am a little wary of Mathews' claim that 'the environmental crisis… is *the result of* an anthropocentric outlook that permeates the Western tradition' (2020, p. 131, italics added; *cf.* Goff's quotation of Naomi Klein on pp. 188–9, about the 'corrosive separation between mind and body… *from which* both the Scientific Revolution and the Industrial Revolution sprang', italics also added). I suspect that anthropocentric theories are more a result than a cause of environmentally destructive practices, whose ultimate origin lies more with the practical advantages that various individuals and societies were able to get from them.

people's lip service to materialism covers up an unconscious dualism. More generally, respect for empirical science, not just as a social institution over there in the universities but as a disciplining factor for one's own most heartfelt beliefs, is less widespread than we might hope. In this context, grouping panpsychism together with dualism, against materialism, seems to me a mistake. Much better would be, so to speak, a 'Monist United Front': despite their differences in the seminar room, panpsychists and materialists should be largely in agreement on most worldly matters. They can agree on the importance of following scientific consensus on empirical questions. And they can agree that, metaphysically, there's just us, and other things made of the same stuff as us. Nothing outside the world, and no deep divides or sharp boundaries within it.

One of the great virtues of *Galileo's Error* is to bring contemporary panpsychist ideas to a wider audience. And one part of that value is that, for some people, panpsychism may offer a more satisfying and appealing environmental philosophy than materialism does. My hope would be that this needn't be a point of conflict between panpsychists and materialists; the broader culture is so full of both outright science denialism and explicit or implicit dualism that the points of agreement between panpsychists and materialists are more important than their points of disagreement. This is especially so in environmental matters, where respect for scientific consensus and recognition of fundamental kinship between humans and nature are so desperately needed.

References

Bird-David, N. (1999) 'Animism' revisited: Personhood, environment, and relational epistemology, *Current Anthropology*, **40s**, pp. 67–91.

Buchanan, J. & Roelofs, L. (2019) Panpsychism, intuitions, and the great chain of being, *Philosophical Studies*, **176** (11), pp. 2991–3017.

Cartwright, N. (1983) *How the Laws of Physics Lie*, New York: Oxford University Press.

Goff, P. (2013) Orthodox property dualism + linguistic theory of vagueness = panpsychism, in Brown, R. (ed.) *Consciousness Inside and Out: Phenomenology, Neuroscience, and the Nature of Experience*, New York: Springer.

Goff, P. (2019) *Galileo's Error: Foundations for a New Science of Consciousness*, London: Penguin Random House.

Harvey, G. (2006) Animals, animists, and academics, *Zygon*, **41** (1), pp. 9–19.

Isaacson, W. (2008) *Einstein: His Life and Universe*, New York: Simon and Schuster.

Leidenhag, J. (this issue) Why a panpsychist should adopt theism: God, Galileo and Goff, *Journal of Consciousness Studies*, **28** (9–10).

Mathews, F. (2003) *For Love of Matter: A Contemporary Panpsychism*, Albany, NY: State University of New York Press.

Mathews, F. (2020) Living cosmos panpsychism, in Seager, W. (ed.) *The Routledge Handbook of Panpsychism*, pp. 131–143, London: Routledge.

Moran, A. (this issue) Grounding the qualitative: A new challenge for panpsychism, *Journal of Consciousness Studies*, **28** (9–10).

Nagel, T. (1986) *The View from Nowhere*, Oxford: Oxford University Press.

New York Times (1929) Einstein believes in 'Sponoza's God', *New York Times*, 25 April, [Online], https://www.nytimes.com/1929/04/25/archives/einstein-believes-in-spinozas-god-scientist-defines-his-faith-in.html.

Papineau, D. (2020) The problem of consciousness, in Kriegel, U. (ed.) *The Oxford Handbook of the Philosophy of Consciousness*, pp. 24–36, Oxford: Oxford University Press.

Rose, D.B. (1996) *Nourishing Terrains: Australian Aboriginal Views of Landscape and Wilderness*, Canberra: Australian Heritage Commission.

Rose, D.B. (2003) Val Plumwood's philosophical animism: Attentive interactions in the sentient world, *Environmental Humanities*, **3**, pp. 93–109.

Skrbina, D. (2005) *Panpsychism in the West*, Cambridge, MA: MIT Press.

Vetlesen, A. (2019) *Cosmologies of the Anthropocene: Panpsychism, Animism, and the Limits of Posthumanism*, Abingdon: Taylor & Francis.

Annaka Harris

A Solution to the Combination Problem and the Future of Panpsychism

Abstract: This paper supports the scientific position that panpsychism is a valid category of possible resolutions to the hard problem of consciousness, and it focuses on a solution to the 'combination problem' in panpsychism. I argue for a new way of thinking about consciousness in which consciousness is not viewed in reference to subjects, and that the concept of a 'subject' is borne of the illusion of self. Therefore, we don't face a combination problem if the notion of a subject is superfluous and consciousness itself is pervasive in the form of a field. The paper is also a more general discussion about the importance of pursuing this scientific question in the twenty-first century: is consciousness a more fundamental aspect of the universe than we have previously assumed?

1. Introduction

The great mystery of consciousness is why matter lights up with felt experience. After all, we are composed of particles indistinguishable from those swirling around in the Sun; the atoms that compose your body were once the ingredients of countless stars in our universe's past. They travelled for billions of years to land here — in this particular configuration that is you — and are now reading these words. Imagine following the life of those atoms from their first appearance in space-time to the very moment they became arranged in such a way as to start *experiencing* something.

Correspondence:
Email: contact@annakaharris.com

Many assume there is probably no felt experience associated with the microscopic collection of cells that make up a human blastocyst. But over time these cells multiply and slowly become a human baby, able to detect changes in light and recognize its mother's voice, even while in the womb. And unlike a computer, which can also detect light and recognize voices, this processing is accompanied by an *experience* of light and sound. First, as far as consciousness is concerned there is nothing, and then suddenly, magically... *something*. The mystery lies in the transition. However minimal that initial something is, experience apparently ignites in the inanimate world, materializing out of the darkness.

People often use the word 'consciousness' to refer to higher-order functions, such as self-awareness and thought. But I'm addressing consciousness in the most fundamental sense — as felt experience, regardless of the content. In my 2019 book *Conscious*, I use Thomas Nagel's distinction — whether or not it's *like something* to be an organism — to get to the heart of what I'm referring to. Is it *like something* to be you in this moment? Presumably your answer is yes. Is it *like something* to be the chair you're sitting on? Your answer will (most likely) be an equally definitive no. We can all use this simple difference as a reference point for what I mean by the word 'consciousness'. Is it *like something* to be a grain of sand, a bacterium, an oak tree, an ant, a mouse, a dog? At some point along the spectrum the answer is yes, and the great mystery lies in why 'the lights turn on' for some collections of matter in the universe.

Despite the many advances in neuroscience over recent decades, the hard problem of consciousness persists, and this has caused more scientists and philosophers to wonder whether we've been thinking about the problem backward. Is it possible that consciousness, rather than arising when non-conscious matter behaves a particular way, is an intrinsic property of matter — that it was there all along? This is the type of question Philip Goff explores in depth in his recent book, *Galileo's Error* (2019). It is also the direction my own thinking has led me.

Because our only evidence of consciousness comes from self-report and can never be confirmed first-hand by an outside observer, we are forced to make one of two assumptions: (1) felt experience arises out of non-conscious processing, or (2) consciousness is a property of matter itself (or of something even more fundamental) — a category of views typically referred to as 'panpsychism'. Cases of split-brain patients, locked-in patients, and anaesthesia awareness should give us pause when we assume that islands of consciousness cannot possibly lie beyond the awareness of the 'reporter' in a human brain. But we have labelled the very distinction between conscious and unconscious brain processes solely on the basis of what can be reported on and what cannot. This seems unavoidable, of course, but it's important to remember that all our starting assumptions about consciousness are based on reportability (and to remember that they are in fact *assumptions*). I assume that the processes taking place in my liver are unconscious because 'I' am not conscious of them, but this doesn't necessarily rule out the possibility that liver processing entails consciousness. The fact that a pregnant woman doesn't experience any of the things her unborn baby is experiencing is obviously no reason to believe the baby is unconscious.

We have a deeply ingrained intuition that systems that act like us are conscious and those that don't are not. This is reasonable, because the only source of reporting or behaviour we can interpret, and therefore have access to, comes to us from organisms most similar to us. As a result, the sciences have led with assumption #1: that consciousness arises out of complex processing (namely in brains). But is this actually the most rational starting place? The idea that consciousness emerges out of non-conscious material, in fact, represents a kind of failure of the typical goal of scientific exploration: to arrive at as simple an explanation as possible. The celebrated biologist J.B.S. Haldane, for example, argued that the notion of the 'strong emergence' of consciousness is 'radically opposed to the spirit of science, which has always attempted to explain the complex in terms of the simple... If the scientific point of view is correct, we shall ultimately find them [signs of consciousness in inert matter], at least in rudimentary form, all through the universe'.[1] The truth is that we simply don't know if consciousness only arises only in brains (or if it is a sign of complexity at all, for that matter), and we are heavily relying on an assumption when we make that case. Each transformative shift in our understanding of the universe has delivered the ego-shattering message that we're not special — Earth is not the centre of the universe, and life, including the human brain, is made up of the same particles as the stars. Perhaps it's time to relinquish our last claim to specialness. If a fundamentally new aspect of matter doesn't

---

[1] See Skrbina (2017).

enter the picture at some point, is it possible it was there to begin with and is simply part of the fundamental stuff that is all around us?

My own sense of the correct resolution of the hard problem of consciousness, whether or not we can ever achieve a true understanding, is split between a brain-based explanation and a panpsychist one. So although I'm not convinced that panpsychism offers the correct answer, I *am* convinced that it is a valid category of possible solutions and cannot be easily dismissed. In recent years, my primary focus has been what I see as a solution to the 'combination problem' in panpsychism. In the second and third sections of this paper, I will lay out my argument for why I think we don't necessarily face a combination problem. But first, I will discuss some of my recent thoughts about panpsychism in general.

As an umbrella term to designate any view of the universe in which consciousness is an intrinsic quality of matter, 'panpsychism' is very useful shorthand for this category of thought. But ultimately, I find it to be an obstacle to fruitful dialogue. I have, therefore, made a change in my writing recently and have chosen to omit the word going forward. I find it more effective to simply pose what is now considered by many scientists to be a legitimate scientific question: is consciousness a more fundamental aspect of the universe than the sciences have previously assumed? If we choose to follow assumption #2, where do we land?

When contemplating this category of hypotheses, it's important to first distinguish between consciousness and thought. We should be careful not to reflexively rail against the idea that rocks and spoons are conscious, which is obviously false when put in terms of a rock's being conscious *as a rock*. If consciousness is fundamental, all matter must entail consciousness by definition; but that doesn't mean it makes sense to specify such things as 'moon consciousness' and 'tree consciousness'. We would expect that the region of space-time occupied by a rock, say, entails consciousness because matter is present there. We can't imagine what that region of particles feels like (or even that it has a unified perspective at all, which seems unlikely). What we can be fairly sure of, however, is that it doesn't contain a human-like experience or even a single 'point of view'. Just as we wouldn't expect (the collection of atoms that make up) a rock to get up and walk or sing — that's not what atoms configured in such a way *do* — we wouldn't expect it to have a single, unified point of view. And we certainly wouldn't expect it to have anything like thoughts or intentions.

We should still assume that information in a complex and integrated form is required to produce experiences like ours. We shouldn't feel compelled to wonder whether there is a specific 'rock consciousness' any more than we're compelled to wonder whether there is a 'rock + the-five-blades-of-grass-the-rock-is-touching consciousness'. That description of consciousness is based on an anthropomorphic view, projecting separateness in isolated packages: me, you, rock, spoon. But perhaps felt experience *is* present — in a form we can't imagine — across the matter in any given area of space-time.

2. Subjects and Selves

We have different ways of using the word 'self'. There's the autobiographical self, which is the story of who I am: *my name is Annaka. I have two daughters. I'm a good swimmer...* If I wake up tomorrow with amnesia and don't remember my name or anything else about who I am, I've lost my sense of being the 'self' called Annaka, and I now feel like a very different person. But the sense of self I will be referring to in my discussion of the combination problem goes deeper than the autobiographical self and isn't necessarily bound up with a specific identity. A sense of self can still persist even if I lose my autobiographical self. It's the 'I' that amnesiacs refer to when they say, 'I don't know who I am! I don't remember my name or where I live!' The deeper sense of self is the experience of being a single, independently existing entity that has a precise centre or location and is doing the experiencing. And this concept of the self is an illusion.

The neuroscientist Anil Seth studies how this 'illusion of self' gets constructed. He writes, 'the predictive machinery of perception when directed at the self makes it seem as though there really is a stable essence of "me" at the centre of everything' (Seth, 2021). But he explains that this does not accurately represent the underlying reality: 'Just as experiences of redness are not indications of an externally existing "red," experiences of selfhood do not signify the existence of an "actual self"' (*ibid.*). In addition to processes such as interoception and memory, which help to construct an illusion of self, the brain suffers a type of change-blindness with respect to the self, further strengthening the illusion. Seth explains, 'We are becoming different people all the time. Our perceptions of self are continually changing — you are a slightly different person now than when you started reading this chapter — but this does not mean that we perceive these changes. This subjective blindness to the changing self has consequences. For one thing, it fosters the false intuition that the self is an immutable entity, rather than a bundle of perceptions' (*ibid.*). While it's admittedly a tough illusion to relinquish, we know that the experience of self does not offer us accurate intuitions about consciousness in general, and we know from people's experiences in meditation and under the influence of psychedelic drugs that consciousness persists when the default mode network quiets down and the illusion of self drops away.[2]

3. The Combination Problem

When considering a view of the world in which consciousness is fundamental, the question arises: if the most basic constituents of matter do indeed have some level of conscious experience, how is it that when they form a more complex system — such as a brain — those small points of consciousness combine to create a new conscious subject? For instance, if the individual atoms and cells in my brain are conscious, how do those separate spheres of consciousness merge to form the consciousness 'I' am experiencing? What's more, do all of the smaller, individual points of consciousness cease to exist after giving birth to an entirely new point of view? This is referred to as the *combination problem*, and according to the *Stanford Encyclopedia of Philosophy*, it's 'the hardest problem facing panpsychism' (Goff, 2001). The entry cites a passage from William James as the primary inspiration for the problem: 'Take a hundred of them [feelings], shuffle them and pack them as close together as you can (whatever that may mean); still each remains the same feeling it always was, shut in its own skin, windowless, ignorant of what the other feelings are and mean. There would be a hundred-and-first feeling there, if, when a group or series of such feelings were set up, a consciousness belonging to the group as such should emerge...' (*ibid.*). The combination problem has kept many scientists and philosophers who are otherwise willing to entertain the idea that consciousness is a fundamental property of matter from fully endorsing this area of study. However, it seems to me that the obstacle one faces here isn't a combination problem but the confusion of *consciousness* with the experience of *self*.

---

[2] For more on the illusion of self, see Chapter 5 of Harris (2019).

When discussing the combination problem, philosophers and scientists tend to speak in terms of a 'subject' of consciousness. My claim is that this is actually another way of pointing to the experience of self in its most basic form. In a paper on the combination problem, David Chalmers writes, 'How could any phenomenal relation holding between distinct subjects... suffice for the constitution of a wholly new subject?' (Chalmers, 2012). Likewise, when Greg Rosenberg discusses the related 'boundary problem' (also referenced by James), he writes, 'I start with the observation that consciousness has inherent boundaries. Only some experiences are part of my consciousness; most experiences in the world are not. Arguably, these boundaries are what individuate me as an experiencing subject in the world' (Rosenberg, 2004).

But I think it's wrong to talk about a 'subject' of consciousness. I believe that this way of framing things is entirely due to the experience of self, which we know to be an illusion. All we truly know is that content (or qualia) appears in the universe. The claim that qualia appear *to a subject* in the universe is an additional (and I think unnecessary and false) step. The most powerful sense in which there is an experience of 'I' or 'me' as a single entity to which certain experiences are presented is through the connection of experienced moments of qualia through memory. If I simply experienced green and then sharp pain and then happiness and then bright light, without memory causing these qualia to trail along, there would be no sense in which we would say this is happening to 'me' or to a subject at all. We would just say that qualia are appearing in the universe — like bubbles in a pot of boiling water. There is no subject that this content is appearing to. It's just appearing. And each quale, by definition, is always limited to that specific quale.

Viewing consciousness through the illusion of self gives rise to other categories of the combination problem that don't necessarily specify a 'subject' *per se*, but which I nevertheless believe still originate with the illusion of self: the *quality combination problem* and the *structure combination problem* (Chalmers, 2012). It is in this context that philosophers often refer to the 'privacy' of conscious experience. But again, this strikes me as a perspective from within the illusion of self. Each individual quale can be considered private, but only in the sense that it is defined by some particular qualities and not others — for changing it or combining it with another quale would, by definition, change the quale. But it is not private *for* any subject. When we look at each individual moment of experience — red, pain, pressure, etc. — we can see that qualia arise in the universe in and of themselves (and, theoretically, there could be areas in space-time where pure consciousness exists with no qualia).

Additionally, qualia are not static, and this can be made evident empirically through introspective training such as meditation. Breaking down the flow of experiences into a stream of 'present moments' can help shift our sense of what we typically call 'subjects' or 'selves'. In each moment, new content appears, but the content is clearly not being experienced by a subject. Some Buddhist teachings more accurately refer to the present moment as the 'passing moment' and, when zeroing in on these passing moments, one notices that the red of the flower (sight) and the whistle of the bird (sound) don't arise simultaneously, nor are they solid or concrete in any real sense. Each quale is experienced sequentially and as a process, not as a static object. Then, through memory, the illusion of a full picture is given. But when one is carefully attending to each passing moment, it becomes clear that those 'memory snapshots' are not an accurate rendering of what the experience actually entailed.

In terms of privacy, when we realize that there is no solid centre we can label 'you' or 'me', it makes no sense to talk about where my consciousness ends and yours begins. Content is arising, and some of that content is shared across time through memory, as yet more content. Your perception of yellow isn't 'yours'. It's simply an experience of yellow arising in the universe, derived from interacting forces and fields. It's not private in the typical sense. There is no self for it to be private *for*.

4. A Solution to the Combination Problem

Rather than an obstacle to theories that place consciousness in a fundamental role, the combination problem may be a reason to favour the proposition that consciousness is a fundamental feature of the universe in the form of a continuous, pervasive field, analogous to space-time. Just as space-time and gravity have an interactive relationship, consciousness might be thought of as a fundamental 'field' that interacts with, and is integral to, matter. We typically don't think of space-time as bits and pieces that build on one another (it's simply *everywhere*), and I don't think we should be tempted to think of consciousness, if it is indeed a pervasive field, as divisible into building blocks either. Rather, it makes more sense to talk about a field that contains a range of *content* — the content depending on the other forces or fields it's interacting with. In the same way that gravity is a two-way street — matter warps space-time and the shape of space-time determines how matter moves — a consciousness field would imbue matter with another property, giving rise to the range of content experienced. In this view, content is perhaps divisible, but consciousness isn't. Therefore, consciousness is also not interacting *with itself*, as it would be in the act of 'combining'. Considering consciousness to be fundamental allows for matter to have a specific internal character everywhere, in all its various forms. It doesn't necessitate a reductionist explanation in which all qualia are made up of little 'quale building blocks'. It also allows for content to overlap in space, similar to the way sound waves at different frequencies coexist in the same space by essentially 'passing through' one another. The science journalist George Musser uses an analogy to sound to help explain the stratification of nature by scale: 'Sounds of long and short wavelengths are oblivious to each other; if you sound a deep bass note and a high treble pitch simultaneously, each ripples through the room as though it were the only sound in the world... These waves overlap in the three dimensions of space through which they propagate, yet they're independent of each other, as if they were located in different places' (Musser, 2015).

If consciousness is fundamental, then the questions that prompt the combination problem are potentially the same as all the other questions we might ask about space-time in which we *don't* anticipate this problem. All matter would entail consciousness, and complex systems, such as human brains, would give rise to certain types of content in those locations in space-time. Even if each individual atom has its own experience, *consciousness itself* is not necessarily isolated. The matter might be isolated, and therefore the *content* associated with the consciousness at that location would be isolated. But consciousness itself would not be said to be isolated. Again, we can think of consciousness as analogous to space-time: how it's affected by matter depends on the matter in question (its mass, in the case of space-time). Similarly, a consciousness field might be 'shaped' by matter in terms of experiential quality or content. This line of thinking yields interesting questions: how does the content that appears in an area of consciousness depend on the configuration of matter present in that location in space-time? Are there sometimes countless experiences of overlapping or merging content in the same location?

In a related conversation I had with the neuroscientist Christof Koch, we discussed what might result from a hypothetical experiment in which two brains were connected as successfully as the two hemispheres of a single brain are connected. Since various experiments with split-brain patients — people whose right and left hemispheres have been surgically disconnected — have shown that the contents of consciousness can be separated, would two brains wired together produce a new, integrated mind? If Christof and I had our brains wired together, for instance, would it create a new Christof-Annaka consciousness — a new single point of view? Would a new mind be produced, with access to all the content that had previously been experienced separately by our brains — all our thoughts, memories, fears, abilities, etc. — constituting a new 'person'?

Even if the answer is yes, I don't think we encounter a combination problem in this thought experiment. We run into problems only if we see my and Christof's conscious experiences as 'selves' or 'subjects' — permanent structures of consciousness with fixed boundaries. In the instance of connecting two brains, we might simply have an example of consciousness changing its content or character, in the same way the content of your consciousness changes predictably along with the properties of the physical world. For example, when your eyes are open and then you close them, the trees and sky are available to your field of view and then they're not (or you could say that an *experience of trees and sky appears, followed by an experience of darkness*). When you dream, your brain states produce an experience of environments quite different from your actual surroundings, and perhaps even the feeling that you are a different person altogether. During both my pregnancies, I found myself experiencing drastic variations in the contents of my consciousness — sensations in my uterus I had never before known were on the menu of experience, an obsession with tomatoes and tomato sauces in every form, feelings of panic and other more amorphous emotions, physical pain, insomnia. I didn't feel like 'myself'. And I expect I wouldn't feel like myself during a mind meld with a 62-year-old male neuroscientist either. But it doesn't necessarily point to a *combination problem* for consciousness. Even in our daily lives, content comes and goes, and consciousness itself can seem to flicker in and out.

In the case of connecting two brains, the part of one brain that gives rise to the memory of mailing a letter yesterday, for example, could now give rise to that memory alongside all the qualia (including memories) that are possible in other areas of this new, merged brain. But in connecting two brains, no selves disappear, nor does a new 'self' appear. When I don't remember something I've done in the past,

I don't tend to think of the 'past me' as someone else. Similarly, am I a new self when I learn something new? The past me had no idea what it was like to ski; the present me does. We're not at all tempted to say that the old me has ceased to exist, giving birth to a new mind — an updated version of me that can ski: 'The old me died and gave birth to a new me who skis!' This is clearly not the correct way to think about it. As long as we're speaking in terms of selves, we are essentially new selves in each new moment. I am a different self now from the self I was 20 minutes ago, with new content appearing because my brain is in a different state than it was in 20 minutes ago (less so than if I've ingested LSD during that time, rather than having just written a few paragraphs, but new just the same). So, too, for merged brains — they would share content and, therefore, some content would appear together (or in new forms) that wasn't possible before. It may seem legitimate to ask, 'Where did those two original subjects go?' But they didn't go anywhere. They weren't there to begin with!

The experience of a human brain is largely one of confusing consciousness with the experience of self, and we run into a combination problem only when we drag the concept of a 'self' or a 'subject' into the equation. The solution to the combination problem is that no 'combining' at all is going on with respect to *consciousness itself*. Consciousness could persist as is, while the character and content changes depending on the arrangement of the specific matter in question. In my analogy to the pot of boiling water, the bubbles are the content, and the water is consciousness. Maybe content is sometimes shared across large, intricately connected regions and sometimes confined to very small ones, perhaps even overlapping. If two human brains were connected to each other, both people might feel as if the content of their consciousness had simply expanded, with each person feeling a continuous transformation from the content of one person's consciousness to the whole of the two, until the connection was more or less complete. It's only when you insert the concepts of 'him', 'her', 'you', and 'me' as discrete entities that the expanding or merging of content becomes a combination problem.

Here I'm reminded of the classic device of characters switching places in a story or a film. But when we look closely at what this actually entails, it becomes obvious that there's no 'self' to transport from one person to another. Being someone else would be no different from what it's already like to be that person. It seems paradoxical, but we end up simply stating the obvious: 'That's what it's like to be over there as that configuration of atoms, and this is what it's like to be over here as this configuration of atoms.' It's analogous to saying: 'The configuration of atoms that compose a leaf result in all its expected leaf properties, but if you take all of those atoms and reassemble them into a liquid, they will take on the expected properties. That's what molecules *do* in that configuration, and this is what they *do* in this configuration. Likewise, that's what molecules *feel like* in that configuration, and this is what they *feel like* in this configuration.'

Before we can move forward with a theory in which consciousness plays a fundamental role in the universe, I think it's imperative that we weed out and untangle the illusion of self from our description of consciousness. If consciousness itself doesn't combine, then we no longer face a combination problem. The experience of consciousness need not be continuous or maintained as an individual self or subject. Nor is it necessarily extinguished when the smaller constituents of matter combine to make more complex systems, like brains. The human sense of being a self, along with an experience of continuity over time through memory (in which trails of previous qualia remain in circulation), may in fact be a very rare form of content. Is it possible that alongside the conscious experience of 'me' is a much dimmer experience of each individual neuron, or of different collections of neurons and cells in my body and beyond, like an orchestra of sound waves? Perhaps the universe is literally teeming with consciousness — with content flickering in and out, connecting through memory, separating, overlapping, flowing, in ways we can't quite imagine — ruled by physical laws we don't yet understand.

References

Chalmers, D.J. (2012) The combination problem for panpsychism, [Online], http://consc.net/papers/combination.pdf.

Goff, P. (2001) Panpsychism, in Zalta, E.N. (ed.) *Stanford Encyclopedia of Philosophy*, [Online], https://plato.stanford.edu/entries/panpsychism/.

Goff, P. (2019) *Galileo's Error*, London: Rider.

Harris, A. (2019) *Conscious*, New York: Harper.

Musser, G. (2015) *Spooky Action at a Distance*, New York: Scientific American/FSG.

Rosenberg, G. (2004) *A Place for Consciousness*, Cambridge: Cambridge University Press.

Seth, A. (2021) *Being You*, New York: Dutton.

Skrbina, D. (2017) *Panpsychism in the West*, Cambridge, MA: MIT Press.

Keith Frankish[1]
Galileo's Real Error

Abstract: *Goff argues that Galileo erred in denying that sensory qualities are present in the physical world and that we should correct his error by supposing that all matter has an intrinsic conscious aspect. This paper argues that we should be open to another theoretical option. Galileo's real error, I argue, was not about the location of sensory qualities, but about their very existence. Like most people, Galileo assumed that sensory qualities are instantiated somewhere. I argue that this is a theoretical assumption which can and should be questioned. If we drop it, we can give a natural account of the function of sensory quality talk and explain how our puzzlement about consciousness arises.*

Galileo's error, according to Philip Goff, concerned the location of sensory qualities, such as colours, sounds, smells, and tastes. The new science Galileo championed could not describe such properties, and he concluded that they were not present in the physical world, but existed only in our minds, produced there by the impact of physical stimuli on our sense organs.

Error or not, Galileo's move was a natural one in the early modern period. Where we encounter a world richly arrayed with sensory qualities, science finds only physical structures and processes, describable in mathematical terms. The mind, which was widely agreed to be an immaterial soul, provided a convenient repository for these qualities, along with other features recalcitrant to mechanistic explanation, such as intellectual thought and free will. The fact that sensory qualities varied with observers made this move even more natural.

Correspondence:
Email: k.frankish@gmail.com

[1] Sheffield University, UK.

The move looked less attractive in the mid-twentieth century, when mind–body dualism was widely rejected in favour of some form of mind–brain identity. If the mind is the brain, then it is a far less hospitable home for sensory qualities. Science does not find qualitative properties inside the brain any more than it does outside it. So the scene was set for the contemporary debate about consciousness.

In response, philosophers divided into two broad camps. The first retreated to a qualified form of dualism, holding that the brain has a soul-like *aspect*, composed of non-physical sensory qualities, which reveal themselves to us in a primitive way. The second camp maintained a materialist line, insisting that, despite appearances, sensory qualities are brain properties, which are known to us through mechanisms of introspection. As Goff explains, both camps face serious problems — the dualists that of explaining the relation between the brain's physical and non-physical properties, the materialists that of persuading us that sensory qualities are nothing more than states of soggy pink-grey brain tissue.

Goff proposes an alternative, which involves returning qualities to the physical world. He doesn't return them straightforwardly, however, by locating colours on the surfaces of objects, tastes in food, sounds in the air, and so on. He takes it as a datum that the sensory qualities with which we are acquainted are mental ones — they are forms of conscious experience, which constitute our subjective life and are known to us with more certainty than anything else (Goff, 2019, pp. 3–5). Rather, he proposes a form of panpsychism, according to which conscious experience is the intrinsic nature of all matter and the qualities we experience are constructed from the primitive qualities of the particles that constitute our brains. In effect, Goff keeps qualities in the mind but distributes minds throughout the world.

It's an ingenious idea, and Goff argues for it clearly and powerfully. The view faces its own problems, however, particularly in explaining how primitive sensory qualities combine to form complex ones. Moreover, as I've argued elsewhere, it cannot explain why sensory qualities have psychological and ethical significance (Frankish, 2021). Still, I agree with Goff that it is a mistake to treat sensory qualities as either identical with, or emergent from, neural ones.

Is there another way of resisting the Galilean relocation? Let us begin by imagining an alternative history in which Galileo responded differently to the problem of sensory qualities. In this alternative timeline, people have long believed in a form of panpsychism, though one different from Goff's. They believe that all objects, including the various parts of their own bodies, have immaterial souls, whose nature they can sense intuitively. However, they do not believe that *they themselves* have a soul — a personal soul, distinct from the souls of their organs and limbs. They believe that they are complex machines composed of soul-possessing parts. So, it does not occur to alt-Galileo to locate sensory qualities in his own soul. Instead, he locates them in the souls of the objects to which they appear to belong — redness in the tomato soul, blueness in the sky soul, pain in the toe soul, and so on. The fact that it is intuitively obvious that sensory qualities are located in the objects around us makes this move even more natural for him. Having consigned sensory properties to the souls of objects, alt-timeline scientists get on with the business of explaining the behaviour of objects and our reactions to them in purely physical terms.

Of course, this view faces problems, such as explaining why objects appear differently to different observers and finding a location for the qualities experienced in dreams. Alt-timeline philosophers come up with ingenious answers, suggesting, for example, that souls change their properties depending on the observer and that there are disembodied souls visible only to dreamers.

Science develops, and by the mid-twentieth century alt-timeline philosophers have ceased to believe in souls and have to decide what to say about sensory qualities now. As in our timeline, there are two camps. The first say that, though objects lack souls, they still have a soul-like aspect, whose character we intuit in an immediate way. The second say that, despite appearances, sensory qualities are physical features of objects — reflectance properties of surfaces, and so on. Both camps face similar problems to those facing their counterparts in our timeline. (Then a brilliant philosopher writes a book called *Galileo's Error*, which argues the sensory qualities of objects are compounded from the primitive sensory qualities of their physical elements…)

The point of this story is not that we should adopt alt-Galileo's view instead of Galileo's. It is that the two views are *parallel*. To claim that sensory qualities belong to a mental arena — a soul, or soul-like aspect of a brain — is to make a theoretical proposal every bit as speculative as that of claiming that they belong to the souls of objects. It is not a datum that we are immediately acquainted with mind-located sensory qualities, but a *theory* (*contra* Goff, 2019, pp. 10–11). The alt-timeline philosophers don't conceive of themselves as having an inner world populated with sensory qualities. For them, all the qualitative richness is located in the space around and inside their bodies. (No doubt the same goes for many people in our own timeline, but the alt-timeline philosophers hold the conception explicitly and in spite of scientific and philosophical objections.)

The moral, then, is that the starting point for thinking about consciousness is not an introspective datum — the existence of mind-located sensory qualities (or 'phenomenal' properties). Rather, it is a *problem*: how to reconcile our everyday image of the world as arrayed with sensory qualities with a scientific image of the world that has no place for them. The idea that sensory qualities are located in our minds, like the alt-timeline claim that they are located in object souls, is a theoretical response to this problem, which is shaped by the theorists' intellectual traditions. Each theory has its costs and benefits (the mind version easily explains the observer relativity of sensory qualities but denies their spatial locatedness, while the object-soul theory has the opposite virtues).

Now, since we are in the realm of theory when talking about sensory qualities, maybe we should be open to other theoretical options regarding them. Maybe Galileo's real error — and alt-Galileo's, too — wasn't about the location of sensory qualities at all, but about their very *existence*.[2]

Suppose that instead of puzzling over where sensory qualities are located, Galileo had asked what our *talk* about sensory qualities is doing. This is a much easier question. At a first pass, such talk tracks dispositional features of objects. Each quality concept tracks a worldly feature (often highly disjunctive) which produces a distinctive set of psychological reactions in us — priming effects, beliefs, desires, emotions, behavioural dispositions, etc. Science can describe these features and reactions in complete detail, and evolutionary biology and psychology can explain why we are sensitive to the features and why they evoke the reactions they do.

Things get difficult only if we try to find features within this story, or within some parallel story about non-physical processes, with which we can identify sensory qualities themselves. Like almost everyone else who has thought about it, Galileo assumed that sensory qualities were instantiated *somewhere*, and so had to invent, or co-opt, a suitable substrate for their instantiation. And maybe that's the big

---

[2] This idea is not new, of course, and it has been defended at length by Daniel Dennett, whose work inspires the sketch that follows (see, for example, Dennett, 1991).

mistake — Galileo's Real Error. Maybe sensory qualities are a sort of illusion (Frankish, 2016).

Maybe what's happening is something like this. As well as tracking features of the world, our brains also track the complex reactive patterns these features evoke in us and *misrepresent* these reactive patterns as simple qualitative aspects of the tracked features. Thus, when we conceptualize an object as having a certain sensory quality — redness, say — we are in effect conceptualizing it as affecting us like *this* — where the demonstrative gestures at the complex reactive pattern triggered by red things. We are representing worldly features as ones that have a certain *significance* for us.

Such a view explains why sensory qualities seem to have a dual nature — located in objects but dependent on us. The reason is that sensory quality concepts track features of objects but represent those features as infused with qualities that express the reactions they produce in us. There is nothing mysterious about this, provided we don't ask where the qualities really are.

Am I serious? Am I really denying that the blue of the sky through my window is not real, that it is not instantiated in all its dazzling blueness? I *am* denying it, though there's a sense in which I still can't help *taking* the blueness to be real. It is part of my subjective 'take' on the world — the huge set of automatic psychological reactions to stimuli constructed by brain systems over which I have no control. This take is a psychological condition, and a thing is part of it if my brain produces reactions indicative of the thing's reality. I cannot bypass these reactions and encounter the world raw — though I can, of course, learn to distrust them and reflectively correct my beliefs about what the world is really like. This goes for sensory qualities as much as for any other aspect of the world. They seem undeniably real because our brains produce psychological reactions strongly indicative of their reality, and when we tell ourselves a story about what we are experiencing, they figure in it as peremptory presences.

I should stress that I am not suggesting that it is a *fault* in our brain systems that they construct a quality-suffused take on the world. Far from it. By doing so they enable us to pick out features by their significance for us — to simultaneously express what's happening and what it *means*. Like art, sensory quality reports express important truths through fictional means. It is not a mistake to take such reports seriously; but it is a mistake to take them *literally*. If we ask where sensory qualities are actually located, then the answer is that it's in the same place as Hamlet's indecision and Anna Karenina's intelligence.

This is only a sketch, of course, but I believe it points to a coherent theoretical option, which in turn opens up new lines of scientific enquiry. But if we are to take it seriously, we must stop making Galileo's Real Error.

References

Dennett, D.C. (1991) *Consciousness Explained*, Boston, MA: Little, Brown.

Goff, P. (2019) *Galileo's Error*, London: Pantheon Books.

Frankish, K. (2016) Illusionism as a theory of consciousness, *Journal of Consciousness Studies*, **23** (11–12), pp. 11–39. Reprinted in Frankish, K. (ed.) (2017) *Illusionism as a Theory of Consciousness*, Exeter: Imprint Academic.

Frankish, K. (2021) Panpsychism and the depsychologization of consciousness, *Aristotelian Society Supplementary Volume*, **95** (1), pp. 51–70.

Michelle Liu[1]

Qualities and the Galilean View

Abstract: *It is often thought that sensible qualities such as colours do not exist as properties of physical objects. Focusing on the case of colour, I discuss two views: the Galilean view, according to which colours do not exist as qualities of physical objects, and the naïve view, according to which colours are, as our perception presents them to be, qualities instantiated by physical objects. I argue that it is far from clear that the Galilean view is better than the naïve view. Given the arguments in this paper, the naïve view ought to be taken seriously. The discussion here appeals especially to theorists who, like Goff, are already convinced that the quantitative language of physical science fails to capture all qualities.*

> What is yellow? pears are yellow,
> Rich and ripe and mellow.
> What is green? the grass is green,
> With small flowers between.
> — Christina Rossetti.

1. Introduction

The world around us seems to instantiate sensible qualities. Emeralds are green; lemons are sour; thunderstorms are loud. Our experience of the world not only presents objects as instantiating sensible qualities, but also presents these qualities as having certain qualitative natures.

Correspondence:
Email: y.liu43@herts.ac.uk

---

[1] University of Hertfordshire, Hatfield, UK.

Our perception of a green object, in addition to presenting the object as green, presents what green is like, i.e. its qualitative — as opposed to quantitative — nature (see Kalderon, 2007, p. 563). But Galileo questioned the existence of sensible qualities as properties of physical objects. He writes:

> I think that tastes, odours, colours, and so on are no more than mere names so far as the object in which we place them is concerned, and that they reside only in consciousness. Hence if the living creature were removed, all these qualities would be wiped away and annihilated. (Galileo, 1623/1996, p. 274)

Strictly speaking, Galileo does not deny the existence of sensible qualities. He reduces them to properties that 'reside only in consciousness'. But, in doing so, he denies the existence of sensible qualities as we ordinarily understand them — as properties of physical objects with certain qualitative natures manifested to us in our experiences of them. Here, I shall use the term 'sensible qualities' to refer to properties of this kind. Correspondingly, the term 'colours' is used to refer to qualitative colours, i.e. colours whose qualitative natures are manifested to us in perception. The claim that sensible qualities thus understood do not exist is referred to here as 'the Galilean view'.[2]

In contemporary philosophy of mind, the Galilean view represents a 'scientifically enlightened common sense' (Allen, 2016, p. 176) and is widely held among philosophers working on the hard problem of consciousness. Once the sensible qualities of physical objects are eliminated from our ontology, the only qualities which seem to have been left out by the quantitative language of the physical sciences are the qualia (i.e. phenomenal properties or phenomenal qualities) of conscious subjects. Goff writes:

> Galileo the philosopher created physical science by setting the sensory qualities outside of its domain of inquiry and placing them in the conscious mind. This was a great success, as it allowed what remained to be captured in the quantitative language of mathematics… However, those sensory qualities have come back to bite us, as we now seek a scientific explanation not only of the inanimate world but also of the conscious mind. (Goff, 2019, p. 21)

Like many others, Goff takes qualia in the mind, not sensible qualities in the world, as presenting a challenge to the physicalist worldview.

---

[2] Some theorists refer to this claim as the 'Galilean intuition' (Boghossian and Velleman, 1989; Allen, 2016). I think it is misleading to call this an 'intuition'.

The Galilean view, however, is fundamentally at odds with the *manifest intuition* that sensible qualities, in the sense understood here, exist (see Moran, this issue). The manifest intuition is particularly compelling in the case of colour, which I shall focus on in this paper. The intuition is based on a *phenomenological datum*. Consider the colour *green*. For a normal perceiver, in having a perceptual experience of a green object under a standard condition, it *phenomenally appears* to her that the object is green, and that green has a certain qualitative character. Perception also does not present green as any of the physical properties described by colour science, e.g. surface spectral reflectances (dispositions of surfaces to reflect certain proportions of incident light at certain wavelengths).[3]

The question of whether or not to take the phenomenological datum at face value, and what metaphysical conclusions should be drawn from it, divides those who reject the Galilean view from those who endorse it. In this paper, I shall pitch the Galilean view against what I call 'the naïve view' of colour. The main example of the latter is *colour primitivism*, according to which colours are irreducible, non-disjunctive, intrinsic, qualitative properties of physical objects (e.g. Broackes, 1992; Campbell, 1993; Yablo, 1995; McGinn, 1996; Gert, 2008; Allen, 2016). The main example of the Galilean view is *colour eliminativism*, according to which colours as properties of physical objects simply don't exist (e.g. Hardin, 1993; Maund, 2006; Chalmers, 2006).[4] Galileo himself is best interpreted as an eliminativist about colour (Boghossian and Velleman, 1989). Goff (2019) readily agrees with Galileo and summarily eliminates colours from the world.[5]

---

[3] Some theorists (e.g. McGinn, 1996, p. 542; Tye, 2000, pp. 152–3; Chalmers, 2006, p. 66) take perception to present colours as *intrinsic* features of physical objects (i.e. as independent of any relations to other objects and subjects) in the same way that shape is presented. I think it is less clear that colour phenomenology itself reveals colours as intrinsic. But it is certainly part of our common-sense view that colours are intrinsic — a tree is still green even when the sun goes down or there is no one there to see it (see Roberts, Andow and Schmidtke, 2014, for empirical evidence).

[4] Eliminativist primitivist views (e.g. Pautz, 2020), which take colours to be primitive properties of physical objects but not actually instantiated, and mentalistic views, which take colours to be mental entities, e.g. properties of visual fields but projected onto physical objects (Boghossian and Velleman, 1989), also count as colour eliminativism.

[5] In so far as the Galilean view is supposed to capture Galileo's own view, and by extension Goff's view, of colour (though it is less clear what Goff actually thinks of colour), I shall exclude from the Galilean view *colour physicalism*, which identifies colours with either the surface reflectance properties of physical objects (e.g. Byrne and Hilbert, 2003) or the lower-level microphysical properties that realize the surface

It is far from clear that the Galilean view is better than the naïve view. As I argue, the Galilean view involves significant costs and challenges, and, in contrast, advocates of the naïve view can make plausible manoeuvres in response to common objections. The goal here is not to argue that the naïve view fares better than the Galilean view. Rather, it is to show that the naïve view should be taken seriously instead of being quickly dismissed as is often done.[6]

The structure of the paper is as follows. §2 critically assesses the Galilean view. §3 expounds the naïve view and discusses two objections. §4 concludes by considering the implications of endorsing the naïve view for the problem of consciousness.

2. The Galilean View

The Galilean view commits to an error theory that renders much of our discourse surrounding colours erroneous. Consider:

(i) The jacket is red.
(ii) The redness of the jacket caught Noor's eyes.
(iii) The two jackets are different because one is pink and the other is red.
(iv) The Chinese village looks festive because there are red lanterns everywhere.
(v) Paul Klee's paintings demonstrate masterful arrangements of colour.

(i) is an ordinary attribution of colour; (ii) is a causal explanation that appeals to the colour of an object; (iii) is an explanation of object differences in terms of their colour differences; (iv) is an explanation that appeals to colour symbols; and (v) is a critical statement that appeals to the colours of paintings. On the Galilean view, all these ordinary attributions and explanations are strictly speaking false. In so far as philosophical theories of colour should take our common-sense conception of colour into consideration, an assumption shared by

---

reflectance profiles (e.g. Jackson, 1996); and *colour relationalism*, the view that colours are dispositions to cause certain experiences in perceivers in standard conditions (e.g. Johnston, 1992) or relations held between objects and perceivers under certain conditions (e.g. Cohen, 2009). Colour physicalism and relationalism also face familiar problems (e.g. Boghossian and Velleman, 1989; 1991; McGinn, 1996; Pautz, 2006), such that eliminativism is arguably to be preferred over them and are thus pitched against the naïve view.

[6] Moran (this issue) presupposes the naïve view and explores some of its consequences.

many in the colour debate (e.g. Johnston, 1992; Campbell, 1993), an error theory of colour is unpalatable and counts as a substantial cost of the Galilean view.

In response, Boghossian and Velleman (1989, p. 101) argue that talk of colours is analogous to talk of the sun rising, the falsity of which 'makes no difference to everyday life'. I think taking the Galilean view seriously does make a difference to how we value some of our experiences and their objects. Consider two scenarios. Imagine a world, w_1, in which objects instantiate qualitive colours. People in w_1 admire exotic birds such as crimson rosellas for the beautiful hues of their feathers. Imagine another world, w_2, in which objects are colourless and crimson rosellas are grey. Imagine further that human beings in w_2 are completely colour-blind. But the scientists in that world have invented a harmless pill which affects the visual system in such a way that it can project vibrant colours onto objects, such as red and blue onto colourless crimson rosellas, so they appear just as they do in w_1.

It seems intuitive to say that we would prefer to be in w_1 even though the pill in w_2 guarantees phenomenally identical colour experiences. The intuition here echoes the point made by Nozick's 'experience machine' thought experiment. Nozick (1974) asks us to consider a machine that can simulate all pleasurable experiences. Most of us are not inclined to plug ourselves into such a machine, because the extent to which we value our experiences depends not only on their phenomenology but also on their veridicality, i.e. whether our experiences actually correspond to reality. The experience of admiring a crimson rosella is more valuable in w_1 than in w_2 (after having taken the pill), because subjects in w_1, unlike those in w_2, are actually connected to the colours of the bird. Such a connection, to quote Nozick, 'is valuable *in itself*' (1989, p. 106).

Of course, the world according to the Galilean is not as unappealing as a world where people are all plugged into Nozick's experience machines. One's experiences of friends and family to be projections of one's mind is far worse than for one's colour experiences to be so. Indeed, the point being made here — namely, that the Galilean view affects how we value certain visual experiences — does not apply *equally* to all our colour experiences. It applies especially to a class of our aesthetic experiences, which most of us do value. In aesthetically appreciating a painting or a bird, we often take ourselves to be attending to and appreciating the colours of these objects themselves. If the crimson rosella which I so admire does not in fact instantiate these vibrant colours, it seems that my admiration is simply misdirected, and as a result I might value the aesthetic experience somewhat less. According to the Galilean view, our world is identical to w_2 except that there is no need for us to take the pill since our mind automatically projects colours. Just as aesthetic experiences of crimson rosellas seem less valuable in w_2 than in w_1, the Galilean view would diminish the value of our aesthetic experiences of colourful objects.

Corresponding to this difference in how we value our aesthetic experiences is a difference in how we value the objects of these experiences. One would naturally value crimson rosellas in w_1, which actually instantiate vibrant colours, more than the colourless ones in w_2. Thus it is no surprise that Keats, writing of the colours of the rainbow in *Lamia* (1990, p. 320), laments: 'Do not all charms fly? At the mere touch of cold philosophy?' Goff also laments the disenchantment of nature by the modern scientific worldview which, he points out, 'seems to present us with an immense universe entirely devoid of meaning' (2019, p. 216). Goff argues that his panpsychism 'offers a way of "re-enchanting" the universe' (*ibid.*, p. 217), because 'on the panpsychist worldview, humans have a deep affinity with the natural world: we are conscious creatures embedded in a world of consciousness' (*ibid.*, p. 191). But the world would surely be in less need of re-enchantment were it not stripped of the colours, sounds, and other sensible qualities that ground much of its beauty.

Furthermore, advocates of the Galilean view are faced with the task of supplementing their error theory with an alternative explanation of our ordinary colour discourse. In particular, they are obliged to provide an explanation for the phenomenological datum which underpins this discourse — if colours don't exist as qualities of physical objects, why do they *seem* to exist? Here, advocates of the Galilean view would appeal to the notion of systematic misrepresentation: colour experiences represent physical objects as having colours that they don't possess. Consider an elaboration on this by Chalmers (2006). Chalmers holds the view that colours presented to us in perceptual experience, or what he calls 'perfect colours' (which he takes to be simple, qualitative properties of physical objects), are not instantiated in our world (for reasons I will turn to in the next section). Nevertheless, colour experiences represent uninstantiated perfect colours (see also Pautz, 2020). Chalmers calls this representational content of perceptual experience 'Edenic content', satisfied only in an 'Edenic world' where perfect colours are instantiated.

Chalmers' account raises an initial worry. We can represent in thought uninstantiated properties such as *being a unicorn*, which presumably is a complex property whose representation involves the representation of a horse with a horn. But how do we come to *perceptually* represent seemingly simple but uninstantiated qualities like colours? In response, Chalmers (2006, p. 83) claims that there are other examples where uninstantiated seemingly simple properties are represented in perception. He points to a certain version of the Humean view of causation,[7] on which we perceptually represent the simple property or relation of causation in our experience even though no such causation exists in our world. Chalmers' example appeals to 'phenomenal causality' — the idea that causation is automatically perceived — which was studied and argued for, most notably, by Michotte (1963) using launching events where a moving object contacts a stationary object and the latter starts moving in the same direction upon contact. However, causation certainly does not seem be presented in perceptual experience in the same way that colours are. Many subjects report no causal perception, and whether we perceive causation is also influenced by prior experience and knowledge (e.g. Schlottmann and Anderson, 1993). In contrast, the existence of colour is just as perceptually salient and persistent as that of shape and size. If there are no other robust examples where seemingly simple and uninstantiated properties are perceptually represented, then colour experience, which on Chalmers' picture systematically represents uninstantiated colours, remains mysterious.

In sum, advocates of the Galilean view face the challenge of providing a fully satisfactory explanation for the phenomenological datum that does not render the projection of colours as properties of physical objects mysterious. Moreover, even if such an account is available, the Galilean view still faces the cost of a systematic error theory which potentially diminishes the value of some of our experiences and their objects.

3. The Naïve View

The naïve view takes the phenomenological datum about colour at face value. There are two tenets to the view. First, on this view colours *exist* as qualities of physical objects. Second, this view draws on a

---

[7] On the Humean view, causes and effects are merely constantly conjoined events.

close connection between *colours* and *colour perception*, such that the qualitative natures of the former are *manifested* through the latter.

Let me elaborate on the second tenet. Regarding *colour perception*, the naïve theorist can be either a relationalist or a representationalist. On relationalism, in perceiving a green object, I am related or acquainted with the greenness instantiated by the perceived object. On representationalism, in perceiving a green object, my experience represents the property green. The notion of *manifestation* is closely related to the thesis of *revelation* about colour, according to which, as Johnston (1992, p. 223) puts it, '[t]he intrinsic nature of canary yellow is fully revealed by a standard visual experience as of a canary yellow thing'. Depending on what counts as the 'intrinsic nature' of a colour, *revelation* may turn out to be unnecessarily strong. On a liberal conception of 'intrinsic nature', *manifestation* need not commit to the idea that the intrinsic nature of a colour is fully revealed — only its qualitative nature.

The naïve view preserves the common-sense view about colour and the validity of our ordinary colour judgments. Indeed, one might think that the phenomenological datum provides *prima facie* support for the naïve view. However, it faces objections. I shall discuss two common objections here — the argument from science (e.g. Chalmers, 2006), and the argument from perceptual variation (e.g. Berkeley, 1734; Hardin, 1993). The aim here is to show that these considerations do not give compelling reasons to eliminate colours from physical objects — at least not for a theorist who, like Goff, is already convinced that our quantitative science fails to include all qualities.

According to the argument from science, colours as we naïvely conceive them don't feature in the scientific explanation of colour perception, and if they are irreducible to physical properties then they are causally idle and should not be admitted in our ontology. This objection parallels the causal exclusion argument against non-reductive views of consciousness (Kim, 1993). It seems reasonable to think that, if there is an adequate solution in the latter case, the same solution will also apply to the colour case (Campbell, 1993; Yablo, 1995; Allen, 2016; Cutter, 2018). But, more importantly, if we agree with Goff that quantitative science cannot in principle accommodate phenomenal or sensible qualities in its description of our world, then it is no surprise that colours do not feature in our scientific explanations. If qualia are not eliminated simply because they may be causally idle, nor should colours be.

The argument from perceptual variation builds on the fact that an object's colour can appear in different ways. Such variations come in three categories (see Allen, 2016): (i) intrapersonal variations, where an object might appear to have different colours to the same perceiver under, say, different illumination conditions; (ii) interpersonal variations between colour-blind subjects and 'normal' subjects, as well as between 'normal' subjects, e.g. an object might appear unique green to one perceiver and yellowish green to another under the same conditions; and (iii) interspecies variations where the same object may appear to have different colours to members of different species due to differences in photoreceptors, mechanisms of visual processing, and sensitivities to different ranges of the electromagnetic spectrum.

Let C_1 and C_2 be two determinate colours whose qualitative natures are revealed by their veridical appearances. S_1 and S_2 can be two perceivers from the same or different species, or the same perceiver under different circumstances. Given perceptual variation, an object x can appear C_1 to S_1 and C_2 to S_2. This raises the question of which colour x in fact has; that is, which of x's colour appearances is veridical. Consider the following argument:

(1) If x's C_1 appearance and C_2 appearance are both veridical, then x has both C_1 and C_2 all over at the same time.
(2) Nothing is both C_1 and C_2 all over at the same time [INCOMPATIBILITY].
(3) It is not the case that x's C_1 appearance and C_2 appearance are both veridical.
(4) Either only one of the colour appearances is veridical or neither is.
(5) It is not the case that only one of the colour appearances is veridical.
(6) Neither of the colour appearances is veridical.
(7) Colour appearances are systematically non-veridical, i.e. objects never have the colours they appear to have. (Adapted from Kalderon, 2007, p. 567.)

(1) is unproblematic — if an object's colour appearance is veridical, then it has the colour it appears to have. (2) — INCOMPATIBILITY — is intuitive. Colours stand in exclusion relations — if something is red all over, then it cannot be green all over at the same time. (3) follows from (1) and (2). (4) is entailed by (3). Suppose that either x has C_1 or C_2 but not both. The question of which colour is instantiated, or which appearance is veridical, might seem hard to settle on non-arbitrary

grounds. For instance, if x appears to be C_1 with respect to one species and C_2 with respect to another, it is not clear which species should be prioritized in deciding x's colour. (5) is then drawn. (6) follows from (4) and (5). Cases then generalize, and we arrive at (7), which entails that the naïve view is false.

However, it is far from clear that the above argument is persuasive. How one should respond will plausibly depend on what type of perceptual variation is at issue. For instance, (5) can plausibly be denied with respect to cases of intrapersonal and interpersonal variations (see Allen, 2016, chapter 3; for discussion on colour-blindness, see Broackes, 1992, p. 216). Due to limited space, I shall focus on interspecies variations. There are two responses here. One is to reject (5) by denying that non-human animals perceive colours and embracing the view that there are only *human* colours. But such a response seems unjustifiably anthropocentric. The second response, then, is to reject (2), INCOMPATIBILITY, and adopt colour pluralism, according to which objects simultaneously instantiate multiple colours (Mizrahi, 2006; Kalderon, 2007). Colour pluralism denies colour monism, on which there is only one family of colours (where a family of colours is defined as a group of colours which stand in relations of chromatic similarities, differences, determinations, and exclusions) (Kalderon, 2020).[8]

Colour pluralism, when first encountered, is likely to provoke an 'incredulous stare'. Indeed, proponents of the Galilean view might point out that their view was criticized precisely for contradicting our common-sense view of colour, but that the naïve view, in embracing colour pluralism, ends up contradicting our common sense to at least the same extent. While a full defence of colour pluralism is beyond the scope of this paper, points can be made that diminish its counterintuitiveness.

First, there are many properties of objects that are not detected by our visual systems (e.g. fingerprints in crime scenes). By analogy, it should not be too odd to think that a surface might have other colours than the ones we can see. We know that flowers like marsh marigolds have surface features that reflect varying amounts of ultra-violet light, forming patterns detectable by some birds and insects, but invisible to

---

[8] Colour pluralism has also been appealed to in dealing with interpersonal variation of fine-grained colours (Kalderon, 2007) and even intrapersonal variations (Mizrahi, 2006).

humans (Primack, 1982). It should not be counter-intuitive to suppose that objects like marsh marigolds have multiple colours visible to different species.[9]

Second, the intuitiveness of INCOMPATIBILITY (nothing is both C_1 and C_2 all over at the same time), which contradicts colour pluralism, seems to derive from the following intuitive claim (Harman, 2001, p. 661):

> INCOMPATIBILITY$_{-A}$: Nothing *appears* to be C_1 and C_2 all over at the same time to the same perceiver.

While INCOMPATIBILITY$_{-A}$, which concerns *colour appearances*, seems to be a truism, it does not entail INCOMPATIBILITY, which concerns *colours* themselves. Indeed, as Harman remarks, 'something could be both red all over and green all over at the same time without looking both red all over and green all over at the same time' (*ibid.*). The counter-intuitiveness of colour pluralism seems to diminish once we distinguish INCOMPATIBILITY from INCOMPATIBILITY$_{-A}$. On colour pluralism, colour perception grants a species access to only some, but not all, colours of objects. A pluralist can further hold that different visual systems *select* different families of colours for perception, determining which colours *appear* to the viewer perceptually (see Kalderon, 2007; 2020; Allen, 2016).[10]

Let's now take stock. I have argued that the Galilean view is committed to a systematic error theory and faces the task of explaining how our mind projects colours onto physical objects. I have also shown that, while the naïve view faces objections, responses can be made to diminish the force of these objections. As it stands, it seems far from clear that the Galilean view should be the default view, as it is often assumed in the literature on consciousness. Given the

---

[9] It is worth noting that all theories that treat colours as mind-independent features of objects, not only the naïve view, face the problem presented by perceptual variations in colour experiences and will potentially have to embrace colour pluralism as a result. Indeed, colour pluralism is compatible with various metaphysical theories of colour including colour primitivism, physicalism, and relationalism.

[10] Given the number of potential visual systems, one might worry that colour pluralism leads to the unacceptable consequence of 'colour explosion', such that each object simultaneously has infinite colours (e.g. Chalmers, 2006, p. 68). Note that such an explosion is quantitative not qualitative (Allen, 2016, p. 67). While qualitative parsimony concerning *types* of entities is generally favoured, there is no clear reason to maintain quantitative parsimony, which concerns the number of entities within the same type (Lewis, 1973).

arguments in this and the last sections, the naïve view ought to be taken seriously. In the next section, I shall explore implications which the naïve view of colour has for the problem of consciousness.

4. The Problem of Consciousness

As several theorists have noted, the problem of colour — how colours arise from colourless microphysical properties of objects — is structurally parallel to the problem of consciousness (e.g. Shoemaker, 2003; Byrne, 2006; Kalderon, 2007; Moran, this issue).[11] Just as antiphysicalists have raised the knowledge argument (Jackson, 1982), conceivability argument (Chalmers, 1997), and revelation argument (Goff, 2017) against physicalist theories of consciousness (see also Goff, 2019), one can raise parallel arguments against physicalist theories of colour. One could say that what Jackson's Mary learns upon leaving her room are non-physical facts about what colours are like; that a minimal physical duplicate of our world devoid of colours is conceivable and also possible; and that the essence of a colour is revealed in standard visual experience of that colour and is not revealed as physical (Johnston, 1992; Campbell, 1993).

Those who take the problem of colour seriously often claim that the problem of consciousness 'derives from a particular response to' the former (Kalderon, 2007, p. 594), and it disappears 'once we recognize the source of the puzzlement' which lies with sensible qualities presented or represented by experience (Byrne, 2006, p. 243; Allen, 2016). Does the naïve view of colour make the problem of consciousness disappear?

Regarding the problem of consciousness, we should distinguish between two questions (Pautz, 2010):

> QUALITY QUESTION: Why does a particular conscious state have the phenomenal character it has?
> GENERAL QUESTION: Why is a conscious state conscious at all?

The quality question asks why the phenomenal character of seeing a red apple is the way it is, as opposed to, say, the phenomenal character

---

[11] In solving the problem of consciousness, Goff motivates his panpsychism by appealing to the idea that science fails to reveal the intrinsic nature of fundamental entities. Taking the latter idea seriously, the naïve view of colour opens up a Russellian monist view of colour, according to which colours are grounded in the intrinsic natures of fundamental entities and their causal dispositions (Cutter, 2018; see also Moran, this issue).

of seeing a green apple. The question becomes particularly perplexing if we focus on the brain states that underlie our experiences. Thus, Levine expresses the quality question as a question about how the former can give rise to the latter:

> Let's call the physical story for seeing red 'R' and the physical story for seeing green 'G'... When we consider the qualitative character of our visual experiences when looking at ripe Macintosh apples, as opposed to looking at ripe cucumbers, the difference is not explained by appeal to G and R. For R doesn't really explain why I have the one kind of qualitative experience — the kind I have when looking at Macintosh apples — and not the other. (Levine, 1983, pp. 356–7)

According to the naïve view of colour, as we saw, colours are qualitative properties of physical objects, and their qualitative natures are manifested in perception. With this view in mind, the qualitative differences between the colours themselves — red and green — would naturally feature in an explanation for the phenomenal difference between seeing a red apple and seeing a green apple.[12]

There is a long philosophical tradition that treats colours as less objective and real than shapes. Empirical evidence suggests that philosophers are less likely than non-philosophers to treat colours as being as objective as shapes (Roberts, Andow and Schmidtke, 2014). But suppose that we were never enthralled by the Galilean view; suppose Galileo and other Enlightenment scientists and philosophers never proposed to eliminate colours from the physical world, and instead treated them as something real but beyond the reach of physical science. It then seems that the question of why my experience is the way it is — that is, why my experience is like *this* (pointing inwardly) when seeing a red apple as compared to *that* (again pointing inwardly) while seeing a green apple — doesn't seem to be particularly interesting. Naturally, one would point to the colours of the apples themselves, as well as viewing conditions, in explaining the phenomenal differences between the two experiences. In this sense, Levine's formulation of the quality problem of consciousness indeed 'derives from a particular response to' the problem of colour (Kalderon, 2007, p. 594). The quality problem becomes pertinent once

---

[12] Some theorists take the phenomenal character of a colour experience to be wholly determined by the colour presented or represented by the experience (Campbell, 1993; Byrne, 2006). Here I am only committed to a weaker claim that colours *partially* determine corresponding colour phenomenology.

we eliminate colours from the world and focus only on the physical states that underlie our conscious experiences.

Nevertheless, unlike those who think the source of the problem of consciousness lies with the problem of colour (e.g. Byrne, 2006; Kalderon, 2007; Allen, 2016), I do not believe that the admission of sensible qualities like colours into our ontology makes the problem of consciousness completely disappear. To begin with, the *quality question* concerning non-sensory experiences arguably remains. It is far from clear that the phenomenal characters of many emotions, moods, or *je-ne-sais-quoi* experiences which we have no words for (Camp, 2006) can be explained by making reference to qualities or values presented or represented by experience. Unlike colours, these qualities or properties do not *seem* to exist out in the world. More importantly, the *general question* remains. Why are we conscious rather than not conscious? Why is it that we are *conscious of* colours, for example?

Given that the problem of consciousness does not completely disappear, advocates of the Galilean view, including Goff, who think that only consciousness is physically irreducible, are likely to point to the consideration from simplicity to argue that an ontology which admits irreducible colours as well as consciousness is bloated and inelegant. But the consideration from simplicity never exists in a vacuum. We should not prioritize simplicity if there are good reasons against eliminating a certain type of entity which seems to exist. Moreover, while a worldview that includes colours is ontologically less simple than a worldview without them, it is far from obvious that the latter is theoretically simpler or more elegant overall. Eliminating colours from the world, as we saw, has significant costs.

Acknowledgments

Thanks to Sam Coleman, Philip Goff, Jakub Mihalik, and Alex Moran for their helpful comments.

References

Allen, K. (2016) *A Naïve Realist Theory of Colour*, Oxford: Oxford University Press.

Berkeley, G. (1734) *Three Dialogues between Hylas and Philonous*, Oxford: Oxford University Press.

Boghossian, P. & Velleman, D. (1989) Colour as secondary quality, *Mind*, **98** (389), pp. 81–103.

Boghossian, P. & Velleman, D. (1991) Physicalist theories of colour, *Philosophical Review*, **100**, pp. 67–106.

Broackes, J. (1992) The autonomy of colour, in Charles, D. & Lennon, K. (eds.) *Reduction, Explanation, and Realism*, pp. 421–465, Oxford: Clarendon Press.
Byrne, A. (2006) Color and the mind–body problem, *Dialectica*, **60**, pp. 223–244.
Byrne, A. & Hilbert, D. (2003) Color realism and color science, *Behavioural and Brain Sciences*, **26**, pp. 3–21.
Camp, E. (2006) Metaphor and that certain 'je ne sais quoi', *Philosophical Studies*, **129** (1), pp. 1–25.
Campbell, J. (1993) The simple view of colour, in Haldane, J.J. & Wright, C. (eds.) *Reality, Representation and Projection*, pp. 257–268, Oxford: Oxford University Press.
Chalmers, D.J. (1997) *The Conscious Mind: In Search of a Fundamental Theory*, New York: Oxford University Press.
Chalmers, D.J. (2006) Perception and the fall from Eden, in Gendler, T. & Hawthorne J. (eds.) *Perceptual Experience*, pp. 49–125, Oxford: Oxford University Press.
Cohen, J. (2009) *The Red and the Real*, Oxford: Oxford University Press.
Cutter, B. (2018) Paradise regained: A non-reductive realist account of the sensible qualities, *Australasian Journal of Philosophy*, **98** (1), pp. 38–52.
Galileo, G. (1623/1996) *The Assayer*, in Drake, S. (trans. & ed.) *The Discoveries and Opinions of Galileo*, pp. 228–280, New York: Doubleday.
Gert, J. (2008) What colors could not be: An argument for color primitivism, *Journal of Philosophy*, **105** (3), pp. 128–155.
Goff, P. (2017) *Consciousness and Fundamental Reality*, Oxford: Oxford University Press.
Goff, P. (2019) *Galileo's Error*, New York: Vintage Books.
Hardin, C.L. (1993) *Colour for Philosophers: Unweaving the Rainbow*, Indianapolis, IN: Hackett Publishing.
Harman, G. (2001) General foundations versus rational insight, *Philosophy and Philosophical Research*, **63**, pp. 657–663.
Jackson, F. (1982) Epiphenomenal qualia, *Philosophical Quarterly*, **32**, pp. 127–136.
Jackson, F. (1996) The primary quality view of color, *Philosophical Perspectives*, **10**, pp. 199–219.
Johnston, M. (1992) How to speak of the colours, *Philosophical Studies*, **68**, pp. 221–263.
Kalderon, M. (2007) Colour pluralism, *Philosophical Review*, **116** (4), pp. 563–601.
Kalderon, M. (2020) Monism and pluralism, in Brown, D. & MacPherson, F. (eds.) *Routledge Handbook of the Philosophy of Color*, pp. 327–341, Oxford: Routledge.
Keats, J. (1990) *The Major Works*, Oxford: Oxford University Press.
Kim, J. (1993) *Supervenience and Mind: Selected Philosophical Essays*, Cambridge: Cambridge University Press.
Levine, J. (1983) Materialism and qualia: The explanatory gap, *Pacific Philosophical Quarterly*, **64**, pp. 354–361.
Lewis, D. (1973) *Counterfactuals*, Oxford: Blackwell.
Maund, B. (2006) The illusory theory of colours: An anti-realist theory, *Dialectica*, **60** (3), pp. 245–268.
McGinn, C. (1996) Another look at colour, *Journal of Philosophy*, **93**, pp. 537–553.

Michotte, A. (1963) *The Perception of Causality*, London: Methuen.
Mizrahi, V. (2006) Color objectivism and color pluralism, *Dialectica*, **60** (3), pp. 283–306.
Moran, A. (this issue) Grounding the qualitative: A new challenge for panpsychism, *Journal of Consciousness Studies*, **28** (9–10).
Nozick, R. (1974) *Anarchy, State, and Utopia*, New York, New York: Basic Books.
Nozick, R. (1989) *The Examined Life: Philosophical Medications*, New York: Simon & Schuster.
Pautz, A. (2006) Can the physicalist explain colour structure in terms of colour experience?, *Australasian Journal of Philosophy*, **84** (4), pp. 535–664.
Pautz, A. (2010) Do theories of consciousness rest on a mistake?, *Philosophical Issues*, **20**, pp. 333–367.
Pautz, A. (2020) How does color experience represent the world?, in Brown, D. & MacPherson, F. (eds.) *Routledge Handbook of the Philosophy of Colour*, pp. 367–389, Oxford: Routledge.
Primack, R. (1982) Ultraviolet patterns in flowers, or flowers as viewed by insects, *Arnoldia*, **42** (3), pp. 139–146.
Roberts, P., Andow, J. & Schmidtke, K. (2014) Colour relationalism and the real deliverances of introspection, *Erkenntnis*, **79**, pp. 1173–1189.
Schlottmann, A. & Anderson, N.H. (1993) An information intergration approach to phenomenal causality, *Memory & Cognition*, **21** (6), pp. 785–801.
Shoemaker, S. (2003) Content, character, and colour, *Philosophical Issues*, **13**, pp. 253–278.
Tye, M. (2000) *Consciousness, Color, and Content*, Cambridge, MA: MIT Press.
Yablo, S. (1995) Singling out properties, *Philosophical Perspectives*, **9**, pp. 477–502.

Alex Moran[1]

Grounding the Qualitative

A New Challenge for Panpsychism

Abstract: This paper presents a novel challenge for the panpsychist solution to the problem of consciousness. It advances three main claims. First, that the problem of consciousness is really an instance of a more general problem: that of grounding the qualitative. Second, that we should want a general solution to this problem. Third, that panpsychism cannot provide it. I also suggest two further things: (1) that alternative kinds of Russellian monism may avoid the problem in ways panpsychists cannot, and (2) that a kind of neo-Aristotelian or ground-theoretical physicalism fares just as well here if not better.

> *For myself, I think that the only plausible way that a Materialist can deal with the secondary qualities is completely to reverse the whole programme started by Galileo, a programme that has persisted for so long. What we should do is put these qualities back into the physical world again.* — D.M. Armstrong (1999, p. 124)

1. Introduction

One central thesis of *Galileo's Error* is that the problem of consciousness arises because of an apparently fundamental difference between the nature of the physical world and the nature of the conscious mind. On the one hand, it would appear that physical reality can be exhaustively described in quantitative terms. On the other

Correspondence:
Email: alexander.moran@philosophy.ox.ac.uk

---

[1] University of Oxford, UK.

hand, however, it seems that consciousness is an essentially qualitative phenomenon. Accordingly, there seems to be no possibility of locating consciousness within the physical world. That is, it seems that we cannot find a place for consciousness in nature due to the fact that, while consciousness is fundamentally qualitative, physical reality is fundamentally quantitative (see esp. Goff, 2019, chapter 1).

The problem of consciousness, as Goff conceives of it, can thus be represented by the following argument:

1. Physical properties are quantitative properties.
2. Mental properties are qualitative properties.
3. Qualitative properties are not identical to quantitative properties.
4. Qualitative properties are not grounded in quantitative properties.

∴ Mental properties are neither identical to nor grounded in physical properties.

Premise 1 follows from the conception of the physical that Goff believes we have inherited from Galileo, on which physical reality can be described entirely in mathematical or quantitative terms (*cf. ibid.*, p. 21). As for premise 2, this is supported by the familiar idea that phenomenally conscious states are qualitative states, such that there is 'something that it is like' to be in them (*cf.* e.g. Nagel, 1974; Levine, 1983; see also the extensive discussion in Goff, 2017; 2019). As for premises 3 and 4, the basic idea is that there is such a radical difference between merely quantitative properties on the one hand, and qualitative properties on the other, that it is hard to see how qualitative properties could either be identical to, or even metaphysically grounded in, merely quantitative properties (*cf.* Section 5).

The conclusion of the argument, however, is essentially property dualism. (It is also consistent with substance dualism.) After all, the conclusion states that mental properties are neither identical to nor grounded in physical properties. But this means that mental properties are something 'over and above' physical properties, in just the kind of way that traditional property dualists maintain. If we wish to avoid dualism, therefore, we must resist one of the premises.

According to Goff, we should indeed be looking to resist dualism (as many of us will agree). But what are the options? We can distinguish three moves that traditional physicalists might make. The more radical kind of physicalist might reject premise 2. Eliminativists might deny that there are any mental properties at all. Illusionists, meanwhile, might say that, while there are mental properties, they do not really have the qualitative nature that they seem to have. Less radical physicalists, by contrast, who are 'realists' about the qualitative nature of the mental, have two further options. Reductive physicalists will reject premise 3, arguing that mental properties, with just the qualitative natures that they seem to have, are identical to physical properties after all. Non-reductive physicalists, meanwhile, will reject premise 4. On this view, mental properties are qualitative properties that are distinct from but metaphysically grounded in underlying physical properties.

My own view is that something like this last position represents the best hope for physicalists, and I return to this idea below. First, however, I bring Goff's panpsychist response into view. Drawing on ideas from Bertrand Russell (1927) and Arthur Eddington (1928), Goff first points out that physics characterizes matter only in terms of its relational properties, leaving open its intrinsic nature. Goff then argues that we know the intrinsic nature of at least some matter; namely, the matter inside our brains. The idea then is that we can combine these points in order to conjecture about the intrinsic nature of matter in general. In particular, the conjecture is that the intrinsic nature of matter is constituted by consciousness. As Goff puts it:

> All we get from physics is this big black-and-white abstract structure, which we must somehow fill in with intrinsic nature. We know how to color in one bit of it: the brains of living organisms are colored in with consciousness. How to color in the rest? The most elegant, simple, sensible option is to color in the rest of reality with the same pen. (2019, p. 133)

That is, the idea is that the intrinsic nature of matter is constituted by the (rudimentary) kind of consciousness that particles instantiate. It thus emerges, on Goff's panpsychism, that 'consciousness *is* the intrinsic nature of matter' (*ibid.*, p. 132).[2]

Goff refers to this as the *simplicity argument* for panpsychism. Notably, the argument has nothing especially to do with the mind–body problem. Rather, the argument is that (i) physics does not tell us about the intrinsic nature of matter and yet (ii) we do know that the intrinsic nature of the matter making up human brains is constituted

---

[2] The idea is that the kind of consciousness that constitutes the intrinsic nature of fundamental particles is a rudimentary kind of consciousness, distinct from the kind of consciousness that we human beings instantiate. One way to think of this is to suppose that human consciousness and the kind of rudimentary consciousness possessed by particles are determinates of the same determinable.

by consciousness, so (iii) it is reasonable to think that consciousness constitutes the intrinsic nature of all matter (*cf. ibid.*, p. 134).[3] What Goff argues is that, while we have this as an independent argument for panpsychism, we can also draw on panpsychism as a means of avoiding the above argument for dualism.

In particular, Goff argues that panpsychists can challenge premise 1. To help see this, we can borrow a useful distinction from Chalmers (2015). On the one hand, there are *narrowly physical properties*, i.e. the relational properties of matter as described by physics. On the other, however, within a panpsychist framework, there are *broadly physical properties*, which include not only narrowly physical properties, but also those rudimentary conscious properties that make up the intrinsic nature of matter. With this distinction drawn, we can then point to an ambiguity in premise 1:

1a. Narrow physical properties are quantitative properties.
1b. Broadly physical properties are quantitative properties.

The panpsychist then reasons so. If the argument employs 1a, then, while this premise holds, the conclusion does not establish dualism. For it leaves open that mental properties are either identical to or grounded in broadly physical properties. Whereas, if the argument employs 1b, then this premise is false, and so again fails to establish dualism. This is because broadly physical properties include those rudimentary conscious properties that constitute the intrinsic nature of matter, whereby these rudimentary conscious properties are qualitative properties.

Moreover, there is a further aspect to the panpsychist response. On that view, broadly physical properties are in fact (at least partly) qualitative properties. Arguably, this makes it easier to see how mental properties could be either identical to or else grounded in underlying (broadly) physical properties. Hence, panpsychism appears to promise us a way of locating mental properties in the physical world.

Questions remain as to how exactly the mental properties that human persons instantiate relate to the broadly physical properties postulated by the panpsychist. One sort of panpsychist will say that

---

[3] Of course, it might be added that part of what makes this conjecture plausible is that it helps to solve the mind–body problem. So perhaps the two issues are not entirely independent.

such mental properties are identical to broadly physical properties. Another sort will say that such mental properties are grounded in underlying broadly physical properties. For present purposes, we needn't engage further with this question (for relevant discussion see Goff, 2019; ms; Chalmers, 2015). The argument to follow will put pressure on both panpsychist views.

2. A Parallel Argument

The challenge that I want to press turns on the thought that there are qualitative properties in nature besides conscious properties. In particular, the properties I have in mind are the familiar 'secondary qualities', as I shall call them, following philosophical tradition; that is to say, the qualities of objects including colours, sounds, smells, and tastes. It is a familiar point that such qualities have a distinctive sensuous or qualitative nature. Consider the following passage from Chalmers:

> Phenomenologically, it seems to us as if visual experience presents simple intrinsic qualities of objects in the world, spread out over the surface of the object. When I have a phenomenally red experience of an object, the object seems to be simply, primitively, *red*. The apparent redness does not seem to be a microphysical property, or a mental property, or a disposition, or an unspecified property that plays an appropriate causal role. Rather, it seems to be a simple qualitative property, with a distinctive sensuous nature. We might call this property perfect redness: the sort of property that might have been instantiated in Eden. (Chalmers, 2006, p. 67)

Two ideas emerge from this passage. First, that the secondary qualities each have a distinctive qualitative character.[4] Second, that in virtue of this, it is hard to see how they could be reducible to physical properties. In other words, it seems hard to see how such properties could be explicable just in terms of the familiar physical properties and relations that fundamental physics speaks of, just as in the case of mental properties, which also seem to be irreducible.[5] Both points, it

---

[4] There is a case for saying that some primary qualities, too, at least as they are presented to us in sensory experience, have a distinctive qualitative nature. I don't press this point here, but for relevant discussion see Broad (1923), Johnston (ms), Moran (ms-a), Strawson (1979). Nb. Sometimes, following Goff (2019), I will just speak of the sensory qualities in general.

[5] *Cf.* Armstrong (1964, pp. 173–4): '[T]he secondary qualities seem to be, in some sense, *simple* qualities, with the consequence that we are unable to give an account of them in

seems, are compelling. Taken together, however, they suggest that we can run the following argument for thinking that the secondary qualities are non-physical properties:

1. Physical properties are quantitative properties.
2'. Secondary qualities are qualitative properties.
3. Qualitative properties are not identical to quantitative properties.
4. Qualitative properties are not grounded in quantitative properties.

∴ Secondary qualities are neither identical to nor grounded in physical properties.

Plausibly, whatever reasons we might have for wanting to deny that mental properties are non-physical properties, there will be analogous reasons to deny that secondary qualities are non-physical properties. Accordingly, we have the same sort of motivation for wanting to resist the kind of 'secondary quality dualism' that this argument leads to as we have for wanting to resist the more traditional kind of dualism about mental properties. What I want to argue now, however, is that panpsychists lack the resources to defuse this argument. I'll then explain why I believe that this poses a problem for the panpsychist view.[6]

The first point is straightforward. Panpsychists urge that the intrinsic nature of matter is constituted by (rudimentary) consciousness. And this is meant to make it easier to see how mental properties could be identical to or grounded in broadly physical properties. As Goff explains:

> The challenge for the materialist is to bridge the gap between the *objective quantities* of physical science and the *subjective qualities* of conscious experience. But... this project is of dubious coherence, and... not something we have made the slightest progress on. [By contrast, the panpsychist faces the] more tractable... challenge of getting from *simple* subjective qualities to *complex* subjective qualities. (2019, p. 146)

---

terms of anything else. They seem to be "intractable", there seems to be no prospect of reducing them to anything else, or exhibiting them as constructions out of simpler elements.'

[6] Some other authors who discuss the above kind of argument, and who have noted the parallel with the initial mind–body problem, include Byrne (2006), Fish (2013), Kalderon (2007), Johnston (1996), Liu (this issue), Moran (ms-b), Pautz (2013), and Shoemaker (1996).

This move, however, does nothing to help us explain instantiations of non-mental qualitative properties such as the colours. If the challenge is to see how the instantiation of redness by an apple, for example, is somehow reducible to the instantiation of certain physical properties, then the claim that physical properties include rudimentary conscious ones does not help. Put differently, if it is hard to see how redness could be reducible to physical features when these are narrowly construed, it is just as hard to see how redness could be reducible to physical features when these are broadly construed, given that this construal just amounts to thinking of broadly physical features as including rudimentary conscious properties.

One way to emphasize the point is to contrast the following passages. First, Wittgenstein:

> If I turn my attention in a particular way on to my consciousness, and astonished, say to myself: 'THIS is supposed to be produced by a process in the brain!' — as it were clutching my forehead. (Wittgenstein, 1958, 1. 412)

Perhaps we could make progress lessening the astonishment here by supposing that the intrinsic nature of matter is constituted by a rudimentary kind of consciousness. But now consider the following parallel worry articulated by Shoemaker:

> I look at a shiny red apple and say to myself 'THIS is supposed to be a cloud of electrons, protons, etc. scattered through mostly empty space.' And focusing on its color, I say 'THIS is supposed to be a reflectance property of the surface of such a cloud of fundamental particles'. (Shoemaker, 1996, p. 248)

The trouble is that supposing that the relevant electrons, protons, etc. have consciousness as their intrinsic nature does nothing to explain or help clarify how the instantiation of redness by the apple could be constituted by the instantiation of physical properties and relations by the cloud of particles making it up. In other words, the supposition that the particles are conscious simply does not help us to see how the secondary qualities get to be instantiated by ordinary physical things.

Suppose we grant this point. Why is this a problem for panpsychists? After all, panpsychists are exclusively addressing the mind–body problem. And for all we've said, the panpsychist can solve *that* problem. So where is the issue?

The central point is that we should be looking for a *unified* solution to these problems. The arguments I have outlined are really instances of *a general argument*, which represents what we might refer to as the *problem of grounding the qualitative.* What I want to suggest at this juncture is that the mind–body problem, as well as the problem of accounting for the secondary qualities, would seem to be merely two instances of this more general concern (*cf.* Moran, ms-b):

1. Physical properties are quantitative properties.
2″. Some properties (mental and non-mental) are qualitative properties.
3. Qualitative properties are not identical to quantitative properties.
4. Qualitative properties are not grounded in quantitative properties.

∴ Some properties are neither identical to nor grounded in physical properties.

What we should be looking for, therefore, is a general explanation as to how qualitative properties, whether mental or non-mental, can be understood in physical terms. Panpsychism, however, cannot provide this. And therein lies the problem.[7]

3. Projectivist Panpsychism

There is an important line of response that panpsychists might make at this juncture. So far, I have been assuming a certain kind of realism about the secondary qualities, which we can refer to, following Chalmers (2006), as the *Edenic view*. The idea is that external things really do instantiate the qualitative properties that they appear, in conscious experience, to have; that things really are coloured, and make the sounds, and have the smells and tastes, that perception presents them as having. It has been common for philosophers, however, to deny this. On that view, there is something it is like to see a rose. But the rose itself is not really red as it appears to be. Rather, the rose's red appearance is a mere function of how sensory experience represents things to be.[8] In fact, this is precisely the view that Galileo took of the sensory qualities:

---

[7] It is sometimes argued that the *real* problem is that of grounding the secondary qualities in the physical, and that the hard problem of consciousness is secondary at best (see e.g. Allen, 2016; Byrne, 2006; Cutter, 2018; Fish, 2008; 2009; 2013; Johnston, 1996; Kalderon, 2007). My own view, by contrast, is that we have one general problem here, of which both the problem of consciousness and the problem of secondary qualities are instances. For more details see Moran (ms-b).

[8] This expression is intended neutrally. There are various ways to make sense of how exactly, on a non-Edenic view, sensory experience nevertheless portrays the world as

> I think that tastes, odours, colours, and so on are no more than mere names so far as the object in which we place them is concerned, and that they reside only in consciousness. Hence if the living creature were removed, all these qualities would be wiped away and annihilated. (1623/1996, p. 274)

The idea is that such qualities are not really features of external things, but are instead properties of experience (or 'qualia', as they are sometimes called), which exist only in the mind. Common sense is therefore guilty of a fundamental *projectivist* error: we mistake the qualitative properties of (or otherwise involved in) sensory experience for qualitative properties of external items (*cf.* Johnston, 1996; Kalderon, 2007).[9]

Suppose, just for the sake of argument, that a projectivist view could be motivated. My argument against panpsychism would then be undercut. The problem of consciousness would pose a genuine problem. But there would be no analogous problem involving secondary qualities. Moreover, the only qualitative properties in reality would be mental properties. Accordingly, my criticism of the panpsychist, namely as posing a too narrow solution to a general problem, will not succeed.

There is some evidence that Goff accepts this kind of projectivist panpsychism. At the beginning of Goff's book, we're told that Galileo first took the radical step of viewing the sensible qualities of external objects as having existence only in the mind. Notably, moreover, Goff seems to accept this aspect of the Galilean programme. For, Goff sees the philosophic challenge in precisely Galilean terms: as that of explaining how the sensible qualities, *reimagined, on Galilean lines, as mental properties rather than features of external things*, fit into the external world. Consider:

> ...Galileo took the secondary qualities (sounds, smells, tastes, odours) out of its domain of inquiry: by reimagining them as forms of consciousness... The fact that physical science has been extremely successful when it ignores the sensory qualities gives us no reason to think that

---

containing the secondary qualities, without having to presuppose representationalism in the contemporary sense.

[9] For discussion of how considerations of simplicity may seem to motivate this view, see Liu (this issue). Goff (personal correspondence) argues that we have a kind of privileged access to our qualia, of a sort that we do not have to the qualitative features of external things, and that this too might motivate projectivism. To my mind, however, this gets the order of explanation the wrong way around — both epistemically (see Martin, 1998) and metaphysically (see Moran, ms-b).

it will be similarly successful if and when it turns its attention to the sensory qualities, reimagined as forms of consciousness. (2019, p. 136)

Here, Goff seems to grant that the sensible qualities, which *seem to be* features of external things, should be (re)conceived as properties of mental states. Goff then sees the challenge as that of making sense of the place of these qualitative mental properties within the rest of the physical world (*cf.* Goff, 2017, chapter 1; 2019).

But this brings out an important point. One might have hoped, when reading the initial pages of *Galileo's Error*, that we would be provided with materials for reversing the whole Galilean programme: that is, for putting the sensible qualities of external objects back into the external world, as well accounting for the place of mental qualitative properties within nature. Goff even suggests at one point that this is something that his panpsychism will help with:

In 1623 Galileo took the sensory qualities out of the physical world. Three hundred years later in 1927 Russell and Eddington finally found a way to put them back. (2019, pp. 137–8)

What I have brought out here, however, is that this is not so. Granted, for all that I have said, panpsychism may help us to locate mental qualitative properties within the physical world. But it does nothing to help us locate the sensible qualities of external things. It follows that Goff's panpsychism needs to be projectivist panpsychism, which denies the reality of Edenic qualities conceived as features of external things. It has to deny, that is, just like Galileo, that external things really have the various qualitative properties, the various sounds and tastes and colours, that they seem to have. But this means that Goff's panpsychism cannot reverse the Galilean programme in the way that it initially seemed to promise. In fact, for those of us who are sympathetic to the idea that we live in an Edenic world, it seems that Goff's panpsychist view remains beholden to a fundamental Galilean mistake; namely, the mistake of 'mentalizing' the sensible qualities — of taking what are in fact qualities of external items and treating them as mere creatures of the mind.[10] In turn, this suggests that Goff's framework does not in fact allow us to fully reverse the Galilean programme in the way that *Galileo's Error* seemed to promise.

---

[10] Recent advocates of the Edenic view include Allen (2016), Cutter (2018), and Liu (this issue). It is, of course, controversial whether or not Galilean projectivism constitutes a *mistake*.

4. Russellian Monism

There is another response Goff might by sympathetic to, and that in any case is worth exploring. In *Galileo's Error*, Goff distinguishes panpsychism in particular from the more general position known as 'Russellian monism' (of which panpsychism is just one instantiation). The difference may be characterized by again utilizing the distinction between narrow and broadly physical properties. Narrow physical properties are dispositional properties such as mass and charge. They tell us what physical things do; not what they are. But broadly physical properties also include what I will call *quiddities*: the intrinsic properties that characterize the intrinsic nature of the entities that physics deals in.[11] Panpsychists are Russellian monists for whom the quiddities are to be thought of as conscious properties. However, *some* kinds of Russellian monist do not suppose the quiddities are conscious properties (although they still insist that whatever nature such quiddities have, they will help us to resolve the mind–body problem). What I want to suggest is that certain non-panpsychist forms of Russellian monism may be better off than panpsychism when it comes to meeting the challenge that I set out above. For, such views are in a position to claim that qualitative properties in general are ultimately explained by the intrinsic nature of matter, whatever that turns out to be. Panpsychists, meanwhile, as we have seen, are not in a position to make this claim.

One way to illustrate this is as follows. Traditional versions of Russellian monism, just like panpsychism, are geared towards resolving the problem of consciousness with which we started. In recent work, however, Cutter (2018) has argued for *sensible quality Russellian monism*. On that view, the quiddities that partially characterize the broadly physical properties are not conscious properties but rather properties of another kind. Leaving open their nature entirely leaves us with neutral monism proper: the idea that matter has an intrinsic nature unknown to us. But there are various other non-panpsychist options for specifying the intrinsic natures or the quiddities of matter (*cf. ibid.*; Liu, this issue). Among them is a variation of an idea, prominently defended in recent times by Coleman (2015), known as 'panqualityism'. On this view, the quiddities are not conscious properties, but rather qualitative properties whose nature is

---

[11] Goff (2019) refers to these as 'intrinsic natures' throughout *Galileo's Error*.

not further specified, besides the claim that they are non-experiential. According to Cutter (2018), this kind of view is well-suited to accounting for the secondary qualities, though it is not well-suited for handling the mind–body problem for which it was originally designed. However, it is not obvious that this last is right. Instead, it seems, one could argue that higher-level qualitative properties in general are ultimately derived from the more basic qualitative properties that constitute the intrinsic nature of matter.

What the above brings out, I submit, is that if our ambition is to deal, not just with the mind–body problem, but with the more general problem of accounting for the full range of qualitative properties that nature contains, then, if we are to be Russellian monists, we must endorse a kind of Russellian monism other than panpsychism. As panpsychists, we may be able to handle the mind–body problem. However, we will be unable to account for the place of the secondary qualities in nature, and hence the more general problem will remain unsolved. As non-panpsychist Russellian monists, by contrast, we have a chance at solving the more general problem. So, the result appears to be that, if we are to be Russellian monists, we should not be panpsychists, but rather Russellian monists of some other kind.[12]

My own view, however, is that to answer the more general problem we need not adopt any form of Russellian monism. To end the paper, then, I wish to outline a different proposal of my own; a kind of neo-Aristotelian physicalism that has been gaining traction in the recent literature (*cf.* Dasgupta, 2014; Schaffer, 2017).

5. Grounding Physicalism

Recall the following argument:

1. Physical properties are quantitative properties.
2″. Some properties (mental and non-mental) are qualitative properties.
3. Qualitative properties are not identical to quantitative properties.
4. Qualitative properties are not grounded in quantitative properties.

---

[12] Note that Goff is sympathetic to this kind of Russellian monist position and is open to non-panpsychist versions of this sort of Russellian view. See e.g. Goff (2017; 2019, p. 137). However, notice that if Goff retreats to Russellian monism then the simplicity argument from earlier for panpsychism can no longer be relied upon. To that extent, Russellian monism of the non-panpsychist kind looks considerably less well-motivated. I leave open how much of a worry that is.

∴ Some properties (mental and non-mental) are neither identical to nor grounded in physical properties.

The premise I would want to resist here is:

4. Qualitative properties are not grounded in quantitative properties.

If we make this claim then, in relation to the pro-dualist argument of Section 2, we can deny that mental properties are not grounded in physical ones. And in relation to the anti-physicalist argument about the sensible qualities of Section 3, we can deny that non-mental qualitative properties like the redness of the rose are not grounded in underlying fundamental physical properties. All of these qualitative properties thus end up being non-fundamental physical features of reality that are conceived as being grounded in more fundamental physical features.[13]

Now, Goff (2019; ms) is unsympathetic to this kind of view, doubting that it constitutes a genuine variety of physicalism (*cf.* Pautz, ms; Schaffer, 2017, offers a nice reply to this charge; see also Moran, ms-c). Here is his main concern:

> Many people take materialism to be the view that the brain *produces* consciousness, as though consciousness were some peculiar kind of gas that the physical workings of the brain bring into being. However, such a view would not be materialism, as it implies that consciousness is something over and above the physical workings of the brain... In fact, materialism is the view that experiences and feelings are identical with states of the brain... (Goff, 2019, pp. 92–3)

The idea seems to be that any view on which we have anything less than identity between the physical and the mental implies that mental properties are 'something over and above' physical properties, meaning that we would not have genuine form of materialism.[14]

---

[13] Note that this view also implies that premise 1 is false, since it implies that some higher-level (non-fundamental) physical properties are in fact qualitative properties, namely certain mental properties and certain secondary qualities. Thus, instead of claiming that all physical properties are quantitative properties, we should say that all fundamental physical properties are quantitative properties. This then leaves room for non-fundamental physical properties to be both qualitative and ultimately grounded in the fundamental quantitative properties.

[14] That said, Goff (2017) is somewhat more permissive. On that view, mental properties need not be identical to physical features, although they do have to be what Goff calls 'constitutively grounded' in such features. Presumably, Goff would argue that it is problematic to suppose that mental properties are grounded in this way in fundamental physical properties, whereas I would wish to deny precisely this claim.

However, the phrase here 'something over and above' is open to interpretation. If x is something over and above y just because x and y are distinct, then the smile of the cat is something over and above the smiling cat. This, however, is implausible, given that the smile of the cat is grounded in and dependent on the smiling cat itself. (That this is so, of course, is what constitutes the philosophical underpinnings of a lovely joke in *Alice in Wonderland*.) Thus, there is room to deny that x cannot be over and above y unless x is identical to y. And, hence, there is room to claim that, in the relevant sense, mental properties are nothing over and above their ultimate physical grounds, despite being distinct therefrom.[15]

But there are further concerns. First, there is a modal argument. As Cutter (2018, p. 50) rightly emphasizes, 'there is some intuitive plausibility to the idea that no collection of non-qualitative properties could be sufficient for the instantiation of a qualitative property' (*cf.* Chalmers, 2015, p. 268; Coleman, 2015, p. 76). If that is right, however, then we can appeal to the idea that *grounds necessitate* to generate a problem for the kind of 'grounding physicalism' I am trying briefly to motivate. Let us suppose that qualitative property Q is grounded in physical properties $P_1, P_2 \ldots P_n$. If no collection of physical properties is *sufficient* for the a qualitative one, then we can have $P_1, P_2 \ldots P_n$ without Q. Yet it is widely held that grounding does not allow for that. If *what makes it the case* that some object O is Q is that some particles have $P_1, P_2 \ldots P_n$, i.e. if the particles having $P_1, P_2 \ldots P_n$ is what metaphysically explains why O has Q, then one might think that necessarily, whenever those particles have $P_1, P_2 \ldots P_n$, then O must have Q. So the argument is:

1. Since $P_1, P_2 \ldots P_n$ are quantitative, they are not sufficient for the instantiation of Q.
2. If any properties $G_1, G_2 \ldots G_n$ are not sufficient for the instantiation of some property F, then F is not grounded in the instantiation of $G_1, G_2 \ldots G_n$.

∴ The instantiation of Q is not grounded in the instantiation of $P_1, P_2 \ldots P_n$.

---

[15] One way to further develop this thought is to point out that, in general, grounded items derive their natures and identities from the more fundamental items in which they are grounded. *Cf.* Moran (ms-c).

While I lack the space to develop this idea here, I think the premise to deny here is premise 2, i.e. the principle stating that if G grounds F then, necessarily, whenever G occurs G must ground F. I argue against this in Moran (2018; 2021); others have done so elsewhere. But think of it this way. It was once a dogma that causation must necessitate: that if a causes b then a has to suffice for b. However, many philosophers are now happy to deny that this is so (dispositional essentialists aside). Perhaps, therefore, the same thing holds in the case of grounding. In other words, it may be that constitutive determination, no less than causal determination, does not require necessitation. Granted, in some cases, the ground will be sufficient for what it grounds: for example, when an item is scarlet, this is sufficient for the item to be red. However, consistently with this, there may be counter-examples to the more general idea that grounds always necessitate. For example, it is plausible that the general fact that <all swans are white> is grounded in a range of particular facts to the effect that each actual swan is white. However, since there could have been an additional non-white swan, there is a possible world in which all of those particular facts that act as grounds obtain, despite the corresponding general fact failing to obtain. The particular facts, therefore, do not necessitate the general fact. Arguably, however, this does not undermine the plausible idea that, as things stand, the general fact is grounded in the various particular facts (*cf.* Bader, ms; Sider, 2020). Moreover, this is just one possible example of contingent grounding. Perhaps, then, when qualitative properties are grounded in underlying quantitative properties, we have another example of contingent grounding.

I will consider one further argument. I said earlier in Section 2 we can motivate the idea that qualitative properties cannot be grounded in quantitative ones by appeal to familiar explanatory gap type concerns. One might just think: *how could a qualitative property be grounded in a quantitative one, given that these properties are so different?* But then, if there is no intelligible connection between these types of properties, one might wonder how the one set could be grounded in the other. That is, the presence of an explanatory gap would seem to preclude the qualitative properties in nature from being grounded in the fundamental quantitative ones.

This worry, I think, really takes us to the heart of the matter. That is, it brings out the real nature of the problem of grounding the qualitative in the fundamental physical. Fundamentally, the problem is analogous to several other 'location' problems.[16] One well-known example concerns locating the abstract within a fundamentally concrete world. Another concerns the place of asymmetric relations in a world with only symmetric relations at the fundamental level.[17] A third concerns how we can derive dispositional properties from a world that is fundamentally categorical in nature. The central challenge, in all these cases, is to get from one from kind of property to another even when the properties are of radically different sorts. Now I submit that one cannot do this when one looks only at the nature of the properties. Hence the importance of a kind of physicalism that also allows us to look at the relationship between them. What we need is a grounding relation that can act as a bridge to take us from the more fundamental property of kind K to the radically heterogeneous and comparatively more derivative property of kind K*.

Even once we posit a grounding relation, however, explanatory gaps may remain. But perhaps grounding relations admit of explanatory gaps (Schaffer, 2017). Indeed, we should arguably expect such gaps, if the grounding relation is, as I have claimed, able to connect properties with radically different natures. The view I recommend, therefore, is that while qualitative properties and their underlying physical grounds are radically different in nature, the former derive from, and are dependent on, the latter. When a person is conscious, or when a rose is red, these facts obtain in virtue of more basic facts involving quantitative physical properties at the fundamental level. If such a view can be developed, then we can answer the general problem of grounding the qualitative without having to speculate about the intrinsic nature of matter. And that, it seems to me, is a view well worth exploring.

Acknowledgments

A version of this paper was presented to the Philosophy of Mind work in progress group at Oxford; my thanks to the participants on that occasion. Special thanks to Philip Goff for helpful comments, and to Dominic Alford-Duguid for discussion.

---

[16] I borrow the idea of a 'location problem' from Jackson (1998).

[17] There is reason to think that, at the fundamental level, reality contains only symmetric relations. So how does reality at the higher levels contain asymmetry? One way of trying to solve this puzzle is developed in detail in Bader (2020).

References

Allen, K. (2016) *A Naïve Realist Theory of Colour*, Oxford: Oxford University Press.
Armstrong, D.M. (1964) *Perception and the Physical World*, New York: Humanities Press.
Armstrong, D.M. (1999) *The Mind–Body Problem: An Opiniated Introduction*, London: Routledge.
Bader, R.M. (2020) Fundamentality and non-symmetric relations, in Glick, D., Darby, F. & Marmodoro, A. (eds.) *The Foundation of Reality: Fundamentality, Space, and Time*, Oxford: Oxford University Press.
Bader, R.M. (ms) Conditional grounding, draft.
Broad, C. (1923) *Scientific Thought*, London: Routledge and Kegan Paul.
Byrne, A. (2006) Color and the mind–body problem, *Dialectica*, **60**, pp. 223–244.
Chalmers, D.J. (2006) Perception and the fall from Eden, in Gendler, T.S. & Hawthorne, J. (eds.) *Perceptual Experience*, Oxford: Clarendon Press.
Chalmers, D.J. (2015) Panpsychism and panprotopsychism, in Alter, T. & Nagasawa, Y. (eds.) *Consciousness in a Physical World: Perspectives on Russellian Monism*, New York: Oxford University Press.
Coleman, S. (2015) 'Neuro-cosmology', in Coates, P. & Coleman, S. (eds.) *Phenomenal Qualities: Sense, Perception, and Consciousness*, Oxford: Oxford University Press.
Cutter, B. (2018) Paradise regained: A non-reductive realist account of the sensible qualities, *Australasian Journal of Philosophy*, **98** (1), pp.38–52.
Dasgupta, S. (2014) The possibility of physicalism, *Journal of Philosophy*, **111** (9–10), pp. 557–592.
Eddington, A. (1928) *The Nature of the Physical World*, London: Macmillan.
Fish, W. (2008) Relationalism and the problems of consciousness, *Teorema: Revista Internacional de Filosofía*, **27** (3), pp. 167–180.
Fish, W. (2009) *Perception, Hallucination, and Illusion*, Oxford: Oxford University Press.
Fish, W. (2013) Perception, hallucination, and illusion: Reply to my critics, *Philosophical Studies*, **163** (1), pp. 57–66.
Galileo, G. (1623/1996) *The Assayer*, in Drake, S. (trans. and ed.) *The Discoveries and Opinions of Galileo*, New York: Doubleday.
Goff, P. (2017) *Consciousness and Fundamental Reality*, Oxford: Oxford University Press.
Goff, P. (2019) *Galileo's Error*, London: Penguin, Random House.
Goff, P. (ms) How exactly does panpsychism help explain consciousness, draft.
Jackson, F. (1998) *From Metaphysics to Ethics: A Defence of Conceptual Analysis*, Oxford: Clarendon Press.
Johnston, M. (1996) A mind–body problem at the surfaces of things, *Philosophical Issues*, **7** (Perception), pp. 219–229.
Johnston, M. (ms) *The Manifest*, draft.
Kalderon, M. (2007) Colour pluralism, *Philosophical Review*, **116** (4), pp. 563–601.
Levine J. (1983) Materialism and qualia: The explanatory gap, *The Philosophical Quarterly*, **64** (October), pp. 354–361.
Liu, M. (this issue) Qualities and the Galilean view, *Journal of Consciousness Studies*, **28** (9–10).

Martin, M.G.F. (1998) *Setting Things Before the Mind*, Cambridge: Cambridge University Press.

Moran, A. (2018) Kind-dependent grounding, *Analytic Philosophy*, **59** (3), pp. 359–390.

Moran, A. (2021) Living without microphysical supervenience, *Philosophical Studies*, [Online], https://doi.org/10.1007/s11098-021-01664-7.

Moran, A. (ms-a) Naïve realism, illusion, and the phenomenal principle, draft.

Moran, A. (ms-b) The mind–body problem reconfigured, draft.

Moran, A. (ms-c) On why non-reductive physicalists need contingent grounding laws, draft.

Nagel, T. (1974) What is it like to be a bat?, *The Philosophical Review*, **83** (October), pp. 435–450.

Pautz, A. (2013) Do the benefits of naïve realism outweigh the costs? Comments on Fish, Perception, Hallucination and Illusion, *Philosophical Studies*, **163** (1), pp. 25–36.

Pautz, A. (ms) How to achieve the physicalist dream-theory: Identity or grounding, draft.

Russell, B. (1927) *The Analysis of Mind*, London: Routledge.

Schaffer, J. (2017) The ground between the gaps, *Philosopher's Imprint*, **17** (11), pp. 1–26.

Sider, T. (2020) *The Tools of Metaphysics and the Metaphysics of Science*, Oxford: Oxford University Press.

Shoemaker, S. (1996) Self-knowledge and inner sense lecture III: The phenomenal character of experience, in *The First-Person Perspective and Other Essays*, New York: Cambridge University Press.

Strawson, P.F. (1979) Perception and its objects, in McDonald, G. (ed.) *Perception and Identity: Essays Presented to A.J. Ayer*, London: Macmillan.

Wittgenstein, L. (1958) *Philosophical Investigations*, Oxford: Blackwell.

Alyssa Ney[1]

Panpsychism and the Limits of Physical Science

Abstract: *This essay critically engages with two themes in Philip Goff's book Galileo's Error regarding, first, the limits to what we can learn from physical science and, second, the comparative metaphysical and ethical implications of panpsychism and physicalism. I argue that the instrumentalist and structuralist theses Goff uses to prop up his claims about the limits of physical science are unmotivated and, even if they were motivated, would not support the sort of panpsychism Goff recommends. The second part of the essay shows why physicalism is, contrary to Goff's claims, no worse than panpsychism when it comes to providing accounts of free will, objective value, and meaning.*

1. Introduction

As Philip Goff (2019) rightly notes in his recent book, *Galileo's Error: Foundations for a New Science of Consciousness*, the debate between physicalism and dualism is not an abstract metaphysical debate, but one with normative consequences about the proper aims for science. In his work, including this book, Goff is focused on these views' implications about the correct way to study mental phenomena scientifically, especially the qualitative features of experience (hereafter, phenomenal consciousness).

Physicalists believe that phenomenal consciousness should be studied using the tools of physical science. And many physicalists are

Correspondence:
Email: aney@ucdavis.edu

---

[1] University of California, Davis, CA, USA.

further optimistic that physical science will be able to provide a comprehensive understanding of phenomenal consciousness. On the other hand, dualists believe that physical science cannot provide a comprehensive understanding of phenomenal consciousness. Consciousness is a non-physical phenomenon and, as such, its scientific study must make use of a distinct set of methods from those of physical science. Goff defends a third view, an alternative to both dualism and physicalism. This is a version of panpsychism. According to Goff's view, phenomenal consciousness cannot be understood using the tools of physical science, and so should not be investigated that way. However, this is not because phenomenal consciousness is distinct from those phenomena studied by physical science. Rather, the qualities constitutive of phenomenal consciousness (hereafter, phenomenal properties) are the intrinsic natures of the entities studied by physical science. There are phenomenal properties that are the intrinsic natures we associate with mass and charge, perhaps more complex phenomenal properties that are the intrinsic natures we associate with chemical properties like acidity and the resonance properties of certain molecules, and so on up the chain of complex systems studied by physical science. Goff's panpsychist argues that, even though physical objects have phenomenal properties as their intrinsic natures, these intrinsic natures cannot be studied by physical science because physical science has limits barring it from providing insight into intrinsic natures. However, the new post-Galilean science of consciousness Goff advocates can give us insight into these intrinsic natures, thus providing a complete model of reality.

In this critical essay, I will be primarily interested in questioning Goff's assertions about the limits of physical science. In his book, Goff makes use of two different views about what these limits are. However, as I will argue, neither view is plausible, nor are these views that are commonly defended in contemporary philosophy of science. Although my focus for the majority of this essay will be on how one's view about the mind–body problem impacts one's view about the limits and proper aims of physical science, I was also struck while reading *Galileo's Error* by Goff's statements about physicalism's and panpsychism's implications for other issues such as free will, ethics, and what he sees as a scarily looming demise of human civilization. And so, I will conclude with comments on his discussion of those topics.

2. Physical Science and Intrinsic Natures

The central argument for panpsychism that Goff presents in *Galileo's Error*, what he calls 'the simplicity argument', may be formulated as follows:

1. The entities of physical science have intrinsic natures.
2. But physical science tells us absolutely nothing about these intrinsic natures.
3. We do, however, learn from our conscious experiences about intrinsic natures. These intrinsic natures are the qualitative features of our experiences (phenomenal properties).
4. And from (3) we may infer that some entities of physical science, i.e. the matter inside our brains, have intrinsic natures that are these phenomenal properties.
5. The simplest hypothesis that accommodates the previous premises is panpsychism: the view that the intrinsic nature of all of the entities of physical science, not just those of the matter inside our brains, involves phenomenal properties.
6. Thus, we should be panpsychists (Goff, 2019, p. 134).

Goff attributes this argument to Arthur Eddington. As mentioned, in Goff's picture, the qualitative features of conscious experience provide the intrinsic natures of entities across the physical sciences. We only have knowledge of the qualitative features of our own experiences. And so we only have knowledge of the intrinsic natures of certain complex biological systems (brains). But since entities across the physical sciences must have intrinsic natures, the simplest hypothesis states that the intrinsic natures of all of these entities are qualitative features of conscious experiences. This is the 'panpsychist move': since physical science posits entities but fails to describe their natures, we infer that it is phenomenal properties that constitute these natures. There is much to be sceptical of in this argument, however, in what follows, I focus on Goff's premise 2: that physical science tells us absolutely nothing about intrinsic natures.

There is a technical and less technical way of understanding what is meant by 'intrinsic natures'. The less technical way is that intrinsic natures concern 'what things are' or, more precisely, 'what things are in themselves and fundamentally'. The 'in themselves' part is why we say 'intrinsic' and the 'fundamentally' part is why we say 'natures'. The technical way starts from a metaphysical distinction between intrinsic and extrinsic properties. Intrinsic properties are properties that concern how an object is in itself, not how it is with respect to other things. Extrinsic properties concern not how an object is in itself, but how it with respect to other things (Lewis, 1986). Consider a tennis ball. The ball's shape is an intrinsic property of it, since it concerns just how it is in itself. But the ball's ability to bounce high on a hard court is an extrinsic property of it, since it concerns how it is with respect to other things. I don't know if tennis balls have intrinsic natures, and if they do whether being roughly spherical is part of their intrinsic natures.[2] But the intrinsic natures of objects concern not just what they are accidentally like, but what they are like essentially or fundamentally. Perhaps it is part of the essence of a tennis ball to be roughly spherical. If so, then this property will constitute part of the object's intrinsic nature. How to understand essences and fundamentality is highly contested in contemporary metaphysics, and so I won't discuss this issue further (but see Tahko, 2018).

Goff presents two philosophical views about science in order to support his claim that physics tells us nothing about intrinsic natures. The first we may call *the way of instrumentalism*, the second, *the way of structural realism*. Here are some passages in which Goff pursues the way of instrumentalism:

> If physics is not telling us the nature of physical reality, what is it telling us? The crucial point Eddington is trying to convey… is that *physics is a tool for prediction*. (2019, p. 125, italics in original)

And:

> What is being exposed here is a certain popular perception of physics, according to which the purpose of physics is to hold up a mirror to reality. This is simply not what physical science is in the business of… The quantitative sciences of physics and chemistry have had great success, but this is in part because they were designed to fulfill a specific and limited goal: the prediction of behavior. (*ibid.*, p. 129)

If scientific realists believe that the claims of our best scientific theories are true or approximately true, that they aim to provide faithful representations of how our world is, even when we are not making observations, and that the objects posited by our best scientific theories exist in roughly the way they say they do, instrumentalists

---

[2] Although I won't press this point in the remainder of this note, one reasonable critique of Goff's argument would begin by questioning the utility of the antiquated concept of intrinsic natures for the current debate about the proper aims and limits of contemporary physical science.

deny all of the above. Instrumentalists deny that our best scientific theories even aim at truth; they hold that our best scientific theories are not intended to give faithful representations of our world, and that we should not believe in entities corresponding to the terms of even our best scientific theories. Rather, these theories should just be understood as tools, tools for better manipulating the world around us, tools for more reliably predicting the results of these manipulations or other observations we may make.

Instrumentalism is certainly not a mainstream view in philosophy of science. Although some of the physicists associated with the Copenhagen interpretation of quantum mechanics made claims that are naturally interpreted in an instrumentalist light, such an interpretation of physics has fallen out of favour in mainstream philosophy of physics (Lewis, 2016; Becker, 2018). Even those physicists today who defend serious successors to the Copenhagen interpretation, the 'ψ-epistemicists', don't adopt such a crude instrumentalism, adopting instead scientific realism (see Leifer, 2014). Given instrumentalism is so far out of the mainstream, if Goff wants to use it to support premises in his argument for panpsychism, he owes us some kind of defence of this position.

I wonder though whether instrumentalism is too blunt a tool to get Goff to his desired conclusion. For if instrumentalism tells us that physical science doesn't provide faithful representations of the world, and so can't tell us about intrinsic natures, then it seems neither can Goff's new science of consciousness. For Goff's science of consciousness is supposed to incorporate not only entities with intrinsic natures, but presumably also the robust causal structures that the entities with these intrinsic natures are capable of instantiating. If Goff wants to advocate a wholesale instrumentalism about physical science, then how can post-Galilean science include an account of these causal structures?

More promising is what I see as Goff's second way to the claim that physical science tells us nothing about intrinsic natures: this is the way of structural realism. Structural realism, by contrast with instrumentalism, is a mainstream view in the philosophy of science. It comes in two forms. The first, epistemic structural realism, is the view that science only teaches us about structures, or relations between things. It cannot teach us about individual objects and their intrinsic properties. The second, ontic structural realism, is the view that science only teaches us about structures or relations, because there are only structures and relations. There are no objects with intrinsic properties. In an especially hardline version of ontic structural realism, this is because, fundamentally, there aren't objects at all. Since ontic structural realists reject the intrinsic natures that play such a central role in Goff's panpsychist argument, he can't adopt the stronger ontic version of scientific realism. However, he can adopt the weaker epistemic version, and it seems that he does so when he makes claims like, 'physics tells us not what matter *is* but only what it *does*' (2019, p. 125).

Although Goff makes these claims, in an appendix, he distances himself from the way of structural realism, and moves back to the more contentious way of instrumentalism:

> [W]hen I said that the equations of physics 'tell us what matter does,' this was really just a loose way of saying that they are a tool for prediction. In fact, careful reflection reveals that physical science doesn't even tell us what matter does. (*ibid.*, p. 176)

The reason Goff backs off from structural realism (what he calls 'causal structuralism') is that he thinks it is wrong that physical science tells us what objects do.

According to Goff, physical science neither tells us what objects are nor what they do. It doesn't say what objects are because this would require giving an account of the properties of things. This, for Goff, would require telling us what properties like mass are in themselves. Both Goff and the structural realists think physical science doesn't do this. However, structural realists think physical science does tell us what objects do, by saying, for example, that when objects have mass, this causes them to move in certain ways; that is, to change their distances from one another. Goff denies this is telling us what objects do, since this, he claims, would require physical science to have an account of what distance is and it doesn't do this anymore than it tells us what mass is. But, this seems to be a confusion. Physical science doesn't need to tell us what distance is. We already know what it is by seeing objects separated all around us. We learn about distance from perception. We certainly don't need a new post-Galilean panpsychist science to tell us what distance is.

Although I myself am not sympathetic to structural realism of either form, epistemic structural realism is the position Goff needs for his simplicity argument to be sound.[3] And so we should ask whether it is

---

[3] I argue that physical science does characterize the intrinsic natures of things in Ney (2015). The basic idea is that our best physical theories give very different mathematical

sustainable. For most structural realists today find epistemic structural realism to be an unstable position, and take the appeal of epistemic structural realism to provide motivation to adopt the stronger ontic variety of the position. It was famously the argument of James Ladyman in his now classic paper 'What is Structural Realism?' that 'structural realism gains no advantage over traditional scientific realism if it is understood merely as an epistemological refinement of it, and that instead it ought to be developed as a metaphysical position' (1998, p. 411). The argument here is that, for structural realism to be a version of realism, it must be telling us that the picture of the world we get from our best scientific theories is approximately correct and that the world is roughly the way these theories say it is. Thus, if physical theories only characterize the world in terms of extrinsic properties and relations, then a realist should think this is all there is.

So, when Goff presents the current state of the art in philosophy of science as adopting the view that physical science doesn't describe intrinsic natures, this isn't off the mark, but what he fails to note is that the same state of the art also holds the view that there aren't such things. The intermediate position Goff wants to hold that physical science doesn't describe intrinsic natures but that nonetheless there are such things may be motivated for Goff by introspection, but the impression *Galileo's Error* may leave one with, that this is a common view in contemporary philosophy of science, is questionable.

3. Physicalism and the Demise of Civilization

I've been concerned here mostly with the normative consequences of adopting physicalism or panpsychism for science. But in the last chapter of *Galileo's Error*, 'Consciousness and the Meaning of Life', Goff argues that which position one adopts on the mind–body problem also has consequences for how one may live one's life, and that the widespread adoption of physicalism in the past century has produced negative impacts to human civilization on a massive scale. He begins the chapter with the claim that panpsychism is 'a theory of Reality

---

characterizations of particles that are charged and not charged, that lack spin and have integer spin and half-integer spin, that have mass and are massless. And, in doing so, these theories provide intrinsic characterizations that distinguish these properties, e.g. that distinguish mass from charge from spin, just like the different geometrical characterizations of squares and circles and triangles provide intrinsic characterizations that distinguish them.

somewhat more consonant with human happiness than rival views' (2019, p. 184). However, although the chapter does discuss how Goff thinks that panpsychism is better able to (a) motivate people to act on the climate crisis (than dualism), (b) provide a metaphysics of free will (better than dualism and physicalism), (c) provide a metaphysics that can ground the existence of objective values (better than physicalism), and (d) steer people away from capitalist and nationalist inclinations (better than physicalism), we never see how the view can lead us to greater happiness and it is completely unclear to me how it would. But setting this point about happiness aside, in the space I have remaining, I will run through his complaints about physicalism and how panpsychism is supposed to do better, to show why none of his points (b)–(d) are convincing.

3.1. On free will

Although Goff is officially neutral on the question of whether humans have free will, he argues that if we are to have it then it must be the case that our actions are both not determined, and also not random (*ibid.*, p. 198). He offers us a middle way: if our actions are responsive to our reasons, then our actions can be, in the sense that matters, 'up to us' and hence free. This focus on reasons-responsiveness is something that is familiar in the free will literature and most commonly associated with the work of the compatibilist John Martin Fischer (1994). It is a strategy physicalists commonly appeal to in making sense of free will, even if determinism is true.

But Goff thinks we need to reject physicalism to get a genuine freedom-cum-reasons-responsiveness. This is never clearly expressed; however, the closest he comes is by articulating what he thinks the panpsychist can get us, that the physicalist cannot:

> It could be that past events *pressure* physical entities in the present to behave a certain way but that it is always up to present physical entities whether or not to accept that pressure... What I am trying to capture is how things seem to be when I am making a free decision. It feels like it's genuinely up to me, while at the same time there exist certain pressures in the form of my inclinations... In extreme cases, such as torture, the pressure may overpower me, making it inevitable that I will choose in the direction I am being pressured. But in many ordinary cases, it is up to me whether or not to yield to the pressure of my inclinations, or which inclination I choose to yield to. (2019, p. 201)

The problem with this is that, on the panpsychist model Goff defends in his book, the causal structures of the world are (or better, can be) completely described by physical science. And so on his own view, although it may feel as if it is really only he that is providing the pressure to act that way, these feelings don't — can't — break the physical causal structures. Like other compatibilists about free will, I don't take this to be a problem because among those actions that are determined by the physical laws there are those one may distinguish that are responsive to my reasons and those that are not. And, like Goff, I take the fact that the former are reasons-responsive to make it the case that they are free. What I am pointing out is simply that there is no room for anything more in Goff's proposed metaphysics.

3.2. Objective value

In the discussion of objective value and all that follows for the remainder of the book, Goff develops further arguments for panpsychism. These rely on an account of mystical experiences. It should be noted that these are experiences Goff says he himself has never had, and so the arguments discussed in the remaining sections are explored in a speculative spirit. He isn't convinced there are such experiences, but neither is he 'persuaded by the arguments that have been advanced to show that mystical experiences must be delusions' (*ibid.*, p. 207).

According to the account Goff is interested in exploring, mystical experiences reveal a formless consciousness that is a part of all experiences (*ibid.*, p. 206). These arguments for panpsychism rely in addition on a metaphysical inference from this putative fact about our experiences to the claim that in some sense 'we are all one' (*ibid.*, p. 207). Note that this 'we' doesn't mean 'we humans', but we everything. According to the panpsychist view Goff entertains for the last sections of the book, everything has experience and so everything whatsoever is one: you, me, every non-human animal, the trees, the forest, everything. I think it is obvious that, even if multiple entities may have experiences that are qualitatively similar in some respect, this does not imply that these entities are numerically identical. But Goff's arguments do rely on this inference.

So, suppose that there is a formless consciousness that forms a backdrop to all of our experiences, and suppose from that it is reasonable to infer we are all one. Still, I cannot see how anything Goff says in the remainder of the book about the benefits of panpsychism over physicalism follows.

Goff complains that physicalists are not capable of providing a foundation for facts about objective value. He discusses Sam Harris's proposal that we can learn from the sciences of well-being and happiness what produces well-being and happiness, and that this can in turn tell us what has objective value and what does not. Goff argues that, to provide an account of objective value, these sciences would also have to tell us that well-being or happiness has objective value:

> It is of course true that empirical investigation is a good way of working out how to make people happy, but it doesn't follow that empirical investigation can shed light on the truly foundational ethics claim here, which is that human happiness or well-being is of objective moral value. (*ibid.*, p. 211)

To this, Goff considers a response Harris gives to this objection:

> Even if each conscious being has a unique nadir on the moral landscape, we can still conceive of a state of the universe in which everyone suffers as much as he or she (or it) possibly can. If you think we cannot say this would be 'bad,' then I don't know what you could mean by the word 'bad' (and I don't think you know what you mean by it either)... It seems uncontroversial to say that a change that leaves everyone worse off, by any rational standard, can be reasonably called 'bad,' if this word is to have any meaning at all. (Harris, 2010, p. 39)

Goff replies that this may be uncontroversial, but the question is what grounds this moral value. I think Goff is wrong to characterize Harris's reply as merely being that it is uncontroversial that happiness and well-being have objective moral value. Harris's claim isn't merely that it is uncontroversial. It is that it is *analytic* (true by definition) that well-being and happiness have moral value. It is uncontroversial because it is analytic. This is clear from the quote even though Harris doesn't use this philosophical jargon. Now, there is reason to be sceptical that today's sciences of well-being and happiness can tell us what has objective value and what does not.[4] But, if it is an analytic truth that happiness has objective value, and these sciences can discover what makes people happy, then we can make inferences to what more specific actions have or lack moral value.

Let me be clear. A physicalist does not have to accept that facts about objective moral value are analytically grounded in facts about

---

[4] See especially Anna Alexandrova (2017) which argues *inter alia* that scientific theories of well-being tend to presuppose rather than argue for a particular theory of well-being and that such theories need philosophy to be more convincing accounts of their targets.

happiness. There are other naturalist models for the grounding of objective moral value available (e.g. Railton, 1986; Jackson, 1998). However, the view that it is analytic that well-being or happiness has objective moral value and that the physical sciences can discover empirically which things make for happiness or well-being is at least more compelling than what Goff is exploring, namely that since 'the most basic element of *my* mind — the formless consciousness which forms the backdrop of each experience — is identical with the most basic element of *your* mind', there is objective moral value (2019, p. 213). It is just not clear how this identity claim is supposed to get us to any value claim. I think the idea is supposed to be something like: if I care about me, then I care about you too, if you are identical to me. But we still don't have an account of why I care about me in the first place, why I or anyone else has moral value. So, I can't see how adopting panpsychism makes us any better off.

3.3. Capitalism and nationalism

Finally, I think it is important to address Goff's accusations that the embrace of physicalism is partly to blame for what he views as two ills of modern society, capitalism and nationalism, and why he thinks the world would be a better place if we would instead adopt panpsychism.

We should start by acknowledging that adoption of the physicalist attitude by scientists across disciplines has significantly transformed our world for the better. Adoption of the physicalist attitude has led to dramatic developments in science, technology, medicine, and public policy that would not have otherwise taken place. To suggest that physicalism has been bad for humanity and that we as a society steer away from it is a very problematic claim without a strong argument to support it.

Following Max Weber and others, Goff traces the problems with physicalism to its disenchantment of the world. Earlier in the book, he discusses the fact that new research is showing that actually not just non-human animals but also plants have some kind of inner mental life (*ibid.*, pp. 191–5). This is supposed to offer support for panpsychism and the idea that by adopting panpsychism, but not physicalism, we can once again see the world around us as enchanted, as conscious. But there is nothing about this research on plant consciousness that is incompatible with physicalism. Indeed, the research proceeds not by the scientists introspecting what these plants feel, but rather by using standard tools of physical science. So, if we need more enchantment to feel a connection with the environment, we can have that and be physicalists too.

Goff's main assertions in this section are that the disenchantment brought by the rise of physicalism has led to a sense of a loss of meaning that in turn has led to 'consumerism and the endless quest for economic growth to make sense of our lives' and, worse, to nationalism. By contrast, Goff says that if panpsychism is true then 'we *belong in*' the universe, that 'we can live in nature, in the universe', that 'we are at home in the cosmos' (*ibid.*, p. 217). And this is supposed to lead us away from capitalist and nationalist impulses.

I am very sceptical of the idea that physicalism leads to a loss of meaning for our lives. Physicalism is in no way incompatible with the pursuit of life projects and goals, the enjoyment of nature, the joys of connecting with family and friends. Physicalism too is a humanism. And of course the physicalist (even a physicalist who does not believe that the being is one) also believes we belong in, live in, and are at home in the cosmos. If that is enough to give life meaning or steer one away from consumerism and nationalism, then physicalism has those consequences as well. We don't need panpsychism to make the world a better place.

References

Alexandrova, A. (2017) *A Philosophy for the Science of Well-Being*, New York: Oxford University Press.
Becker, A. (2018) *What is Real? The Unfinished Quest for the Meaning of Quantum Physics*, New York: Basic Books.
Fischer, J.M. (1994) *The Metaphysics of Free Will*, Oxford: Blackwell.
Goff, P. (2019) *Galileo's Error: Foundations for a New Science of Consciousness*, New York: Vintage.
Harris, S. (2010) *The Moral Landscape: How Science Can Determine Human Values*, New York: Free Press.
Jackson, F. (1998) *From Metaphysics to Ethics: A Defence of Conceptual Analysis*, Oxford: Clarendon.
Ladyman, J. (1998) What is structural realism?, *Studies in History and Philosophy of Science*, **29** (3), pp. 409–424.
Leifer, M. (2014) Is the quantum state real? An extended review of ψ-ontology theorems, *Quanta*, **3** (1), pp. 67–155.
Lewis, D.K. (1986) *On the Plurality of Worlds*, Oxford: Blackwell.
Lewis, P. (2016) *Quantum Ontology*, New York: Oxford University Press.
Ney, A. (2015) A physicalist critique of Russellian monism, in Alter, T. & Nagasawa, Y. (eds.) *Consciousness in the Physical World*, pp. 346–369, Oxford: Oxford University Press.
Railton, P. (1986) Moral realism, *Philosophical Review*, **95** (2), pp. 163–207.

Tahko, T. (2018) Fundamentality, in Zalta, E.N. (ed.) *The Stanford Online Encyclopedia of Philosophy*, [Online], https://plato.stanford.edu/archives/fall2018/entries/fundamentality/ [16 June 2021].

Damian Aleksiev[1]

Missing Entities

Has Panpsychism Lost the Physical World?

Abstract: Panpsychists aspire to explain human consciousness, but can they also account for the physical world? In this paper, I argue that proponents of a popular form of panpsychism cannot. I pose a new challenge against this form of panpsychism: it faces an explanatory gap between the fundamental experiences it posits and some physical entities. I call the problem of explaining the existence of these physical entities within the panpsychist framework 'the missing entities problem'. Space-time, the quantum state, and quantum gravitational entities constitute three explanatory gaps as instances of the missing entities problem. Panpsychists are obliged to solve all instances of the missing entities problem; otherwise, panpsychism cannot be considered a viable theory of consciousness.

1. Introduction

There is a lot to like about Philip Goff's *Galileo's Error*. The book is a concise introduction to the philosophy of consciousness. Without a loss of rigour, Goff brings the academic discussions of consciousness from the ivory tower to the broader public. Moreover, there is a lot to like about Goff's preferred metaphysical theory — *panpsychism*. Panpsychism brings a breath of fresh air to the stale debates between physicalists and dualists. It promises an account of human

Correspondence:
Email: aleksiev.damian@gmail.com

[1] Central European University, Budapest, Hungary.

consciousness compatible with both the data of physics and introspection. Goff and fellow panpsychists aspire to solve the mystery of consciousness with a worldview shift. They posit that human consciousness and the physical world are of the same kind, are both essentially experiential. Although substantial, this rethinking of the nature of reality seems justified if panpsychism can indeed deliver on its promises.

I use this occasion to pose a new challenge against panpsychism. Panpsychists are standardly challenged for whether they can account for human consciousness. However, it has so far been neglected whether they can account for the *physical world*. I set out to explore this question by analysing whether the entities entailed by some of our best theories of physics are compatible with panpsychism. I do this by analysing aspects of Goff's take on panpsychism both from *Galileo's Error* and from his wider academic work.

I argue that, if panpsychism were true, the existence of at least some *physical entities* would be left unexplained. I call this *the missing entities problem* for panpsychism. I define *three explanatory gaps* between the hypothetical fundamental experiences (that panpsychists posit) and different physical entities as instances of this problem. Panpsychists are obliged to solve all instances of the missing entities problem. Otherwise, the worldview shift proposed by panpsychists is unwarranted, and panpsychism cannot be considered a viable theory of consciousness.

2. Missing Entities

Goff argues that Galileo's error was to think that the *quantitative* vocabulary of physics fully captures the essences of physical entities. As Galileo himself puts this, in a famous passage:

> [The book of the universe] is written in mathematical language, and its characters are triangles, circles, and other geometrical figures. Without these it is humanly impossible to understand a word of it, and one wanders around pointlessly in a dark labyrinth. (2008, p. 183)

Galileo was a mind–body dualist. He thought experiences are real, yet quite unlike physical entities. He saw experiences as essentially not only quantitative but moreover *qualitative*. He took this to entail that physical entities and experiences are different in kind.

Contemporary philosophers of mind are, by and large, physicalists. Physicalists reject Galileo's dualism due to theoretical considerations based on the current empirical evidence.[2] Nonetheless, many physicalists follow Galileo in his putative error. They embrace the Galilean conception of physical reality (what Goff calls the '*purely physical*' conception). According to these physicalists, the fundamental entities are purely physical. In their view, the purely physical facts *ground* the experiential facts. 'Grounding', as I understand it, is a relation that holds between the more fundamental facts (as grounds) and the less fundamental facts (as groundees). Grounds *determine* and *explain* the obtaining of their groundees.

Pure physicalism is often contested because it faces an *explanatory gap* between the pure physical and the experiential facts. An explanatory gap, simply put, means that there is *no intelligible connection* between a ground and a groundee.[3]

To locate explanatory gaps, as a good heuristic, think about what an ideal reasoner could deduce about reality from its fundamental elements. For example, think of Laplace's fictional demon. If Laplace's demon could never deduce a groundee from its fundamental ground, then there is an explanatory gap between fundamental reality and that groundee. Plausibly, Laplace's demon could deduce the properties of H_2O (and, in general, of all purely physical groundees) from the fundamental, purely physical grounds. In contrast, plausibly, not even Laplace's demon could deduce what red feels like (and, in general, what any experience feels like) from the purely physical facts. If so, pure physicalism faces the above-mentioned explanatory gap.

Explanatory gaps matter because they might reveal mistakes in our conception of reality. They might indicate the falsity of the grounding claims they involve. Goff (2017, pp. 100–3) argues that there should be *no* explanatory gaps in true cases of grounding when the ground and groundee are thought under '*transparent*' concepts. In Goff's usage, a transparent concept reveals the *full essence* of its referent; it reveals '*what it is for that entity to be part of reality*' (*ibid.*, p. 15).

Goff argues that both pure physical and phenomenal concepts are transparent. Pure physical concepts refer to purely physical entities.

---

[2] Many physicalists argue that dualism violates the 'causal closure' of the physical. See Papineau (2001).

[3] Readers seeking a more rigorous definition should think of explanatory gaps in terms of a *lack of a priori entailment* between grounds and groundees. See Chalmers and Jackson (2001).

As Goff argues, they reveal that purely physical entities are essentially quantitative: their essences are pure physical structures (*ibid.*, p. 101). Phenomenal concepts, in contrast, are the concepts we use in introspection when thinking about experiences in terms of what they feel like. In Goff's view, phenomenal concepts reveal that experiences are essentially qualitative: their essences are their phenomenal characters, are what experiences feel like (*ibid.*, pp. 107–8).[4]

I summarize Goff's ideas in the following two theses:

> *No Explanatory Gaps*: There are no explanatory gaps in true cases of grounding where both the ground and the groundee are thought under transparent concepts.
>
> *Transparency*: Pure physical concepts and phenomenal concepts are transparent.

Given *No Explanatory Gaps* and *Transparency*, the explanatory gap between the pure physical and the experiential facts entails that *pure physicalism is false*.[5]

Goff aspires to fix Galileo's error in the light of *No Explanatory Gaps* and *Transparency*. He and fellow panpsychists *redefine* the Galilean conception of fundamental physical reality. According to Goff's preferred version of panpsychism, it is implausible that pure quantities could exist autonomously without having some deeper qualitative ground. Goff and fellow panpsychists posit that experiences are the perfect candidates for such a ground. In Goff's own words:

> All we get from physics is this big black-and-white abstract structure, which we must somehow fill in with intrinsic nature. We know how to color in one bit of it: the brains of living organisms are colored in with consciousness. How to color in the rest? The most elegant, simple, sensible option is to color in the rest of reality with the same pen. (2019, p. 135)

The resulting view is panpsychism. Or, more precisely, *Russellian pure panpsychism*. The adjective 'Russellian' is in honour of the philosopher and mathematician Bertrand Russell. It designates the

---

4   I take this to imply that the contents of an experience *are* the properties of the experience itself. As Goff puts it: 'Arguably, the qualities in our experience just are, in their essential nature, experience-characterizing properties' (2017, p. 161).

5   It is worth noting that this argument does not apply to all versions of physicalism. Some physicalists are happy to accept explanatory gaps between grounds and groundees. See Schaffer (2017) for a defence of such a view.

conviction (that Russell likewise held) that the fundamental physical structure of the cosmos has a qualitative ground. The adjective 'pure' designates that fundamental reality is *entirely* experiential. The structure of the fundamental experiences is objective; it obtains independently of human observation.[6] Physics accurately describes this structure. If panpsychism is true, the fundamental physical structure of the cosmos *is* the structure of the fundamental experiences.

As Goff points out, panpsychists typically conceive of the fundamental experiences as 'unimaginably simple' (*ibid.*, p. 113). In their view, these simple hypothetical experiences ground both human experiences and all physical entities.

Many contemporary panpsychists are moreover *reductive* panpsychists.[7] They posit that the above-mentioned fundamental experiences are the building blocks of everything. I illustrate this idea with a theological metaphor: if such reductive panpsychism is true, all that God would need to do to create the cosmos is to create the fundamental experiences; everything else would follow metaphysically 'for free'.[8] Moreover, epistemically, in line with *No Explanatory Gaps*, reductive panpsychism promises a cosmos *without* explanatory gaps. If reductive panpsychism (of the above kind) is true, the knowledge of the fundamental experiences would, in principle, entail knowledge of all other facts for an ideal reasoner.

Throughout this paper, unless otherwise specified, I use 'panpsychism' to refer to *pure Russellian reductive panpsychism* as I have defined it above. Goff is sympathetic to this version of panpsychism, both in *Galileo's Error* and throughout his academic work.[9] Panpsychism, so construed, seem to be the most cohesive, parsimonious, and widely accepted version. It is a form of experiential monism that aspires to explain human consciousness better than physicalism.

---

[6] And, in general, independently of the observations of any non-fundamental subject.

[7] More rigorously defined, 'reductive panpsychism', as I use the term here, is panpsychism that is both metaphysically and epistemically reductive; it is the conjunction of constitutive and type-A panpsychism. The kind of grounding at play here is 'grounding by analysis', in Goff's (2017) terminology.

[8] I stress this is only a metaphor and a heuristic: panpsychists are *not* necessarily committed to the existence of God.

[9] Goff also explores and defends other versions of panpsychism such as: consciousness+ panpsychism, emergentist panpsychism, and hybrid panpsychism. I briefly discuss consciousness+ panpsychism in Section 4. The rest of these views are beyond the scope of this paper.

It is not clear whether panpsychism can fulfil its explanatory promise. The panpsychist framework has *two essential metaphysical seams*:

(a) between the fundamental experiences and human experiences, and
(b) between the fundamental experiences and all physical entities.

Panpsychists must be wary of any unclosable explanatory gaps at either of these seams. Unclosable explanatory gaps violate *No Explanatory Gaps*; they obtain even for reasoners with all the required transparent concepts. Unclosable explanatory gaps entail cracks in the elegant panpsychist framework: they entail that panpsychism — as I have defined it — is false.

Case (a) corresponds to '*the combination problem*'.[10] The combination problem is a serious and well-discussed problem for panpsychism. However, the combination problem is not the only serious problem that panpsychism faces. Given *No Explanatory Gaps*, panpsychists are obliged to resolve both (a) and (b). Case (b) is mostly ignored in the literature. I use this occasion to bring (b) into the spotlight.

Against panpsychism, I argue that if fundamental reality were purely experiential, some physical entities might lack an intelligible explanation in terms of fundamental reality. If so, were we to rebuild the cosmos from pure experiences, these physical entities would be *missing* from our reconstruction of reality. I call the problem of explaining the existence of these physical entities *the missing entities problem*. I express this problem as a broad explanatory gap as follows:

> *Missing Entities*: There is an explanatory gap between the fundamental experiences (as grounds) and some physical entities (as groundees).

In the rest of the paper, I show how to use *Missing Entities* to argue against panpsychism.

A note on terminology. I use the term 'entity' in a broad sense. In my usage, directly observable entities (such as tables and planets), indirectly observable entities (such as space-time and micro-particles), and purely theoretical entities (such as the quantum state and spin networks) all count as entities. The challenge I raise against

---

[10] For more on the combination problem see Chalmers (2016).

panpsychism is that experiences do not have the right structure to be the grounds of all physical entities.

3. Three Explanatory Gaps

The three principles referred to above — *No Explanatory Gaps*, *Transparency*, and *Missing Entities* — work together as premises in an argument against panpsychism that I call 'the missing entities argument':

P-1. *No Explanatory Gaps* is true.
P-2. *Transparency* is true.
P-3. *Missing Entities* is true.
C. Panpsychism is false.

If sound, the missing entities argument entails that at least some physical entities are *not grounded* in the fundamental experiences. If so, going back to the theological metaphor I used earlier, it would not be enough that God creates the fundamental experiences to create the cosmos. Instead, God would *need to do more* to bring the missing entities into existence. If so, experiences alone are not enough to recreate the cosmos. Thus, panpsychism is false.[11]

Goff (2017, pp. 100–3) explicitly defends P-1 and P-2. I expect these two premises to be uncontroversial for reductive panpsychists. Rejecting P-1 and P-2 undermines a key motivation for panpsychism. Without *No Explanatory Gaps* and *Transparency*, panpsychists would struggle to reject physicalism based on explanatory gap worries. Thus, panpsychists are better off accepting P-1 and P-2.

With P-1 and P-2 out of the way, I dedicate the rest of this section to the defence of P-3. I defend P-3 by defining *three explanatory gaps* as instances of it. They involve space-time, the quantum state, and quantum-gravitational entities. All three of these gaps involve experiences (as grounds) and pure physical entities (as groundees). All three gaps express the same idea — the idea that the structure of the cosmos is essentially different from the structure of experiences.[12] Beyond

---

[11] As I stated already, the missing entities argument only targets pure Russellian reductive panpsychism. *Impure* (and, thus, more complex) forms of panpsychism positing both experiences *and* physical structure as fundamental are not its target.

[12] Thus, *Missing Entities* is like the hard problem of consciousness in reverse. In the role of P-3, *Missing Entities* resembles the hard problem turned upside-down and put to use against the panpsychist.

these three gaps, there might be other ways to specify P-3 involving different physical entities or properties. It only takes *one* instance of *Missing Entities* to challenge panpsychism. The panpsychist is obliged to close all known instances of *Missing Entities*.

In the case of all three explanatory gaps, I set out to establish *Missing Entities* by first analysing *human experiences* and then inferring that the hypothetical fundamental experiences are similar. I base this inference on the following thesis:

> *Good Model*: Human experiences are a good model for the fundamental experiences posited by panpsychists.

Good Model might appear controversial. It entails that human experiences can be used as proxies for the putative fundamental experiences. It entails that we are justified to use the available transparent concepts of human experiences — referenced in *Transparency* — in place of the unavailable transparent concepts of the putative fundamental experiences. If so, the phenomenal concepts involved in *Missing Entities* are covered by *Transparency*. This consequence of *Good Model* might appear overly strong, given that the fundamental experiences are unknown. Panpsychists broadly agree that humans do not directly access the fundamental experiences.

In defence of *Good Model*, I point out that fundamental experiences are a theoretical posit. Their primary role is to explain human experiences. Human experiences are our *only guide* to the fundamental experiences. Thus, I take it that it is reasonable to assume the fundamental experiences must be *similar* to human experiences. How similar? Similar enough to ensure there is *no explanatory gap* between them (as grounds) and human experiences (as groundees).

Rejecting *Good Model* comes at a high price for the panpsychist. First, it shrouds panpsychism in mystery. If the fundamental experiences are a complete mystery, it is impossible to investigate how they might ground the rest of reality. Second, and more importantly, it turns the combination problem into a potentially unclosable chasm. If the fundamental experiences are not similar to our experiences, it is unreasonable to expect they could intelligibly explain our experiences. In light of this, *Good Model* presents the panpsychist with a dilemma: either (1) reject *Good Model* but shroud panpsychism in mystery and face a potently unsolvable combination problem, or (2) accept *Good Model* but deal with the missing entities argument. The missing entities argument targets panpsychists willing to accept horn (2) of this dilemma. I count Goff among these panpsychists, given his persistent ambition to solve the combination problem in his academic work.

3.1. The Space-time Gap

The first gap involves *space-time* as a groundee.

> *Space-time Gap*: There is an explanatory gap between the fundamental experiences (as grounds) and space-time (as a groundee).

The theory of relativity is our best current theory of space and time.[13] The theory of relativity — in its most straightforward ontological interpretation — entails the existence of *space-time*. The theory of relativity, so understood, is a substantivalist theory of space-time. If space-time substantivalism is true, space-time is a ubiquitous and dynamic entity with no further physical ground. In this section, I assume that space-time substantivalism is true, and I base *Space-time Gap* on it.

Space-time has an *essential geometric structure*. It is a *four-dimensional manifold* structured by the *space-time metric*. In geometry, metric structures determine *distances*. The space-time metric determines the space-time distance between any two events in space-time.[14] Space-time distances are invariant; they are the same for all observers. If space-time's metric structure were different, the theory of relativity would no longer apply to space-time; space-time would no longer be the same entity, it would no longer *be* space-time. If so, the space-time metric is likely essential to space-time.[15]

Visual experiences, more than any other experiences, are associated with distances. I can clearly see and measure the distance between two points on a line, and I can clearly tell that some objects are close to me while others are further away from me. Thus, it appears that at least some visual experiences might be essentially metric. If so, could some visual experiences, in some contexts, instantiate the space-time metric?

---

[13] By 'the theory of relativity', I have in mind the conjunction of the special and the general theory of relativity.

[14] What I call the 'space-time distance' is often called the 'space-time interval' in the physics literature (standardly designated as '$(\Delta s)^2$'). Here, I use the term 'space-time distance' both for simplicity and to emphasize that this quantity is the space-time analogue of the Euclidean distance.

[15] In the literature, this metaphysical position is known as *metric essentialism*; its *locus classicus* is Maudlin (1989).

The space-time metric entails facts of *positive* distances but also of *negative* and *null* distances. The distinction between positive, negative, and null space-time distances constitutes space-time's causal structure. These three kinds of space-time distances obtain *everywhere* in space-time: at all scales and in all regions. There are perfectly natural reference frames in every cell of our bodies where space-time events are at positive distances but also at negative and null distances from one another.

Human experiences, in contrast to space-time, seem to instantiate only positive distances. Our ordinary concept of distance is the concept of a positive quantity separating some entities. Although negative and null distances are coherent, they appear to be *neither perceivable nor imaginable*. If so, given that phenomenal concepts are transparent, it seems reasonable to assume that no human visual experience essentially has the space-time metric.

Beyond visual experiences, the same point seems to apply to all other human experiences. To the best of my introspective ability, I cannot find any experience that exhibits the properties of space-time's metric structure. To the best of my introspective ability, no human experience instantiates negative and null distances. If so, again, given that phenomenal concepts are transparent, *no* human experience appears to have the space-time metric.

Putting the above together: space-time essentially has the space-time metric, while all human experiences lack the space-time metric. If so, via *Good Model*, the putative fundamental experiences (posited by the panpsychist) likely lack the space-time metric. If so, it seems that not even an ideal intellect (like Laplace's demon) could deduce space-time's existence from the putative fundamental experiences. Thus, *Space-time Gap* is true.

If panpsychism is true, space-time is a dependent entity grounded in the putative fundamental experiences. Thus, although space-time has no further physical ground, it has an experiential ground. However, *Space-time Gap* — in the role of P-3 of the missing entities argument — entails that, if panpsychism were true, space-time would lack an explanation in terms of fundamental reality. Thus, if panpsychism were true, there might be no space-time. Thus, if space-time substantivalism is true, *Space-time Gap*, as a premise of the missing entities argument, entails that panpsychism is false.

3.2. The Quantum State Gap

The second gap involves the *high-dimensional quantum state*, posited by some proponents of realist quantum theories, as a groundee.

> *Quantum State Gap*: There is an explanatory gap between the fundamental experiences (as grounds) and the high-dimensional quantum state (as a groundee).

Quantum theory is our best current theory of matter. Quantum theory's predictive power is undeniable, yet its ontological implications are an enigma. The wave function is the central mathematical device of quantum theory. According to proponents of *realist* quantum theories, the wave function is not only a useful mathematical device; instead, it represents a real entity.[16] Philosopher of physics Tim Maudlin (2019) calls this entity the '*quantum state*'.

The properties of the quantum state are a matter of heated debate. There is a lot of disagreement about how much of the wave function's mathematical structure corresponds to the quantum state's structure. Specifically, one question of key importance is whether the quantum state has the same number of dimensions as the wave function.

The wave function's domain is standardly called a *configuration space*. It is typically characterized as *3N-dimensional*, where N stands for the number of particles it describes. The universe is estimated to have at least 10^{80} particles. If so, the configuration space of our universe's wave function has at least *3×10^{80} dimensions*.

Proponents of (what I call) *high-dimensionalism* argue that the quantum state is 3N-dimensional, just like our universe's wave function. In their view, the 3D space aspect of space-time is grounded in the 3N-dimensional quantum state. *Quantum State Gap* involves this high-dimensional conception of the quantum state.

Contributors to this volume, Alyssa Ney (2012) and Sean Carroll (2019) are notable proponents of high-dimensionalism. It is worth noting that Goff (forthcoming) argues against high-dimensionalism based on the worry that it cannot account for the reality of human consciousness. Thus, this section is not directly an attack on Goff's view but is aimed more generally at panpsychists sympathetic towards high-dimensionalism.

---

[16] Realist quantum theories include Bohmian, Everettian, and spontaneous collapse quantum theories.

Given high-dimensionalism, the quantum state is essentially 3N-dimensional. The complexity of the quantum state, so defined, is staggering. It is impossible to imagine (or otherwise experience) anything even remotely close to this high-dimensional structure. If so, and given the transparency of phenomenal concepts, clearly, no human experience is essentially 3×10^{80} dimensional.

By now, my defence of *Quantum State Gap* should be obvious. If panpsychism is true, the quantum state is grounded in some fundamental experiences. However, via *Good Model*, it is reasonable to assume that none of the putative fundamental experiences is 3×10^{80} dimensional. Based on this, were panpsychism true, not even Laplace's demon could deduce the existence of the quantum state from the fundamental experiences. Thus, *Quantum State Gap* is true.

Quantum State Gap — in the role of P-3 of the missing entities argument — entails that, if panpsychism were true, the high-dimensional quantum state would lack an explanation in terms of fundamental reality. If panpsychism were true, there might be no high-dimensional quantum state. Thus, if high-dimensionalism is true, *Quantum State Gap*, as a premise of the missing entities argument, entails that panpsychism is false.

3.3. The Quantum Gravity Gap

The third gap involves the *timeless quantum-gravitational entities*, posited by some theories of quantum gravity, as groundees.

> *Quantum Gravity Gap*: There is an explanatory gap between the fundamental experiences (as grounds) and timeless quantum-gravitational entities (as groundees).

Theories of quantum gravity aspire to explain gravity in a way that is compatible with quantum theory. Some theories of quantum gravity aspire to accomplish this by *quantizing* space-time. Contributor to this volume Carlo Rovelli (2004) is a leading figure in the development of one such theory: loop quantum gravity. These theories of quantum gravity posit that space-time, as a whole, is grounded in fundamental structures that lack a spatial and temporal metric. These structures seem to be essentially *neither spatial nor temporal*.

I use the term 'timeless quantum-gravitational entity' as a blanket term for any fundamental entity that lacks an essential temporal metric and is posited by a theory of quantum gravity.[17] The postulated existence of timeless quantum-gravitational entities gives rise to *Quantum Gravity Gap*.

Time appears to be an essential property of all human experience.[18] No human experience seems possible unless it has some temporal duration, unless it lasts some time. This should be plain to everyone who has reflected on any ordinary experience. If human experiences are essentially temporal, via *Good Model*, it follows that the putative fundamental experiences are likewise temporal. If so, *Quantum Gravity Gap* is true.

Quantum Gravity Gap is not a new challenge for panpsychism. Susan Schneider (2018) has already raised an analogous challenge. Goff acknowledges her challenge and offers a tentative solution (2019, pp. 209–10). Based on Miri Albahari's (2019) work, he suggests that one kind of human experience might be essentially timeless. This is *mystical experience*. Mystical experiences are typically induced either by deep meditation or by the use of psychedelics. They have been reported across many different cultures and religions. People who have undergone mystical experiences often describe them as experiences of pure oneness beyond space and time.

Given that phenomenal concepts are transparent, mystical experiences might be essentially timeless. However, this alone is not sufficient to close *Quantum Gravity Gap*. Closing *Quantum Gravity Gap* requires experiences that both (a) are essentially timeless and (b) have the *right kind* of timeless structure. I argue that even if mystical experiences satisfy (a), they most likely fail to satisfy (b). The 'right kind of timeless structure' required in (b) is the structure of the specific quantum-gravitational entities we are trying to ground. I illustrate this by reference to *Space-time Gap*.

To close *Space-time Gap*, it is not sufficient that some experience has *some* metric. Instead, some experience must have the space-time metric. Likewise, to close *Quantum Gravity Gap*, it is not sufficient that some experience has *some* timeless structure. Instead, some experience must have the structure of the specific quantum-gravitational entities they are expected to ground (or, at least, a structure sufficiently similar to allow for an intelligible connection).

---

[17] In loop quantum gravity, the timeless quantum-gravitational entities are called 'spin networks'.

[18] For an independent defence of this claim, see Phillips (2014).

It is highly unlikely that mystical experiences have the structures of quantum-gravitational entities. Reports indicate that timeless mystical experiences are experiences of pure oneness, *without any differentiation*.[19] Thus, plausibly, they lack any definable structure. If so, via *Good Model*, it follows that the timeless fundamental experiences are likely to lack any definable structure. Moreover, given *Transparency*, if mystical experiences had a structure even roughly resembling that of quantum-gravitational entities, mystics would have told us about it by now. Yet, they have not. Instead, it was physicists who developed quantum gravity. Thus, again, via *Good Model*, it follows that even if there are timeless fundamental experiences, their structure is likely not similar to that of quantum-gravitational entities.

In summary, timeless quantum-gravitational entities have specific timeless structures. The reports of mystical experiences indicate that some human experiences might be essentially timeless. However, these experiences seem to lack structures similar to the ones posited by theories of quantum gravity. If so, via *Good Model*, it follows that the putative fundamental experiences likewise lack timeless quantum-gravitational structure. If so, were panpsychism true, not even Laplace's demon could deduce the existence of these timeless quantum-gravitational entities from the fundamental experiences. Thus, *Quantum Gravity Gap* is true.

Quantum Gravity Gap — in the role of P-3 of the missing entities argument — entails that, if panpsychism were true, timeless quantum-gravitational entities would lack an explanation in terms of fundamental reality. If panpsychism were true, there might be no timeless quantum-gravitational entities. Thus, if any of the theories positing such entities is true, *Quantum Gravity Gap*, as a premise of the missing entities argument, entails that panpsychism is false.

4. The Consciousness+ Response

Missing Entities is true if at least one of the above explanatory gaps is true. If *Missing Entities* is true, given my previous defence of the other premises, the missing entities argument is sound. If so, panpsychism is false.

The missing entities argument refutes a specific version of panpsychism that is pure, Russellian, and reductive. This is one version of

---

[19] See Albahari (2019).

panpsychism that Goff is sympathetic towards. However, he also defends other versions. Notably, in his academic work, Goff (2017, pp. 179–81, 230–5) develops a unique version of panpsychism he calls '*consciousness+ panpsychism*'.

Consciousness+ panpsychism posits that fundamental reality is constituted of consciousness+ properties. These hypothetical properties enfold 'experiential and non-experiential aspects into a single nature' (*ibid.*, p. 180). Consciousness+ panpsychism is impure, Russellian, and reductive. Goff develops it as one potential solution to the combination problem.[20] The non-experiential aspects of consciousness+ are completely mysterious. Goff argues that the addition of these hidden aspects to experiences might solve the combination problem.

Goff envisions consciousness+ panpsychism as metaphysical monism. Consciousness+ properties are supposed to be unitary. This makes sense in the context of solving the combination problem. Presumably, the experiential aspects of consciousness+ explain human experiences *qua* experiences. In contrast, presumably, their plus-aspects explain how the putative fundamental experiences combine into human experiences. If so, both the experiential and non-experiential aspects of consciousness+ have necessary *explanatory roles* in solving the combination problem.

The missing entities argument fails to refute consciousness+ panpsychism. As I have argued, experiences alone cannot close *Missing Entities*. However, consciousness+ properties might close *Missing Entities* in virtue of their plus-aspects. If so, consciousness+ panpsychism avoids both the combination and the missing entities problem. Nevertheless, I believe, stretching the explanatory role of consciousness+ properties this far might come at a price. The price is the loss of *monism*.

Let us assume — as Goff argues — that Galileo's conception of the physical world is mistaken. Thus, *contra* Galileo, physical entities have some deeper qualitative ground. Moreover, suppose we are justified to posit consciousness+ properties to explain human consciousness (perhaps due to explanatory gap worries). So far, so good. Now, ask yourself: what is the best candidate for the deeper qualitative ground of physical entities?

---

[20] It is worth noting that Goff also proposes other potential solutions to the combination problem. Perhaps the most notable among these is his 'phenomenal bonding' proposal (Goff, 2016).

Missing Entities indicates that experiences alone (as long as they are similar to our human experiences) are *not* apt to ground the physical world.[21] On the other hand, the plus-aspect of consciousness+ is tailor-made for closing explanatory gaps. It seems that, when it comes to grounding the physical world, the plus-aspect can do *all* the explanatory work and do it *alone*. In contrast, experiences can do this same explanatory work *only if assisted* by the plus-aspect. This indicates that the experiential aspect of consciousness+ might be explanatorily *redundant* in closing *Missing Entities*.

If I am right about the above, the plus-aspect of consciousness+ seems to be the best candidate for the qualitative ground in the scenario under consideration. Although the two aspects of consciousness+ come together in solving the combination problem, they come apart in closing *Missing Entities*. But then, is there any strong reason to think consciousness+ is unitary? We might as well be dualists at this point.

Consider the following dualist view.[22] Imagine a cosmos where two distinct kinds of properties are fundamental: non-experiential (physical) qualities and experiences. The physical qualities ground the physical world. The fundamental experiences correlate with them throughout the cosmos: like a ghost in the machine. Yet, these fundamental experiences do not ground any physical entities. The two kinds of fundamental properties come together in complex systems (such as humans). There, they mutually ground higher-order experiences. Although rough, this sketch seems coherent. The onus is on Goff to explain why his monistic consciousness+ panpsychism is preferable to this or a similar form of dualism.

5. Conclusion

Panpsychism promises a lot. It is a well-motivated and elegant metaphysical theory. However, panpsychism faces serious challenges from both the combination problem and, as I argued, the missing entities problem. I used this opportunity to elucidate the missing entities problem as a new problem for panpsychism.

All the instances of *Missing Entities* I outlined express the same underlying idea: the structures of the cosmos are essentially different

---

[21] Assuming, of course, that one of the theories of physics I have examined is true.
[22] Or rather: *dualist panpsychism*.

from the structures of experiences. I based *Missing Entities* on an inference from human experiences to the putative fundamental experiences. Our experiences can ground rich structures. Yet, the structures of the cosmos are richer. The structures of the cosmos are beyond the structures of our experiences. I argued, via *Good Model*, that panpsychism is most likely to work if the fundamental experiences it posits are similar to our experiences. *Missing Entities* is the outcome of this inference. *Missing Entities* — as a premise of the missing entities argument — entails that the physical structure of the cosmos cannot be purely experiential.[23] If so, Galileo might have been right to separate physical structure and experiences.

Acknowledgments

I am deeply grateful to Philip Goff for his comments on the paper and, more importantly, for teaching me about panpsychism and then inviting me to argue against it. Many thanks also to Jamie Elliot, Zhiwei Gu, and Alex Moran for commenting on the paper.

References

Albahari, M. (2019) Perennial idealism: A mystical solution to the mind–body problem, *Philosophers' Imprint*, **19** (44), pp. 1–37.
Carroll, S. (2019) *Something Deeply Hidden: Quantum Worlds and the Emergence of Spacetime*, New York: Dutton Books.
Chalmers, D.J. (2016) The combination problem for panpsychism, in Brüntrup, G. & Jaskolla, L. (eds.) *Panpsychism: Contemporary Perspectives*, New York: Oxford University Press.
Chalmers, D.J. & Jackson, F. (2001) Conceptual analysis and reductive explanation, *Philosophical Review*, **110** (3), pp. 315–361.
Galilei, G. (2008) The assayer, in Finocchiaro, M.A. (ed.) *The Essential Galileo*, Indianapolis, IN: Hackett Publishing Company.
Goff, P. (2016) The phenomenal bonding solution to the combination problem, in Brüntrup, G. & Jaskolla, L. (eds.) *Panpsychism: Contemporary Perspectives*, New York: Oxford University Press.
Goff, P. (2017) *Consciousness and Fundamental Reality*, New York: Oxford University Press.
Goff, P. (2019) *Galileo's Error: Foundations for a New Science of Consciousness*, London: Rider.
Goff, P. (forthcoming) Quantum mechanics and the consciousness constraint, in Gao, S. (ed.) *Quantum Mechanics and Consciousness*, New York: Oxford University Press.

---

[23] At the very least, the fundamental experiences certainly *cannot be unimaginably simple*, as Goff and many other panpsychists posit.

Maudlin, T. (1989) The essence of space-time, *PSA: Proceedings of the 1988 Biennial Meeting of the Philosophy of Science Association*, **2**, pp. 82–91.

Maudlin, T. (2019) *Philosophy of Physics: Quantum Theory*, Princeton, NJ: Princeton University Press.

Ney, A. (2012) The status of our ordinary three dimensions in a quantum universe, *Noûs*, **46** (3), pp. 525–560.

Papineau, D. (2001) The rise of physicalism, in Gillett, C. & Loewer, B. (eds.) *Physicalism and its Discontents*, Cambridge: Cambridge University Press.

Phillips, I. (2014) Experience of and in time, *Philosophy Compass*, **9** (2), pp. 131–144.

Rovelli, C. (2004) *Quantum Gravity*, Cambridge: Cambridge University Press.

Schaffer, J. (2017) The ground between the gaps, *Philosophers' Imprint*, **17** (11), pp. 1–26.

Schneider, S. (2018) Spacetime emergence, panpsychism and the nature of consciousness, *Scientific American*, [Online], https://blogs.scientificamerican.com/observations/spacetime-emergence-panpsychism-and-the-nature-of-consciousness/ [30 May 2021].

Ralph Stefan Weir[1]

Can a Post-Galilean Science of Consciousness Avoid Substance Dualism?

Abstract: *In Galileo's Error, Philip Goff sets out a manifesto for a post-Galilean science of consciousness. Article three of the manifesto says that a post-Galilean science of consciousness should reject dualism. I argue that there is an important sense of 'dualism' in which Goff's arguments are not only compatible with but entail dualism, and not only dualism but substance dualism. I do not suggest that this is bad news for the post-Galilean revolution. But if it is correct then the revolution might equally be considered a restoration.*

1. Introduction

Philip Goff argues that we must stop seeing consciousness as something to be squeezed into the mathematical picture of the physical world that we have inherited from Galileo's conception of science. Instead, we need a new research programme that treats consciousness as a qualitative datum in its own right, and that accounts for human and animal consciousness in terms of basic forms of consciousness that exist as basic properties of matter. The penultimate chapter of *Galileo's Error* sums up this research programme in a 'Post-Galilean Manifesto'. Article three of the manifesto reads:

Correspondence:
Email: ralphsweir@gmail.com

---

[1] University of Lincoln, Lincolnshire, UK.

Anti-Dualism: Consciousness is not separate from the physical world; rather consciousness is located in the intrinsic nature of the physical world. (Goff, 2019a, p. 174)

This essay argues that there is an important sense of 'dualism' in which Goff's arguments are not only compatible with but entail dualism, and not only dualism but substance dualism. Substance dualism, in the sense I have in mind, is the view that (i) there are two sides to reality, a fundamentally mental side and a fundamentally non-mental side; and (ii) the fundamentally mental side consists of mental substances resembling Cartesian souls.

I claim that Goff's arguments entail substance dualism in the sense specified. I do not suggest that this is bad news for post-Galilean science of consciousness, or that it threatens the main advantages of Goff's position which, in its latest form (Goff, ms), is the only theory of consciousness I am aware of that seems totally comfortable with the (putative) empirical data and that allows conscious states to act causally on the physical world without leaving an 'explanatory gap' between the physical and the mental. The aim of this essay is rather to push for a certain view about what a theory of consciousness motivated by Goff's arguments must look like.

2. Substance Dualism

Goff argues that, *contra* materialism, there exist fundamentally mental properties. But he suggests that the fundamentally mental properties are among the properties whose dynamic structure is described by physics. The fundamentally mental properties are not, therefore, 'separate from the physical world' where 'physical world' means the world whose dynamic structure is described by physics, irrespective of its fundamental nature. Hence, Goff's position contradicts dualism where 'dualism' means:

> Dualism 1: there exists a side to reality whose dynamic structure is described by physics and a side whose dynamic structure is not described by physics.

But there remains the question: are *all* of the properties whose dynamic structure is described by physics fundamentally mental, or only some? If the answer is only some it follows that there are two sides to reality, a fundamentally mental side and a fundamentally non-mental side. So, Goff's position is compatible with dualism where 'dualism' means:

> Dualism 2: there exists a side to reality that is fundamentally mental and a side that is not fundamentally mental.

The distinction between Dualism 1 and Dualism 2 cuts across a more familiar distinction between property dualism and substance dualism. Property dualism is the view that in addition to physical things there are non-physical mental properties, but no non-physical substances.[2] Substance dualism is the view that in addition to physical things there are non-physical substances. Classic defences of property dualism include Chalmers (1996) and Kim (2005). Classic defences of substance dualism include Plato's *Phaedo* and Descartes' *Meditations*. Suppose our post-Galilean theory of consciousness is a dualist one, in the sense of Dualism 2. We can then ask: are the fundamentally mental things all properties, or are some substances? If the fundamentally mental things are all properties, our post-Galilean theory will be a kind of property dualism. If the fundamentally mental things include substances, our post-Galilean theory will be a kind of substance dualism.

'Substance' is a term of art. Here, we are interested in the sense of 'substance' that captures the distinction between substance dualism and property dualism. With that in mind, I propose that x is a substance if and only if x could exist by itself. That is, a substance is something whose existence does not necessitate the existence of anything further. Substances, so understood, contrast with properties because properties, on an intuitive view, cannot exist by themselves. As Justin Broackes says, 'nothing can be circular *and nothing else* (e.g. not of a particular diameter); nothing can be red *and nothing else* (e.g. not of any extent or shape)' (2006, p. 156). This is a reasonable account of substance, in the present context, for two reasons. First, the definition of substances as things that can exist by themselves is roughly the definition that is in play in Descartes' paradigmatic defence of substance dualism.[3] Secondly, substance dualism is widely thought of as a view on which the non-physical entities are capable of disembodied existence, and any fundamentally mental thing that can

---

[2] It is generally accepted that for something to be 'physical', in the sense that defines materialism, it must be both fundamentally non-mental and among the things described by physics. So, something can count as 'non-physical' either because it is not described by physics or because it is fundamentally mental.

[3] See e.g. 'Fourth Replies' (AT VII, 226) and *Principles* (AT VIII, 24–5; AT IX B, 147), both Descartes (1967–76). I discuss the nature and history of substances in detail in Weir (forthcoming).

exist by itself will *a fortiori* be capable of disembodied existence in the sense that it will be capable of existing in the absence of anything fundamentally non-mental.

A standard objection to the definition of a substance as something that could exist by itself says that nothing could satisfy this definition because nothing could exist without its own parts or intrinsic properties. But there is a familiar sense in which the statement that *x* could exist 'by itself' does not entail that *x* could exist without its parts and intrinsic properties since the parts and intrinsic properties of *x* are intuitively included in *x* (*cf.* Gorman, 2006, p. 151; 2012; Koslicki, 2018, pp. 173–87; Weir, forthcoming). For example, the statement that there is a coffee cup 'by itself' on your desk does not mean that the cup does not have a shape, size, base, or handle. For these are not 'other things' on your desk in the relevant sense — they are included in the cup. A substance is something that could exist by itself in the sense that it could exist without anything that is not included in it.[4]

I claim that Goff's arguments entail that the fundamentally mental properties belong to mental substances — fundamentally mental things that could exist by themselves.[5] In the case of the mental substances that I will argue Goff must posit, the fundamentally mental properties are all phenomenal properties, i.e. experiential properties of the sort that determine what it is like to be a given conscious subject at a given time. The mental substances to which they belong are therefore what Goff (2010; 2017, pp. 229–30) would call 'pure subjects': beings whose nature is exhausted by what it is like to be them. They are just like Cartesian souls, on a common conception of Cartesian souls.

First, I will explain why Goff's arguments entail that phenomenal properties belong to mental substances. Following this, I will explain why Goff's arguments also entail that there is a fundamentally non-mental side to reality, in addition to these mental substances. If so,

---

[4] A further objection says that nothing could exist without one or more necessary and/or abstract beings. The simplest way to avoid this problem is by restricting the domain of the definition of substance to contingent concreta.

[5] I assume that all properties of mental substances must be fundamentally mental, since even property dualists must posit some substances that have some fundamentally mental properties. Hence I would not count what Schneider (2012, p. 63) calls 'hybrid' substances as mental substances.

3. The Zombie Argument

Materialism is the view that everything is physical, conscious experience included. Materialists usually suppose that consciousness is some kind of brain state. Goff (2019a, pp. 86–96) argues against materialism on the basis that (i) it is possible for brain states to exist without conscious experiences; and (ii) if it is possible for brain states to exist without conscious experiences, then conscious experiences are not identical to or constituted by brain states. Goff defends premise (i) on the basis that zombies are possible, where a zombie is a physical duplicate of an actual human without any conscious experience.

How do we know that zombies are possible? Goff argues that this is essentially a matter of logic (*ibid.*, p. 89). Careful reflection shows that there is no contradiction in the idea of a physical duplicate of a conscious human, without their conscious experiences. But Goff makes one qualification. He grants that the fact that there is no contradiction in the idea of a zombie only shows that zombies are possible *if* we fully understand what it is we are conceiving of.[6] Goff concedes that, if we were ignorant of the nature of matter or of consciousness, the fact that there is no contradiction in the idea of a zombie would not show that zombies are possible.

It is this qualification that opens the way for panpsychism. For Goff goes on to argue that, in so far as we confine our attention to what physics tells us, we *do not know* the nature of matter. 'Physics tells us not what matter *is,* but what it *does*' (Goff, 2019a, p. 125). For this reason, we can make sense of the panpsychist hypothesis that the intrinsic nature of matter involves conscious experience. And, according to Goff, it is this hypothesis that best accounts for the place of consciousness in nature.

Strictly then, Goff only claims that we know that zombies are possible *given the assumption* that matter is fundamentally non-mental. But this is enough to show that materialism is false because materialism says that there is nothing more to consciousness than the

---

[6] In more technical terms: we definitely know that zombies are conceivable/logically possible. But Goff grants that it only follows that they are 'metaphysically' possible if we have a full understanding of what we are conceiving of. See Goff (2019a, p. 91, fn. 1; 2017, pp. 86–105; 2019b).

fundamentally non-mental brain states (or whatever) described by physics: materialism presupposes that matter is fundamentally non-mental.

There is a second way in which one might object to Goff's inference from the fact that there is no contradiction in the idea of a zombie to the conclusion that zombies are possible. Perhaps the inference is invalid because we do not know the nature of our *conscious experiences*. If so, we can reject the zombie argument against materialism while also avoiding panpsychism, on the basis of our ignorance of the nature of consciousness. But Goff argues that this objection to the zombie argument is implausible. For according to Goff (*ibid.*, pp. 106–09) we *do* know the nature of our conscious experiences, simply by reflecting on those experiences when we have them.[7] 'The job of science', Goff (2019a, p. 109) urges, 'is not to tell us what feelings are (we already know what a feeling is when we feel it) but rather to give an account of the place of feelings in a general theory of reality.'

4. Possible Ghosts

Goff's version of the zombie argument rests on two principles. First, if there is no contradiction in the idea of a scenario, and we know the nature of the entities involved, then that scenario is possible. Secondly, we know the nature of our conscious experiences. If these principles are true then the fact that there is no contradiction in the idea of a zombie entails that materialism is false. I suggest that, in conjunction with another argument of Goff's, the same principles entail that phenomenal properties belong to mental substances. I have in mind Goff's argument for the conceivability of 'lonely ghost twins' (see Goff, 2010; 2012; 2014).

A 'ghost' is the phenomenal complement of a zombie, 'a creature whose being is exhausted by its being conscious, by there being something that it is like to be it' (Goff, 2010, p. 123). Someone's 'ghost twin' is 'a creature that has qualitatively identical conscious

---

[7] Elsewhere Goff (2011; 2017, pp. 107–20) expresses this point by saying that our phenomenal concepts are 'transparent': they reveal what it is for phenomenal properties to be part of reality. A 'phenomenal concept' here is a concept of a phenomenal property of the sort that one can acquire simply by undergoing an experience with that property and attending to it, e.g. the concept of what red looks like that you can acquire by looking at a pillar box and reflecting on how it appears colour-wise. Goff (2011) is the best place to start for an understanding of the idea that phenomenal concepts are transparent.

experience' to that person but 'whose nature is exhausted by its conscious experience' (*ibid.*, p. 124). A 'lonely ghost twin' is a ghost twin that exists entirely by itself. Goff argues that the coherence of Cartesian doubt about the external world shows that there is no contradiction in the idea that one's lonely ghost twin:

> To entertain that I am the only thing that exists, and that I exist as a thing with no properties other than my conscious experience, just is to conceive of my [lonely] ghost twin. (*ibid.*, pp. 124–5; *cf.* 2019a, pp. 4–5)

To recapitulate, Goff argues that there is no contradiction in the idea that one's ghost twin should exist by itself with only its phenomenal properties; that we know the nature of phenomenal properties; and that if there is no contradiction in the idea of a scenario, and we know the nature of the entities involved, then that scenario is possible. These three theses jointly entail that our ghost twins *could in fact* exist by themselves with only their phenomenal properties: *they* are mental substances.

So far Goff would agree. But he would emphasize that our ghost twins *are not us*. They are *phenomenal duplicates* of us that do not actually exist. The fact that our ghost twins could exist by themselves, with only their phenomenal properties, shows that our ghost twins are mental substances. But it is a further step to infer that *we* are mental substances. Goff (2010, n. 9, p. 128; 2014, p. 12) rejects this further step. In the locations cited, Goff is considering Descartes' argument from the possibility that *I am* the only thing that exists and that *I exist* as a thing with no properties other than my phenomenal properties to the conclusion that *I am* a mental substance. Goff rejects this argument on the basis that we do not know the nature of the referent of 'I' in the way that we know the nature of our conscious experiences.[8] For this reason:

> There may be corporeal aspects of the 'I' I am conceiving of... which have no conceptual association with the mental properties in terms of which I am conceiving of that 'I'. For this reason, it seems to me that Descartes' argument fails as an argument for substance dualism. (Goff, 2014, p. 12)

---

[8] Goff (2010, p. 127, n. 9, p. 128; *cf.* 2014, p. 12; 2017, p. 98) argues that, in general, concepts that refer to particular individuals are not 'transparent': they do not reveal the nature of the entities they refer to.

If this objection is successful then one can accept the possibility of lonely ghost twins without having to posit any *actual* mental substances.

5. Real Ghosts

Initially, Goff's objection to the disembodiment argument for mental substances looks entirely reasonable. But I believe that it is in fact unsustainable. To see why, take some conscious subject *s*. Now, let us introduce the term '*s*-minus' to name that part or aspect of reality that *s*'s ghost-twin duplicates. That is, '*s*-minus' refers to the subject of *s*'s conscious experience, including all of its phenomenal properties, but *excluding* all non-phenomenal properties. We can now run an argument from the possibility of the scenario in which *s*-minus is the only thing that exists and exists with only its phenomenal properties to the conclusion that *s*-minus is a mental substance. If this argument is unsound, this cannot be because *s*-minus has further aspects beyond the mental properties under which we are conceiving of it, since we have set those aside by stipulation.

Goff might object that it is not clear that *there is* such thing as *s*-minus. For we have introduced *s*-minus by directing our attention at an existing thing, *s*, and setting aside all non-phenomenal properties. And sometimes, this kind of 'systematic setting aside' (Campbell, 1990, p. 3) leaves us with an abstraction like the Cheshire cat's grin or the waiter's manner of walking — things which according to some theorists do not, strictly speaking, exist. However, the natural reason for doubting that, for example, the Cheshire cat's grin exists in its own right *just is* that the idea of the grin without the rest of the cat is incoherent. If, by contrast, the idea of *s*-minus existing by itself contains no contradiction, then the manner in which *s*-minus was introduced gives us no more reason to doubt that it exists than we have in the case of such ordinary objects as the Cheshire cat's tail or the waiter's tie.[9]

---

[9] I assume here that we can reason about the modal properties of *s*-minus, and decide on that basis whether it exists. In the meantime we can be neutral on whether '*s*-minus' names a real object or what Peter van Inwagen (1990) would call a 'virtual object', i.e. something that does not exist in the strict sense required of metaphysical theorizing. One might worry that if *s*-minus exists so will, for example, *s*-minus less any one of its phenomenal properties, with the result that there exist countless distinct subjects where there should be only one, as in the case Peter Unger's (2004) 'mental problem of the many' and Mark Johnston's (2016) 'personite problem' (*cf*. Moran, 2018; forthcoming).

Alternatively, Goff might argue that it is impossible for s-minus to exist by itself, not because it has further properties, beyond the mental properties under which we are conceiving of it, but for some other reason. But this cannot be because s-minus possesses or lacks any property that s's ghost twin does not also possess or lack. The scenario in which s-minus exists by itself and the scenario in which s's ghost twin exists by itself are, *ex hypothesi*, qualitatively exactly the same. So if the scenario in which s-minus exists by itself is impossible, this must be due to s-minus's identity alone, i.e. it must be due to the brute fact that s-minus — although it is exactly like s's ghost twin in every other respect — is simply not the same thing as (is not identical with) s's ghost twin.[10] This would be very different to the case of the waiter's manner or walking or the Cheshire cat's grin. For the reason why these things cannot exist by themselves has to do with their qualitative nature: no qualitative duplicate of the gait or the grin *could* exist by itself.[11]

Could the mere identity of s-minus — it's being what it is and not another thing — be responsible for the fact that s-minus cannot exist by itself? At first, this might seem plausible. For mere identity certainly *does* make *some* differences to what is possible for an individual: i.e. those that that follow from the logic of identity. There is nothing puzzling, for example, about the fact that s's ghost twin can whereas s-minus cannot exist without s-minus, simply because s-minus is s-minus, whereas s's ghost twin is not s-minus. But the thesis that s's ghost twin can whereas s-minus cannot exist by itself requires that mere identity can make non-logical differences to what is possible for qualitative duplicates.[12] For two reasons, I suggest that Goff should reject the thesis that mere identity can result in such differences.

This problem can be avoided either by granting that x is a subject only if there exists no larger group of co-conscious experiences that includes all of the experiences involved in x or by denying that a subject less one of its phenomenal properties could exist by itself. I discuss this further in a future work.

[10] In the technical jargon, since s-minus and s's ghost twin are *qualitatively identical*, any difference in what is possible for them must be due to the fact that they are not *numerically identical*.

[11] It seems to me, at any rate, that gaits and grins cannot exist by themselves. Some might disagree (e.g. Hume, 2007, §1.4.5.5; Williams, 1953, pp. 79–80). If so, substances are more widespread than I suppose and s-minus is all the more likely to be one of them.

[12] By 'non-logical' differences I mean those that do not follow *a priori* from the logic of identity plus 'transparent' concepts of the entities involved.

First, the idea that mere identity can result in non-logical differences to what is possible for qualitative duplicates is inherently implausible. Consider, for example, whether there could be an exact qualitative duplicate of your coffee cup that is incapable of shattering, not because of its superior manufacture, or any difference in its qualitative character, but simply because of its identity — because it is the particular coffee cup that it is. This seems absurd. If the nature of possibility is as mysterious as this example implies then it is hard to see how we can know that *any* non-actual scenario is possible.

Secondly, the thesis that mere identity can result in non-logical differences to what is possible for qualitative duplicates undermines the zombie argument. For the zombie argument is based on the premise that it is possible for physical *duplicates* of actual humans — our zombie *twins* — to exist with only their physical properties without being conscious. It infers that there is more to *actual* human beings than our physical bodies. But if mere identity can result in non-logical differences in what is possible for qualitative duplicates then the inference is invalid. For, in that case, it remains possible that actual human organisms are wholly physical beings that are nonetheless, necessarily, conscious. (I use 'physical' here in the sense that implies being fundamentally non-mental.)

If this response to the zombie argument sounds absurd, I suggest that this is because it is implausible that mere identity can make non-logical differences to what is possible for qualitative duplicates. If it is possible for your zombie twin to exist with only its physical properties and no consciousness, then the same is true of that part or aspect of reality that your zombie twin duplicates. And if it is possible for your ghost twin to exist with only its phenomenal properties and no body, then the same is true of that part or aspect of reality that your ghost twin duplicates. The zombie argument is therefore safe from objections that require that mere identity can make non-logical differences to what is possible for qualitative duplicates. But so is the disembodiment argument for mental substances when that argument is reformulated in terms of '*s*-minus' instead of 'I'.[13]

---

[13] Perhaps a similar approach would work with 'I' or 'I-minus' as well. But 'I' carries philosophical baggage that we can leave behind in the present context. As far as this essay is concerned, Goff's arguments are compatible with the view that I am not a mental substance but, for example, the composite of a mental substance and a body.

6. Idealism or Substance Dualism?

I discuss one way of resisting the disembodiment argument for mental substances in a moment. First, I pause to consider how the argument affects Goff's position if it is accepted. The immediate consequence is that phenomenal properties belong to mental substances resembling Cartesian souls. There are two directions in which a position that starts with this thesis might be developed — one idealist, the other dualist.

The first option is that the mental substances exhaust reality. In Goff's terminology, our world is a 'ghost world'. If so, the result will be a form of idealism where idealism is 'the thesis that the universe is fundamentally mental' (Chalmers, 2019, p. 353). It is then necessary to give a revisionary account of those things that we usually take to be fundamentally non-mental. There are two classic strategies, leading to what Chalmers (*ibid.*, pp. 353–4) describes as 'anti-realist' and 'realist' forms of idealism respectively:

> *Anti-realist idealism* reduces the perceptible world to the perceptual experiences of those conscious subjects that are already part of our non-revisionary view, i.e. humans, animals, and perhaps God. The resulting form of idealism resembles that of the eighteenth-century Irish philosopher George Berkeley.
>
> *Realist idealism* posits additional conscious subjects to fill the causal and structural roles that we usually suppose to be played by unconscious matter, e.g. fundamental particles really exist, but they are conscious subjects. The resulting form of idealism resembles that of the seventeenth–eighteenth-century German philosopher, Gottfried Wilhelm Leibniz.[14]

Realist idealism better coheres with Goff's approach to consciousness. For it has the panpsychist consequence that what we usually suppose to be unconscious matter is in fact conscious.[15]

The second option is that the mental substances do not exhaust reality. Rather, there is *also* a fundamentally non-mental side to

---

[14] I discuss these two forms of idealism further in Weir (2021).

[15] With the right 'localization' and 'thinning' laws, an idealist version of Goff's (ms) hybrid cosmopsychism might resemble Berkeley's idealism in limiting the conscious subjects to human and animal minds and one cosmic mind. It would remain a realist form of idealism in asserting that the existence of what we usually take to be unconscious matter consists in its being conscious — as an aspect of the cosmic mind — rather than in its being perceived.

reality. The result is a form of substance dualism, in the sense defined in this paper. That is, on the resulting view, (i) there are two sides to reality, a fundamentally mental side and a fundamentally non-mental side; and (ii) the fundamentally mental side consists of mental substances resembling Cartesian souls. Given Goff's wider commitments, the resulting substance dualist theory is likely to differ in a number of respects from Descartes'. First, this will be a form of substance dualism that aims to account for mental causation by fitting the mental substances into the dynamic structure described by the physics of what we usually take to be unconscious matter. For example, the mental substances with which our minds are identified will be part of the 'deep nature' of our biological 'bodies'. Secondly, depending on which iteration of Goff's theory we are considering, there may be many more mental substances than just the minds of humans and other sentient animals. Thirdly, although the mental substances with which our minds are identified will be capable of existing in the absence of anything fundamentally non-mental, including any fundamentally non-mental part of their biological 'bodies', they will not be capable of existing without that fundamentally mental part of their biological 'bodies' with which they are themselves identified. Fourthly, there is no reason to expect that the fundamentally non-mental side of reality will resemble Cartesian matter.[16] These differences do not affect the fact that the position is substance dualist in the sense I have defined.

There are, then, two options: idealism or substance dualism. A final argument of Goff's weighs in favour of substance dualism (though this is not, of course, how he intends it). Goff argues that for two reasons there must be more to reality than conscious subjects characterized solely by phenomenal properties. First, Goff (2017, pp. 179–86, 229–30) argues that reductionist panpsychists must posit some non-phenomenal aspect of reality to explain how conscious subjects at the human scale are grounded in more basic conscious subjects. Secondly, Goff (2017, pp. 231–2, 243–53) argues that it is necessary to posit some non-phenomenal side to reality to account for the existence of causal powers and natural laws.[17] Assuming that the

---

[16] Goff's (2017, pp. 183–6) suggestion that the fundamentally non-mental side to reality might include 'the deep nature of spatial relations' would be a step in this direction, given Descartes' identification of matter with space. But the cosmopsychist positions that Goff (2017; ms) ultimately favours can do without that hypothesis.

[17] Goff's (ms) hybrid cosmopsychism is emergentist about subjects. I assume that it still requires a non-phenomenal side to reality to account for the causal powers and natural

non-phenomenal side to reality is not in some other sense 'fundamentally mental', the result is substance dualism.[18]

In summary: the zombie argument relies on the principle that if the idea of a scenario does not contain a contradiction, and we know the nature of the entities involved, then the scenario is possible, and the principle that we know the nature of our conscious experiences. These two principles, combined with the claim that the idea of one's lonely ghost twin does not contain a contradiction, entail that actual phenomenal properties belong to mental substances resembling Cartesian souls. For this reason, proponents of the zombie argument who grant that the idea of one's lonely ghost twin does not contain a contradiction must choose between substance dualism and idealism. Goff's arguments for the existence of a non-phenomenal side to reality weigh in favour of substance dualism.

7. Transcendental Egos and Mental Substances

I have presented the main argument of this paper as a reply to Goff's objection to the disembodiment argument for mental substances. But I am not at all opposed to Goff's arguments in general. On the contrary, I am interested in reflecting on the consequences of those arguments because I find them persuasive. I find the principles underlying the zombie argument hard to resist. If they are accepted then, in my judgment, it only takes the possibility of lonely ghost twins to leave us with the choice between substance dualism and idealism. There is, however, one objection to the possibility of lonely ghost twins which I think we should take seriously.

Goff characterizes a ghost as something whose existence is exhausted by what it is like to be it. The objection I have in mind says that, in fact, there is a subtle entailment from the existence of conscious experience to the existence of a non-phenomenal reality, so that ghosts are not strictly conceivable after all. The non-phenomenal reality would have to be something easily missed, to have gone

---

laws, not least the 'localization' and 'thinning' laws that govern the emergence of subjects.

[18] Goff conjectures that the non-phenomenal and phenomenal sides to reality might be two aspects of a unitary property which he labels 'consciousness+'. But if the fact that the phenomenal side consists of mental substances that could exist in the absence of the non-phenomenal '+' side is inconsistent with the claim that these are two aspects of a unitary property, then the conjecture that they are two aspects of a unitary property is incompatible with Goff's arguments.

unnoticed. The best candidate, I suggest, is the subject of experience, on the hypothesis that subjecthood is necessitated by, but involves more than just, the possession of phenomenal properties.[19]

Galen Strawson anticipates this proposal, in a similar context, when he considers the hypothesis that 'experience (experiential content) is impossible without a subject of experience' and that 'a subject of experience cannot itself be an entirely experiential (experiential-content) phenomenon' (2003, p. 28). However, Strawson has since argued that subjects of experience are entirely experiential phenomena after all (Strawson, 2008). Goff adopts a similar view, according to which subjecthood is 'a determinable of which each conscious state is a determinate' (2017, p. 178). If so, the fact that conscious experience entails the existence of a subject of experience poses no threat to the conceivability of ghosts understood as entirely experiential beings.

Even if subjecthood involves more than the possession of phenomenal properties, this will not in itself undermine the disembodiment argument for mental substances. This depends on whether we know the nature of subjects of experiences as we do that of the conscious experiences they undergo.[20] If we do, then lonely ghost twins understood as things whose existence is exhausted by their conscious experience *and* their non-phenomenal nature as subjects will remain possible and the disembodiment argument for mental substances will survive. The threat to the argument for mental substance arises *only if* conscious experience entails the existence of a subject that: (i) is not entirely experiential, and (ii) whose nature is unknown to us. A subject, so conceived, would resemble the 'transcendental ego' as that term is used by Kant and the phenomenologists (see Priest, 2000). It seems to me that, once the zombie argument against materialism has been accepted, the existence of a transcendental ego, so defined, poses the only serious threat to the disjunction of substance dualism and idealism. If so, this topic deserves greater attention from analytical philosophers of mind than it has thus far received.

---

[19] Another option would be a substratum or 'bare particular' in which the phenomenal properties 'inhere'. This comes to nearly the same thing and can be handled in the same way.

[20] For a helpful recent discussion of this, see Morris (2021).

8. Prospects for the Post-Galilean Revolution

The 'anti-dualism' article of the post-Galilean manifesto demands that consciousness should not be separate from the dynamic structure described by physics. But there is a sense in which this is no departure from traditional substance dualism. Descartes argues that minds are mental substances. But he does not argue that mental substances are separate from the dynamic structure described by physics. On the contrary, Descartes' physics (an important precursor to Newton's) gives a central role to one mental substance: God. And the promissory note at *Principles* art. 40 (AT VIII, 65) suggests that in a future work Descartes intended to extend his physics to include created minds too. It was just this assimilation of minds into the causal dynamics of the physical world that Gilbert Ryle saw as Descartes' most egregious 'category error'. Ryle's principal complaint was that 'the logical mould into which Descartes pressed his theory of the mind... was the self-same mould into which he and Galileo set their mechanics' (1949, pp. 19–20).

The kind of substance dualism that best accommodates Goff's response to the problem of mental causation *will* differ from Descartes' in treating mental substances as part of the dynamic structure described by the physics *of what we usually take to be unconscious matter*. It would be wrong to understate the importance of this difference. Descartes implies that physics will have to be extended to accommodate anomalous events in the brain caused by created minds. In so far as we are confident that no such anomalies exist, it is an immense advantage of Goff's approach that it does not predict any. In his latest work, Goff (ms) has shown convincingly how this can be achieved. But he does not achieve this by rejecting dualism, where dualism is the view that there is a fundamentally mental side to reality and a fundamentally non-mental side. On the contrary, Goff's arguments lead to a substance dualism, akin to Descartes' or — if the case for a non-phenomenal side to reality is dropped — to an idealism akin to Leibniz's. These positions emerged as the leading responses to the mind–body problem in the seventeenth century among the pioneers of the scientific revolution. This is not a problem for a post-Galilean science of consciousness. But if it is correct then the post-Galilean revolution might equally be considered a restoration.

Acknowledgments

I am very grateful to Harry Cleevely, Philip Goff, Alex Moran, and participants in a seminar at the University of Fribourg organized by Martine Nida-Rümelin and Donnchadh O'Conaill for discussions of this article.

References

Broackes, J. (2006) Substance, *Proceedings of the Aristotelian Society*, **106** (1), pp. 133–168.
Campbell, K. (1990) *Abstract Particulars*, Oxford: Blackwell.
Chalmers, D.J. (1996) *The Conscious Mind: In Search of a Fundamental Theory*, Oxford: Oxford University Press.
Chalmers, D.J. (2019) Idealism, in Seager, W. (ed.) *The Routledge Companion to Panpsychism*, London: Routledge.
Descartes, R. (1964–76) *Oeuvres de Descartes*, Adam, C. & Tannery, P. (eds.), revised ed., Paris: Vrin/C.N.R.S.blu.
Goff, P. (2010) Ghosts and sparse properties: Why physicalists have more to fear from ghosts than zombies, *Philosophy and Phenomenological Research*, **81** (1), pp. 119–139.
Goff, P. (2011) A posteriori physicalists get our phenomenal concepts wrong, *Australasian Journal of Philosophy*, **89** (2), pp. 191–209.
Goff, P. (2012) A priori physicalism, lonely ghosts and Cartesian doubt, *Consciousness and Cognition*, **21**, pp. 742–746.
Goff, P. (2014) The Cartesian argument against physicalism, in Sprevak, M. & Kallestrup, J. (eds.) *New Waves in Philosophy of Mind*, London: Palgrave Macmillan.
Goff, P. (2017) *Consciousness and Fundamental Reality*, Oxford: Oxford University Press.
Goff, P. (2019a) *Galileo's Error*, London: Penguin Random House.
Goff, P. (2019b) Essentialist modal rationalism, *Synthese*, [Online], https://doi.org/10.1007/s11229-019-02109-9.
Goff, P. (ms) How exactly does panpsychism help explain consciousness?
Gorman, M. (2006) Independence and substance, *International Philosophical Quarterly*, **46** (2), pp. 147–159.
Gorman, M. (2012) On substantial independence: A reply to Patrick Toner, *Philosophical Studies*, **159** (2), pp. 293–297.
Hume, D. (2007) *A Treatise of Human Nature*, Norton, D. & Norton, M. (eds.), Oxford: Oxford University Press.
Johnston, M. (2016) The personite problem: Should practical reason be tabled?, *Noûs*, **51** (3), pp. 617–644.
Kim, J. (2005) *Physicalism, or Something Near Enough*, Princeton, NJ: Princeton University Press.
Koslicki, K. (2018) *Form, Matter, Substance*, Oxford: Oxford University Press.
Moran, A. (2018) Kind-dependent grounding, *Analytic Philosophy*, **59** (3), pp. 359–390.
Moran, A. (forthcoming) Living with microphysical supervenience, *Philosophical Studies*.

Morris, K. (2021) Phenomenal transparency and the transparency of subjecthood, *Analysis*, **81** (1), pp. 39–45.
Priest, S. (2000) *The Subject in Question*, London: Routledge.
Ryle, G. (1949) *The Concept of Mind*, London: Hutchinson.
Schneider, S. (2012) Why property dualists must reject substance physicalism, *Philosophical Studies*, **157**, pp. 61–76.
Strawson, G. (2003) Real materialism, in Antony, L. & Hornstein, N. (eds.) *Chomsky and His Critics*, Oxford: Blackwell. Reprinted in Strawson, G. (2008) *Real Materialism and Other Essays*, Oxford: Oxford University Press.
Strawson, G. (2008) What is the relation between an experience, the subject of experience, and the content of experience?, *Real Materialism and Other Essays*, Oxford: Oxford University Press.
Unger, P. (2004) The mental problems of the many, in Zimmerman, D.W. (ed.) *Oxford Studies in Metaphysics*, vol. 1, Oxford: Clarendon Press.
van Inwagen, P. (1990) *Material Beings*, Ithaca, NY: Cornell University Press.
Weir, R.S. (2021) Does idealism solve the problem of consciousness?, in Farris, J. & Göcke, B.P. (eds.) *The Routledge Handbook of Idealism and Immaterialism*, London: Routledge.
Weir, R.S. (forthcoming) Bring back substances!, *The Review of Metaphysics*.
Williams, D.C. (1953) On the elements of being II, *The Review of Metaphysics*, 7 (2), pp. 171–192.

Galen Strawson[1]
(and Bertrand Russell)
'Oh You Materialist!'[2]

Abstract: [1] Materialism in the philosophy of mind — materialism^{PM} — is the view that everything mental is material (or, equivalently, physical). Consciousness — pain, emotional feeling, sensory experience, and so on — certainly exists. So materialism^{PM} is the view that consciousness is wholly material. It has, historically, nothing to do with denial of the existence of consciousness. Its heart is precisely the claim that consciousness — consciousness! — is wholly material. [2] 'Physicalism', the view introduced by members of the Vienna Circle in the late 1920s, also has nothing to do with denial of the existence of consciousness. [3] Recently the words 'materialism' and 'physicalism' have come to be treated as synonymous, and as names for a position in the philosophy of mind that does involve denial of the existence of consciousness. They've been used to name a position that (i) directly rejects the heart of materialism (materialism^{PM}) and (ii) is certainly false. This is a pity, because they're good terms for a view that is very likely true.

Correspondence:
Email: gstrawson@austin.utexas.edu

---

[1] University of Texas at Austin, TX, USA.
[2] Darwin (1838/1987, p. 271), addressing himself in his notebook. 'Why is thought, being a secretion of brain, more wonderful than gravity, a property of matter?' (*ibid.*, p. 614; here Darwin follows Cabanis and others, and uses the word 'thought' in the then standard Cartesian way to mean conscious experience of any sort). When I quote, all italics mark my own emphases, unless otherwise stated.

1.

When I reviewed Philip Goff's thoroughly enjoyable book, *Galileo's Error*,[3] I took issue with some of his historical claims, but principally with his definition of materialism, which led to the conclusion that materialism 'involves a contradiction' and falls into 'incoherence' (2019, pp. 65–6, 69). Something had gone wrong, for this was a *reductio ad absurdum* of his account. Materialism may be false as a philosophical or metaphysical doctrine, but it's certainly not incoherent. And yet Philip had cause for his conclusion, given how far some people have twisted the notion of materialism away from its roots.

I want to pursue this matter a little way, not by examining the details of *Galileo's Error*, but by giving an account of what materialism actually is — serious materialism, realistic, time-honoured, non-crazy, 2,500-year-old materialism, materialism that is — for a start — fully realist about something that certainly exists: *consciousness*. It's worth doing because different uses of the term 'materialism' (and 'physical', and 'consciousness') have caused extraordinary confusion. A hideous amount of time has been wasted. It's been particularly hard on the young, generation after generation of students.[4]

I think philosophers of mind should hold a great conference and try as far as possible to establish an agreed terminology. Pending that fabulous event, I'm going to define a number of key terms. I'm also going to use some of them — e.g. 'conscious' and 'physical' — freely before I define them (we're on Otto Neurath's boat).[5] I'm also going to quote Bertrand Russell a lot, since he usually puts things better than I can.

I've given up trying to persuade anyone of anything. Almost no one changes their mind once they've taken up a position on this issue. Abandoning the goal of persuasion liberates one to be blunt when considering views one believes to be folly. One doesn't have to worry that bluntness usually strengthens people's attachment to its target,

---

[3] https://www.theguardian.com/books/2019/dec/27/galileos-error-by-philip-goff-review.

[4] The word 'materialism' dates from the seventeenth century, but the idea is old; it 'aris[es] almost at the beginning of Greek philosophy' (Russell, 1925, p. v). 'Materialism is as old as philosophy' (Lange, 1865–75/1925, p. 3). For discussion of the Cārvāka and Lokāyata Indian traditions, see e.g. Ganeri (2011), Bhattacharya (2017).

[5] 'Neurath has likened science to a boat which, if we are to rebuild it, we must rebuild plank by plank while staying afloat in it. The philosopher and the scientist are in the same boat' (Quine, 1960, p. 2).

and I find it helps me live with my intellectual despair about the current debate. I don't think philosophy has ever sunk lower than it has in the last sixty years in discussion of the so-called mind–body problem — in spite of some beautiful work. I wish I could be as jolly about folly as Erasmus, but my disposition is melancholic.[6]

2.

Materialism is the view that

[1] everything in the universe is wholly material or physical.[7]

That's it. All things — chairs, pains, whisky, colour experiences, explosions, conscious thoughts, clouds, feelings of guilt, marshmallow, feeling sleepy, stainless steel, thirst, sunlight, nausea, plutonium — are made of the same single kind of fundamental stuff: material or physical stuff. We can put it in more dynamic terms, because stuff is best thought of as process, process-stuff: everything in the universe, conscious or not, is wholly a matter of physical process, physical goings-on.

I use 'stuff' as a theoretically uncluttered, entirely general term for concrete existence. All experiences are concrete occurrences, i.e. stuff. All intrinsic qualities of stuff are themselves stuff, for they concretely exist.[8]

Materialism doesn't have the consequence that ordinary people are wrong about what conscious experiences are, considered specifically as such, i.e. considered specifically in respect of their lived experiential character. They're not. There's a primordial respect in which conscious experiences are exactly as they seem. Necessarily so, for their seeming as they seem, in the having of them, is their being what

---

[6] Erasmus, *In Praise of Folly* (1509/1990).
[7] The *Oxford English Dictionary* defines the philosophical use of 'materialism' as follows: 'nothing exists except matter and its movements and modifications', and records a narrower use according to which 'mental phenomena' in particular 'are nothing more than... the operation of material or physical agencies'.
[8] It's impossible to get things right in metaphysics if one starts out from the (initially natural) picture according to which there is on the one hand stuff, and, on the other hand, its qualities, which somehow or other 'flavour' stuff *without themselves being stuff*. If one is tempted by this picture, it can help to ask oneself exactly where the qualities are located. See Strawson (2021).

they are, so far as their experiential character is concerned.[9] Materialism simply says that they are, considered specifically in respect of their experiential character, wholly physical.

To adopt materialism, then, isn't to change anything in the ordinary view of conscious experience. It's only to change one's conception of the physical — *if*, that is, one's conception of the physical says that conscious experience can't be physical. This is precisely what has always been so striking — thrilling, shocking to some — about materialism. It states that *conscious experience*, of all things, is wholly physical!

Some use the term 'materialism' in other ways; I'll consider a few of them in §§8–10, but this must be where we start. This is the fundamental — I'm going to say the *true* — conception of materialism. Margaret Cavendish is an exemplary materialist in this respect, like her contemporary Anne Conway and her near contemporary Thomas Hobbes: 'nature', she says, 'is altogether material'; 'thoughts, ideas, conceptions, sympathies, antipathies... and Soul, are all material', wholly a matter of 'corporeal motions' (1664, pp. 12, 21).[10]

Cavendish and Conway believe in God, and [1] must in their case (and in the case of some others) be modified to:

[1*] everything in the universe except God is wholly material or physical.

They're still part of a great tradition which also includes, or so I propose (to name only a few, and speak only of the West), Leucippus, Democritus, Lucretius, Galen*, the Church Fathers Tertullian and Minucius Felix (and perhaps Origen, at one point), Regius, Spinoza, Toland, Collins, Hume*, Hartley, Priestley, Lawrence (W.), Darwin (E. and C.), Tyndall, Maudsley, Huxley, Fontenelle, De la Mettrie, d'Holbach, Diderot, Leopardi, Vogt, Moleschott, Büchner, Czolbe, Huschke, Du Bois-Reymond, Russell, Eddington, Whitehead, Einstein, Lorentz, Dirac, Schrödinger, Neurath, Schlick, Strong, Drake, Montague (W.P.), Sellars (R. and W.), Williams (D.C.), Feigl, Quine, Davidson, Place, Searle, Nagel (E. and T.). I go no further

---

[9] The core sense in which ordinary people are not wrong is fully compatible with the kinds of errors they — we — make with respect to 'filling in', in 'change blindness', and so on (see e.g. Grimes, 1996; Simons and Levin, 1997; Chun and Marois, 2002; Pessoa and de Weerd, 2003).

[10] Conway states that 'by material and corporeal... I mean something very different from Hobbes' (c. 1673/1996, p. 65).

forward in time, because the terminological chaos intensifies around 1960.[11]

Materialism isn't in fact a radical thesis. The misapprehension stems, again, from our everyday conception of physical goings-on. This everyday conception is fine in its everyday place, but it can't be imported into philosophy, and in particular the metaphysics of mind. It has no philosophical or scientific justification. It has no philosophical or scientific justification precisely to the extent that it makes it seem that conscious goings-on can't be physical — precisely to the extent that it makes materialism seem radical.

Certainly *physics* provides no such justification. Physics, in Stephen Hawking's words, is 'just a set of rules and equations' (1988, p. 174). It provides no reason for thinking that consciousness isn't wholly physical.

This can be a hard idea. To see it clearly, to really get it, is to solve the so-called mind–body problem. Quite a few people have done so. They've solved the mind–body or consciousness–matter problem. I've done it myself. It took me a long time. To solve the mind–body problem is to cease to feel that there's a problem — for the right reason. It's to see there's no good reason to think there's a problem. It's to get into a certain state of mind. It can, as Russell observes, be hard to maintain. Then one has to work to get back into it.

Either way, all one is really doing is fighting a picture, a bad imaginative picture.

3.

Is physicalism the same as materialism? I'm going to take 'materialism' and 'physicalism' to be synonymous and stick with 'materialism', following David Lewis:

---

[11] I've asterisked two who are probably materialists. There is plenty of scope for argument about this list: Democritus is regularly misquoted (see Strawson, 2019, p. 37). Huxley has no doubt that 'consciousness is a function of the brain' (1886, p. 797), but forcibly rejects the view he calls 'materialism'. Du Bois-Reymond was regularly charged with being a materialist, but some question this. Some think Russell can't be counted a materialist; see p. 239 below. Some wrongly think a panpsychist can't be a materialist, in spite of Lewis's observation that 'a thesis that says [that] panpsychistic materialism… is impossible… is more than just materialism' (1983/1999, p. 36). Spinoza's materialism may also be questioned. He is, like Eddington, Whitehead, Strong, Drake, and many other materialists, a panpsychist in every sense in which he's a materialist, but he also accepts [1], on one natural interpretation of 'wholly', given which being wholly physical doesn't exclude being wholly conscious (see e.g. Garrett, 2017/2018).

> [A]ll fundamental properties and relations that actually occur are physical. This is the thesis of materialism... It was so named when the best physics of the day was the physics of matter alone. Now our best physics acknowledges other bearers of fundamental properties: parts of pervasive fields, parts of causally active spacetime. But it would be pedantry to change the name [to 'physicalism'] on that account, and disown our intellectual ancestors. Or worse, it would be a tacky marketing ploy, akin to British Rail's decree that second class passengers shall now be called 'standard class customers'. (1994/1999, p. 293)

I agree with Lewis, except that I'm inclined to replace his use of the word 'fundamental' with something like 'natural, intrinsic, concrete'. The basic position is clear: the stuff of the universe that banged into existence in the Big Bang was all of the same single fundamental kind, in some sense of 'same' which is compatible with the fact that we take there to be different kinds of fundamental entities (e.g. fermions, i.e. leptons and quarks, and bosons) and different physical fields (e.g. electromagnetic, gravitational). This primordial stuff subsequently differentiated in many ways, but it remained of the same single fundamental kind, which we call 'physical'.

Physicalism, then, is materialism. It's a straightforwardly metaphysical thesis; this is how the word 'physicalism' is used today. It meant something completely different when first introduced into philosophy by Carnap and Neurath around 1928–29, as I'll explain in §6. It was then a thesis about language, in particular scientific language. It wasn't any sort of metaphysical thesis (the Vienna Circle wasn't keen on metaphysics). This wouldn't matter if there hadn't been a strange false leakage from the original use into the current use.

4.

By 'consciousness', 'conscious experience', I mean things of a sort already mentioned, experiences of colour, of thinking, warmth, feeling depressed, in love, all the experiences you have as you listen to someone talking or read this. I mean what some call 'experiential what-it-is-likeness', experiential what-it-is-likeness of any sort whatever, however complex, however primitive, whether in human beings or in spiders (assuming they have conscious experience).

Others use 'consciousness' in different ways. 'Well was James Ward advised to call this "a sand-heap of a term"' (Strong, 1934, p. 313). There's so much disagreement about the best use of the term, in fact, that I'm going to introduce a new term — 'ψ' (*psi*, as in 'psych[olog]ical') — to cover what I mean by it: experiential what-it-is-likeness as just characterized, 'phenomenological' quality, 'qualiality': absolutely everything that life is to one, experientially.

If you don't think this is an adequate characterization of what I mean, you're welcome to stop reading. All that needs to be added here is that we know that ψ exists. It's an ancient point: when it comes to concrete reality (as opposed to mathematics, for example) the only absolutely certain thing is the existence of ψ.

5.

I need to say more about what I mean by the word 'physical', for some will already think I'm using it wrongly. But I want to first repeat the central point once again using the new term:

> [2] materialism (physicalism) has *nothing to do* with denial of the existence of ψ.

How could it? We know for certain that ψ exists. No serious theory of anything denies its existence. No materialist worthy of serious consideration, no one who brings Russell's 'robust sense of reality' (1919, p. 170) to philosophy, can deny its existence.

If you find [2] surprising, given what you've heard and read, you've been grievously misled. [2] is just a boring statement of fact about any theory worthy of the name 'materialism'. The idea that materialism might involve outright denial of the existence of ψ (I'll call this *the Denial*) took flight only in the twentieth century.[12] The claim that ψ is wholly physical has always been the true philosophical heart of materialism.

The pre-twentieth-century consensus is hardly surprising. It's not surprising because the Denial is, in C.D. Broad's technical sense of the term, 'silly':

> [B]y a 'silly' theory I mean one which may be held at the time when one is talking or writing professionally, but which only an inmate of a lunatic asylum would think of carrying into daily life. (1925, p. 5)

The Denial is the silliest (or equal silliest) view that has ever been held by any human being.

---

[12] There were flickerings in the nineteenth century, but Mary Calkins is wrong when she claims in 1930 that 'materialism *in the broad old-fashioned sense of the term*' is 'inconsistent with [the] conviction that mental realities exist' (1930, p. 200).

It's also, of course, a radical view, and radicalism can be intensely seductive, even when, like the Denial, it's radicalism of the 2 + 2 = 17 kind. One thing to do, when one comes across the Denial, also known as 'illusionism' or 'eliminativism' about ψ, is to remember that if the Deniers are right then there's no suffering and never has been any (never mind other kinds of experience).

Some Deniers say that suffering is an illusion because it's really just a matter of having a certain *belief*, being in a certain cognitive state, and so not an actual experience of suffering. They must then hold that it is not in any way unpleasant to have such a belief, not in any way a matter of actual pain or suffering — as one gives birth, say, or is tortured or crucified or burnt alive, or endures clinical depression, or is forced to watch as one's family is raped and murdered in front of one's eyes. This is an alarming view. But the strangest thing about it, and the current fashion for it, is that there isn't even any *prima facie* good reason to believe it (see §11).

6.

What do I mean by 'physical'? I'm using it in the way Russell does in 1914:

> The word 'physical', in all preliminary discussions, is to be understood as meaning 'what is dealt with by physics'. Physics, it is plain, tells us something about some of the constituents of the actual world; what these constituents are may be doubtful, but it is they that are to be called physical, *whatever their nature may prove to be*. (1914/1919, p. 150)

This seems exactly the right move to make, given the incompleteness of our knowledge of the nature of the stuff we call 'physical' (in spite of the knowledge we have of its nature in knowing various laws of physics and — I speak as a materialist — in having ψ). We need to recognize our ignorance. This is the first great step we need to take, when it comes to the supposed consciousness–matter problem. We need to keep our ignorance vivid.

It may help to replace the term 'physical' with a new term that doesn't trigger misleading mental reflexes. The Greek letter 'φ' offers itself (*phi* as in 'physical' to match 'ψ' for 'psych[olog]ical'), but I'm going to use 'χ' (*chi*), which resembles 'X', a traditional marker of the unknown. When we're trying to vivify our ignorance, it's better if the term replacing 'physical' doesn't echo it in any way, given the false assumptions many build into their understanding of 'physical'.

From now on, then, 'χ' has the same reference as 'physical' or 'the physical' (adjective or noun) in Russell's use. One can always read '(the) physical' where I use 'χ', as long as one keeps one's sense of ignorance sharp. Russell is right, as already observed, that it can be hard to maintain.

7.

Objection. 'What your χ materialism comes down to, in effect, is simply "stuff monism", the theory according to which everything is one single kind of fundamental stuff. You've deprived "physical" — "χ" — of any positively descriptive general meaning.'

Reply. If you say so. I've followed Russell (and Eddington and many others). That said, I'm also happy to follow Russell in including *spatio-temporality* and *causal connectedness* as among the known features of χ, and one may certainly think of the reference to these things as supplying some positively descriptive meaning to 'χ'. But if one does this one must also follow Russell (and Moritz Schlick) in stressing that the space of physics is not to be confused with the space of perception on which our intuitive everyday notion of space is founded. (This is another hard idea.)

'What is matter? the whole a mystery' (Darwin, 1838/1987, p. 614). There is much more to say. One should probably add that a number of leading physicists deny that spatio-temporality is a fundamental property of χ. For the moment let me just note that it was once a commonplace, and is becoming so again, that physics has nothing to say about the ultimate, non-structural, intrinsic 'stuff nature' of the physical = χ.

8.

We can now say that materialism is the view that

[1] everything in the universe is wholly χ

— from which it follows immediately that

[3] all ψ is χ.

It's an interesting question whether

[4] all χ is ψ,

i.e. whether all-out ('panexperientialist') panpsychism is true. I think, with Goff and others, including several winners of the Nobel Prize for physics, that it is, all things considered, the most natural, most parsimonious, least implausible view of the fundamental nature of χ, given that ψ certainly exists (Lewis reminds us that it is a possible position for a materialist). This, though, is a question for another time. Here I want to say something about some other uses of 'materialism' ('physicalism'). We become deeply habituated to particular uses of words, and this often blocks our ability to take in what other people are saying. I've certainly failed in this way.

(i) Some take materialism to imply commitment to a *mechanistic* physics. This use has no place here, and has largely faded away. Materialism is fully compatible with the latest instar of relativistic quantum field theory.

(ii) Some equate materialism with atheism, but there's no necessary connection between them. Neither implies the other.

One needs to bear uses **(i)** and **(ii)** in mind when reading older writings, but not otherwise.

(iii) Some refrain from calling themselves materialists even though they agree or at least strongly suspect that [1] is true — that everything in the universe is wholly χ (= physical). This is because they think the term 'materialist' suggests adherence to a conception of matter that has been refuted by science.

This seems to be Bertrand Russell's position. He called himself a 'neutral monist' from 1921 on, and when he was interviewed in 1964, forty-three years later, he was 'not conscious of any serious change in my philosophy since I adopted neutral monism' (Eames, 1967, p. 510). At the same time he confirmed that he

> would describe himself as a materialist, if it were not for the fact that, since the concept of solid matter had disappeared from physics, the label 'materialist' had become ambiguous. (*ibid.*, p. 510)[13]

It seems, then, that we may take Russell's neutral monism to be materialist. His 'neutral stuff', I propose, is what I call 'χ'. It's the

---

[13] In 1944 Russell writes, 'I find myself in ontology increasingly materialistic... In ontology I start by accepting the truth of physics' (1944/1946, p. 700). Matter after all 'has become as ghostly as anything in a spiritualist séance' (1927, p. 104), 'it has begun to seem that matter, like the Cheshire Cat, is becoming gradually diaphanous until nothing of it is left but the grin, caused, presumably, by amusement at those who still think it is there' (1950/1956, p. 145).

subject matter of physics, whatever it is in itself. He thinks it most probable that the laws of physics cover everything that exists — which is not of course to deny the reality of ψ.[14]

Russell holds that our knowledge of the intrinsic nature of χ is restricted to knowledge of its causal–spatio-temporal structure, as expressible in the abstract logico-mathematical terms of physics, *except in one vital respect*. 'We know nothing of the intrinsic quality of physical [χ] phenomena', he writes,

> *except when they happen to be sensations* [i.e. instances of ψ]... there-fore there is *no reason to be surprised that some are sensations, or to suppose that the others are totally unlike sensations*. The gap between mind and matter has been filled in, partly by new views on mind, but much more by the realisation that physics tells us nothing as to the intrinsic character of matter [χ]. (1927, p. 154)[15]

Here in 1927 Russell solves the mind–body or consciousness–matter problem. He holds the same view in 1950: 'We know nothing about the intrinsic quality of physical events except when these are mental events that we directly experience' (1950/1956, p. 153). In his late book *My Philosophical Development*, he records his conviction that he has 'completely solved... the problem of the relation of mind and matter'. I think he's right. 'It is true', he continues, rather touchingly, 'that nobody has accepted what seems to me the solution, but I believe and hope that this is only because my theory has not been understood' (1959, p. 15).[16]

Objection. 'This can't be right. ψ is essentially mental, so Russell's χ isn't genuinely neutral. He's committed to some of it being mental.'

Reply. It seems a strong point. It fails, on Russell's terms, because he denies that ψ is intrinsically mental. He never wavers in his definition of mind — mentality — as something essentially complex that essentially involves cognition, intentionality, deployment of

---

[14] There's a great debate about Russell's view, which I put aside here. See e.g. Wishon (2015), Stubenberg (2016).

[15] Russell doesn't count a thing's structural nature as part of its intrinsic nature. The point about our necessary ignorance was a commonplace at the time, and was built into the Vienna-Circle conception of physicalism: 'physicalism... maintain[s]... that whatever is [scientifically] *knowable* in *any* field of inquiry is *structure*' (Nagel, 1936, p. 41).

[16] Schlick solves it too, and Eddington, and Whitehead, and C.A. Strong, and Durant Drake, and Herbert Feigl, Grover Maxwell, and others. It's arguable that Kant's solution (see e.g. Kant, 1781–87/1933, A358–60, A379–80, A391, B427–8) is marred only by his doctrine that ψ is 'mere appearance'. I can't speak for Mach.

memory and all the causal connection that that entails. An isolated bare sensation isn't a mental occurrence, for Russell. Ignorance of this terminological point has led to a lot of misunderstanding.

I'll end with a little more about this. Before that, let me record some more uses of 'materialism'.

(iv) Some think materialism involves the belief that 'the inner subjective world of experience is to be *explained* in terms of the chemistry of the brain, in something like the way the wetness of water is explained in terms of its molecular structure' (Goff, 2019, p. 53). This use seems to have become widespread in the nineteenth century. Certainly it's clear in Tyndall:

> [Y]ou cannot satisfy the human understanding in its demand for logical continuity between molecular processes and the phenomena of consciousness. This is a rock on which materialism must inevitably split whenever it pretends to be a complete philosophy of the human mind. (1868/1877, p. 334)

But Tyndall remains a materialist in the classical sense — in the sense of [1]. The same goes for Emil Du Bois-Reymond.[17]

There is, certainly, an 'explanatory gap'[18] between the terms we use to denote ψ phenomena and the terms we use to denote phenomena that we ordinarily classify as physical (whether in physics or in everyday life). Tyndall, Huxley, and Du Bois-Reymond are the poets of this point. The gap is unbridgeable — it's *obviously* unbridgeable. But to think that this constitutes an objection to materialism is to be philosophically lost. Sensible materialists know the gap can't be bridged, and they know this is no objection to materialism.

(v) The version of materialism described in **(iv)** is close to the version according to which **(v)** 'reality can [in principle] be *exhaustively described* in the objective vocabulary of physical science' (Goff, 2019, p. 68).

(v) isn't as bad as it may seem. On one reading it's true (if materialism is true). **(v)** is true (if materialism is true) if one takes 'exhaustive' to mean that physical science can in principle give a description, in its

---

[17] See Tyndall's once famous but now forgotten Belfast Address (1874), and Du Bois-Reymond's equally famous but now mostly forgotten Leipzig Address (1872/1874). Du Bois-Reymond is certainly a monist, but some doubt (see note 11) that he is a materialist.

[18] See Levine (1983).

own highly specialized descriptive terms, of absolutely everything that exists — every part and aspect of what exists.[19] **(v)** is false only if one takes 'exhaustive' to mean that it can say — convey — *everything there is to say* about every part of what exists, i.e. that the description it gives is *descriptively* exhaustive.

'Obviously', you say. 'There's no such thing as exhaustive description of reality; reality outruns — infinitely — all possible description.'

Yes, but the abstract numerico-structural descriptions of physical science are profoundly limited even when one puts this point aside. There are, to begin, no terms for emotions or sensory experiences.

This is not to say that physics can't give an exhaustive *physics* description of the ψ/χ phenomena that constitute emotions and sensations. It can — if materialism is true. But it can't say anything about what emotions and sensations are like, considered specifically as such. The mistaken move is from the true claim that

[a] physics can (in principle) give an exhaustive description *in physics terms* of everything that exists

(a crucial component of original — Vienna-Circle — physicalism) to the false claim that

[b] physics can say everything there is to say about everything that exists.

9.

Actually, there's a way to interpret [b] given which it too comes out as true (and also as part of what the Vienna-Circle physicalists had in mind). All you have to do is to start by endorsing the doctrine of the 'incommunicability of content', a doctrine, intensely fashionable in the 1920s and beyond, about what can be truly meaningfully be said — genuinely, fully intersubjectively communicated — in language.

Why does this make [b] come out true? Because, according to the doctrine of the incommunicability of content, very little of what we ordinarily think can be truly meaningfully said can really be truly meaningfully said. Certainly nothing can be said (genuinely fully

---

[19] This is what the members of the Vienna Circle who introduced the term 'physicalism' had in mind. Physicalism (to repeat) was a thesis about language, not any sort of metaphysical thesis.

communicated) about the nature or essence of any ψ. And this is so even though we are each individually 'directly acquainted' with ψ, in Schlick's words, and indeed derive our 'concept of reality' from it (1918–25/1974, p. 234).

The idea is old and simple. If I say 'I'm experiencing pillar-box red', you can't know for sure what my experience is like. My words can't nail it down. They can't package it up for lossless transmission to you; and your experience of pillar-box red may be different. When they introduced the doctrine of physicalism, the Vienna-Circle philosophers were specifically concerned to mark the limits on the genuine content of specifically scientific language, but the point generalizes, and we can see it in action — spectacularly, paradoxically — in §304 of Wittgenstein's *Philosophical Investigations*. His imaginary interlocutor speaks first:

> 'But you will surely admit that there is a difference between pain-behaviour with pain and pain-behaviour without pain.' — Admit it? What greater difference could there be? — 'And yet you again and again reach the conclusion that the sensation itself is a Nothing.' — Not at all. It's not a Something, but not a Nothing either! The conclusion was only that a Nothing would render the same service as a Something about which nothing could be said. (c. 1944/1953, §304)

This may seem fairly strange, but one thing is clear. The denial of the existence of ψ (in this case pain) is of course held to be false ('What greater difference could there be?'). We're not pushed into denying its existence, in the way Wittgenstein's interlocutor supposes, if we endorse the thesis of the incommunicability of content.[20] The basic claim is simply that there's much about reality that can't be captured in language.[21] It hardly licenses a move to saying that the language of the physical sciences can fully capture everything. The transmogrification of the physicalism of the Vienna Circle into the present-day view that many call 'physicalism', and that I call *physics-alism* — the Denial-entailing view 'that the nature or essence of all concrete reality can in principle be *fully captured* in the terms of *physics*' (Strawson, 2006, p. 4), to which we may add the terms of everyday physical

---

[20] Should we endorse it? In *Mental Reality* I argue that we can in fact (and of course) communicate about ψ although there is (strictly speaking) a kind of hazard in it: 'language leaps without looking and lands on its feet' (1994, p. 230).

[21] There are some striking further complications. See Strawson (1994, chapter 8).

description — is one of the great philosophical aberrations of our time.

10.

(vi) The point about **(v)** made in §8 is open to misinterpretation. One way to restate it is to consider Goff's variation of **(v)**, according to which **(vi)** 'reality can be exhaustively *captured* in the quantitative language of physical science' (Goff, 2019, p. 68).

There's *no* reading of this claim according to which it comes out true — for reasons just given: 'exhaustively capturing the nature of reality' implies fully expressing its nature, and of course physical science can't do this (that's not its job). It can't describe what it is like experientially to see the Ka'ba, or Carcassonne, or 'attack ships on fire off the shoulder of Orion' (to quote Roy Batty in *Bladerunner*). It can't capture — convey — the (experiential) nature of any ψ at all. It really is 'just a set of rules and equations'.

(vii) The various confusions that animate **(iv)**–**(vi)** culminate in **(vii)**. In the last 100 years or so, a very small number of people who have called themselves 'materialists' or 'physicalists' have not only denied the existence of ψ. They've also thought **(vii)** that this is part of materialism — that materialism requires them to do so.

This, again, is the great aberration — the Denial — about which I've said enough. There's irony in the fact that it thinks of itself as the perfection of materialism precisely as, and because, it rejects the central, shining materialist idea: the idea that ψ — *ψ*, no less, ψ, real ψ! — is wholly material.

11.

The confusion that is fed by different uses of 'materialism' is intense. Let me repeat the principal point once again: *materialism, serious, realistic materialism, is and always has been fully realist about ψ*. I call such realistic materialism 'real materialism'. The Deniers have lost touch with reality.

How could they? I'm not sure, but the all-too-human psychological mechanism that underwrites such silliness is the same as one finds in the QAnon and Flat Earth movements. The phenomenon is robustly confirmed by experimental psychologists: 'we know that people can maintain an unshakable faith in *any proposition, however absurd*, when they are sustained by a community of like-minded believers' (Kahneman, 2011, p. 217). *Weh ist mir*! When Russell complained of 'the subjectivistic madness which is characteristic of most modern philosophy' (1948, p. 846), he can hardly have imagined what was to come.[22]

Perhaps the most poignant aspect of the Denial, from a philosophical point of view (to pick up the thread at the end of §5), is the fact that there is nothing in physics or everyday experience — nothing, absolutely nothing — that provides any good motivation for it. All there is is habit — or prejudice. Russell describes the mistake: however much people emphasize the unknown character of the physical cause of sensation,

> they still suppose themselves to know enough of it to be sure that it is very different from a mind. This comes, I think, of not having rid their imaginations of the conception of material things as something hard that you can bump into. You can bump into your friend's body, but not into his mind; therefore his body is different from his mind. This sort of argument persists imaginatively in many people who have rejected it intellectually. (*ibid.*, p. 244)

We can't shrug it off overnight — the everyday distinction between mental and physical which makes it seem that ψ can't be physical. Perhaps we can do a little better when we think instead that ψ can be χ, since 'χ' is a term designed to glow with ignorance, with intense *descriptive* thinness — even while it has a clear *reference* (the subject matter of physics: everything in the universe).

Even so it takes work, philosophical work. Not the work of argument, or the usually much more difficult work of searching theoretical description. Philosophy here is a matter of dwelling with a kind of

---

[22] Objection. 'You *can't* say that the Deniers aren't real materialists. They agree with [1]: they hold that everything that concretely exists is wholly physical, and that's obviously — trivially — a sufficient condition of being a real materialist.' Reply. Afraid not. The trouble is that they deny the existence of a vast part of what certainly (concretely) exists. So they don't hold that everything that actually (concretely) exists is wholly physical. So they're not real materialists — *serious* materialists. They're the ultimate philosophical Procrusteans. They simply lop off a vast part of concrete reality that doesn't fit their theory. Objection. 'You *can't* just assert that ψ exists, for they specifically deny just this. So you're begging the question. People can certainly legitimately argue about what concretely exists.' Reply. When you come up against a view that denies something that is certain (a view that holds, for example, that there has never been any suffering), you're bound to end up begging the question against them. Try arguing against someone who holds that $2 + 2 = 5$.

intense and delicate balance — equanimity — on certain thoughts: the thought that one's conscious experience (one's ψ) right now is wholly a matter of one's neural goings-on. (I put my head in my hands.) It's a matter of bringing the thought back to focus as it slips and slips again. One might try this for a minute every day, until it begins to work. It's an astounding experience when it does — if, that is, you start out as benighted as I was. It is, as Durant Drake mildly says, 'a very considerable mental wrench' (1930, p. 286).

I'll say it again. All one needs to solve the mind–body problem are two things, both of which one has: (1) knowledge of the reality of one's current conscious (ψ) goings-on, knowledge that one has in the moment simply in having the experience one is having; (2) knowledge (it is far beyond reasonable doubt) that that reality is neural — or more generally bodily, or more generally χ — goings-on.

One may periodically hesitate about (2), and go back down into the bowels of the mind–body problem (*facilis est descensus averno philosophico*). What one mustn't do is think that one has any good reason to suppose that one's ψ can't be neural goings-on.[23]

One's ψ, once again, is not other than one ordinarily supposes it to be. One's fundamental *theoretico-imaginative* conception of ψ is essentially correct (we deal almost exclusively with such theoretico-imaginative elements when we study things like the mind–body problem). The trouble doesn't lie here. It lies in one's theoretico-imaginative conception of χ, which is certainly not correct — not unless one has completed a lot of the hard philosophical work just described.

It's certainly not correct if it involves any element that makes it seem in any way puzzling that some of it is ψ:

> [T]he physical world is only known as regards certain abstract features of its space-time structure — features which, because of their abstractness, *do not suffice to show whether the physical world is, or is not, different in intrinsic character from the world of mind*. (Russell, 1948, p. 240)[24]

> Beyond certain very abstract mathematical properties, physics can tell us nothing about the character of the physical world. But there is one part of the physical world which we know otherwise than through

---

[23] I argue for this in several places, e.g. Strawson (2016).
[24] We may take him to be using 'physical world' to mean either simply χ, or, more narrowly, χ as described by physics.

physics, namely that part in which our thoughts and feelings are situated. (1944/1946, p. 706)[25]

Schlick agrees, although he objects to calling our direct acquaintance with our conscious experience 'knowledge'. He also solves the mind–body problem. This tradition of solution stretches back to neo-Kantians like Riehl and von Helmholtz who are (unlike Kant himself) outright realists about ψ. The fundamental move is the same, for all the differences of detail. They too solve the mind–body problem. All one has to do to solve it is to fight a picture that has no scientific nor any other theoretical justification.[26]

To the point that nothing in physics gives us good reason to think that ψ is not χ, we can add the point that nothing in our ordinary everyday conception of χ does either. To think that it does is like thinking that it gives us good reason to hold that rock and air, flesh and steel, can't possibly be made of the same stuff.

12.

There's a simple test for whether people understand what's at issue. If they think there are good reasons for believing that materialism — i.e. [1] — isn't true, they fail the test. They haven't yet done the necessary work. That's not to say that materialism — real materialism, ψ-acknowledging materialism — is true, although I have no doubt that it is.

One doesn't have to call it 'materialism' or 'physicalism'. One can simply call it 'monism', meaning stuff monism, the view that there is just one kind of fundamental stuff, while bearing in mind two points: (i) the stuff in question is the subject matter of physics, about which physics has a very great deal to say; (ii) at least some of it is ψ. We can signal our ignorance by calling it 'χ-ism', so long as we never forget that we know that ψ is real and know what it is. We can even call it 'neutral monism' — treating χ as something neutral between 'mind' and 'matter' — so long as we understand 'mind' and 'matter' in the way Russell does, as products of what he calls 'logical

---

[25] A year later Russell suggests that 'it would be better to substitute the word "physicalism" for the word "materialism"'. 'I should define "physicalism"', he continues, 'as the doctrine that [all] events are governed by the laws of physics' (1945/1997, p. 247). There is no (absurd) implication that physics can fully capture the nature of everything.

[26] See in particular Schlick (1918–25/1974, §§31–35), Riehl (1876–87/1894, Part 2, chapter 2), von Helmholtz (1887/1977).

construction', and, again with Russell, never for a moment waver from the thought that we know something fundamental about the intrinsic nature of the neutral stuff simply in having ψ; nor from the thought that 'we do not know enough of the intrinsic character of events outside us to say whether it does or does not differ from that of "mental" events' (Russell, 1927, p. 222).

References

Bhattacharya, R. (2017) A history of materialism from Ajita to Udbhata, in Ganeri, J. (ed.) *The Oxford Handbook of Indian Philosophy*, pp. 344–359, Oxford: Oxford University Press.
Broad, C.D. (1925) *The Mind and Its Place in Nature*, London: Kegan Paul.
Calkins, M. (1930) The philosophical *credo* of an absolutistic personalist, in Adams, G.P. & Montague, W.P. (eds.) *Contemporary American Philosophy: Personal Statements*, vol. 1, New York: Macmillan.
Cavendish, M. (1664) *Philosophical Letters*, London.
Chun, M.M. & Marois, R. (2002) The dark side of visual attention, *Current Opinion in Neurobiology*, **12**, pp. 184–189.
Conway, A. (c. 1673/1996) *The Principles of the Most Ancient and Modern Philosophy*, Coudert, A. & Corse, T. (eds. & trans.), Cambridge: Cambridge University Press.
Darwin, C. (1838/1987) Notebook C 166, in *Charles Darwin's Notebooks, 1836–1844*, Cambridge: Cambridge University Press.
Drake, D. (1930) The philosophy of a meliorist, in Adams, G.P. & Montague, W.P. (eds.) *Contemporary American Philosophy*, vol. 1, pp. 277–297, New York: Macmillan.
Du Bois-Reymond, E. (1872/1874) On the limits of scientific knowledge, *Popular Science Monthly*, **5**, pp. 17–32.
Eames, E. (1967) The consistency of Russell's realism, *Philosophy and Phenomenological Research*, **27**, pp. 502–511.
Erasmus, D. (1509/1990) *In Praise of Folly*, Adams, R.M. (trans.), New York: Norton.
Ganeri, J. (2011) Emergentisms, ancient and modern, *Mind*, **120**, pp. 671–703.
Garrett, D. (2017/2018) The indiscernibility of identicals and the transitivity of identity in Spinoza's logic of the attributes, in Garrett, D., *Nature and Necessity in Spinoza's Philosophy*, Oxford: Oxford University Press.
Goff, P. (2019) *Galileo's Error*, London: Rider.
Grimes, J. (1996) On the failure to detect changes in scenes across saccades, in Akins, K. (ed.) *Perception*, pp. 89–110, New York: Oxford University Press.
Hawking, S. (1988) *A Brief History of Time*, New York: Bantam Books.
Huxley, T.H. (1886) Science and morals, *Fortnightly Review*, **40**, pp. 788–802.
Kahneman, D. (2011) *Thinking, Fast and Slow*, New York: Farrar, Strauss, Giroux.
Kant, I. (1781–87/1933) *Critique of Pure Reason*, Kemp Smith, N. (trans.), London: Macmillan.
Lange, F.A. (1865–75/1925) *The History of Materialism*, Thomas, E.C. (trans.) with an introduction by Bertrand Russell, London: Routledge and Kegan Paul.

Levine, J. (1983) Materialism and qualia: The explanatory gap, *Pacific Philosophical Quarterly*, **64**, pp. 354–361.
Lewis, D. (1983/1999) New work for a theory of universals, in Lewis, D., *Papers in Metaphysics and Epistemology*, pp. 8–55, Cambridge: Cambridge University Press.
Lewis, D. (1994/1999) Reduction of mind, in Lewis, D., *Papers in Metaphysics and Epistemology*, pp. 291–324, Cambridge: Cambridge University Press.
Nagel, E. (1936) Impressions and appraisals of analytic philosophy in Europe I and II, *Journal of Philosophy*, **33**, pp. 5–24, 29–53.
Pessoa, L. & de Weerd, P. (eds.) (2003) *Filling-in*, New York: Oxford University Press.
Quine, W.V. (1960) *Word and Object*, Cambridge, MA: MIT Press.
Riehl, A. (1876–1887/1894) *The Principles of Critical Philosophy*, Fairbanks, A. (trans.), London: Kegan Paul.
Russell, B. (1914/1919) The relation of sense-data to physics, in *Mysticism and Logic*, pp. 145–179, London: Longmans.
Russell, B. (1919) *Introduction to Mathematical Philosophy*, London: Allen and Unwin.
Russell, B. (1925) Materialism, past and present, introduction to Lange, F.A., *A History of Materialism*, Thomas, E. (trans.), London: Routledge & Kegan Paul.
Russell, B. (1927) *An Outline of Philosophy*, London: George Allen and Unwin.
Russell, B. (1944/1946) Reply to criticisms, in Schilpp. P. (ed.) *The Philosophy of Bertrand Russell*, pp. 681–741, Chicago, IL: Northwestern University Press.
Russell, B. (1945/1997) Mind and matter in modern science, in Slater, J. (ed.) *Collected Papers of Bertrand Russell, Volume 11*, London: Routledge.
Russell, B. (1948) *Human Knowledge: Its Scope and Limits*, London: Allen & Unwin.
Russell, B. (1950/1956) Mind and matter, in Russell, B., *Portraits from Memory*, pp. 145–165, New York: Simon and Schuster.
Russell, B. (1959) *My Philosophical Development*, New York: Simon and Schuster.
Schlick, M. (1918–25/1974) *General Theory of Knowledge*, Blumberg, A.E. (trans.), with an introduction by A.E. Blumberg & H. Feigl, New York: Springer-Verlag.
Simons, D. & Levin, D. (1997) Change blindness, *Trends in Cognitive Sciences*, **1**, pp. 261–267.
Strawson, G. (1994) *Mental Reality*, Cambridge, MA: MIT Press.
Strawson, G. (2006) Realistic monism: Why physicalism entails panpsychism, in Freeman, A. (ed.) *Consciousness and Its Place in Nature*, pp. 3–31, Exeter: Imprint Academic.
Strawson, G. (2016) Mind and being, in Brüntrup, G. & Jaskolla, L. (eds.) *Panpsychism*, pp. 75–112, Oxford: Oxford University Press.
Strawson, G. (2019) A hundred years of consciousness: 'A long training in absurdity', *Estudios de Filosofía*, **59**, pp. 9–43.
Strawson, G. (2021) Identity metaphysics, *The Monist*, **104**, pp. 60–90.
Strong, C.A. (1934) A plea for substantialism in psychology, *Journal of Philosophy*, **31**, pp. 309–328.
Stubenberg, L. (2016) Neutral monism, in Zalta, E.N. (ed.) *Stanford Encyclopedia of Philosophy*, [Online], https://plato.stanford.edu/archives/win2016/entries/neutral-monism/.

Tyndall, J. (1868/1877) Scientific materialism, in *Fragments of Science*, 5th ed., pp. 409–422, New York: Appleton.

Tyndall, J. (1874) *Address Delivered Before the British Association Assembled at Belfast*, London: Longmans, Green, and Co.

von Helmholtz, H. (1878/1977) The facts in perception, *Boston Studies in the Philosophy of Science*, **37**, pp. 115–185.

Wishon, D. (2015) Russell on Russellian monism, in Alter, T. & Nagasawa, Y. (eds.) *Consciousness in the Physical World: Perspectives on Russellian Monism*, pp. 91–118, Oxford: Oxford University Press.

Wittgenstein, L. (c. 1944/1953) *Philosophical Investigations*, Anscombe, G.E.M. (trans.), Oxford: Blackwell.

Joanna Leidenhag[1]

Why a Panpsychist Should Adopt Theism

God, Galileo, and Goff

Abstract: This paper argues that there is a deep level of agreement between panpsychism and theism. Goff's Galileo's Error would have been even more compelling than it already is if Goff had portrayed a panpsychist cosmos as the world created by God, not as a spiritual alternative to theism. First, I critique Goff's assumption of incompatibilism, with regards the relationship between science and religion, and argue that panpsychism provides unique resources for articulating divine action. Second, I argue that most panpsychists endorse either the 'principle of sufficient reason' or a 'causal principle' in their rejection of emergence theory, and that if either of these principles are applied to the universe as a whole this would imply a further endorsement of the cosmological argument for the existence of God.

1. Introduction

There is no opposition or competition between panpsychism and Christian theology. Instead, panpsychists and theists have far more in common than many often realize.[2] As someone who has argued for

Correspondence:
Email: j.leidenhag@leeds.ac.uk

---

[1]   University of Leeds, UK.
[2]   Unless otherwise qualified, by 'theism' I mean the belief in a single transcendent or supernatural God, such as is found in the Abrahamic faiths. Beyond transcendence and the power to create, the other attributes or identity of such a being are not discussed in this paper.

panpsychism within my own field of Christian theology, it is unsurprising that I found much to agree with in Philip Goff's *Galileo's Error*. It is a marvellous introductory text displaying intellectual virtues of clarity, charity, and wit. Indeed, my own beliefs about the mind are indebted to much of Goff's previous work. I wholeheartedly agree with his arguments against materialism, some of the points against dualism, and I consider panpsychism the most promising position that warrants further philosophical, theological, and scientific investigation.

What disagreements Goff and I may have will not be in philosophy of mind, but only when Goff ventures into the neighbouring areas of philosophy of religion and theology. Like Goff, I am excited about 'the possibility that, in a panpsychist worldview, the yearnings of faith and the rationality of science might finally come to harmony' (Goff, 2019a, pp. 216–17). The only word I disagree with in this sentence is 'finally'. After all, theologians have been using (and creating) the rational tools of philosophy and science to investigate their own truth claims, internal coherency, and practical out-workings for centuries. So, one of the first points I want to contest in this paper is Goff's implicit endorsement of the widely debunked 'myth of conflict' between science and religion.[3]

The places where Goff implies that science and religion are in conflict are often occasions when Goff is trying to sell panpsychism as a new *scientific* approach to consciousness. What makes something a 'science', or a particular knowledge-seeking enterprise 'scientific' in character, is, of course, no simple question. Philosophers of science have been struggling with this question in its modern form for a century, and no consensus is forthcoming. It is not uncommon, therefore, for philosophers to cut this gnarly corner by relying on the modern myth that science and religion are inevitably in conflict, or to affirm some form of naturalism in a plea for legitimacy. If science and religion are defined in opposition to one another, then all one has to do to make panpsychism seem scientific is to show that it too is opposed to religion. As such, there are a few places in *Galileo's Error* where Goff positions panpsychism in competition with theism and does not consider the possibility that someone, like me, might be a theist and a panpsychist. Whilst I am used to the task of repositioning

---

[3] There is a very large literature on this now, e.g. Brooke (1991), Harrison, (2015), Lightman (2019), and Hardin, Numbers and Binzley (2018).

panpsychism and theism into a relationship of mutual reinforcement rather than competition, my work has thus far primarily sought to introduce and defend panpsychism to my fellow Christian theologians. Here I want to consider reasons that a panpsychist philosopher, like Goff, might have for accepting theism. Rather than leaving panpsychists vulnerable to Ockham's razor or undermining the scientific status of their position, I suggest that the combination of panpsychism with theism is an eminently consistent and appealing position.

In the two sections below, I take a two-pronged approach to move Goff's position closer to theism. In the first section, I tackle some of the places where Goff mentions God in *Galileo's Error* in order to remove the obstacles that are constructed between panpsychism and theism. My intention in this section is merely to show that, despite the implications of some of Goff's comments, there is no competition or opposition between belief in fundamental minds and belief in God. The second section then gives positive reasons why a panpsychist should extend her position and also adopt theism. It is concluded that the combination of panpsychism and theism is the most consistent and satisfying worldview.

2. Removing Incompatibilism's Obstacles

Most of the places where Goff positions panpsychism and (traditional Western mono-)theism as alternatives are to do with rejecting the explanatory power of divine intervention. For Goff, if science can explain how an event occurs — that is, if we can describe the physical causes that are sufficient for bringing about an event — then this rules out any kind of dualism, God–world dualism or mind–body dualism. This view, whereby physical causes are in competition with mental and divine causes, is known as *incompatibilism*; physical (efficient, mechanical, determined) causes are seen to be incompatible with any other kinds of causation.

Let's consider another example where Goff, by presenting theism as analogous to dualism, implicitly places theism in competition with panpsychism. In the chapter against naturalistic dualism, Goff draws the comparison between divine intervention and the mind–body interaction problem. Goff's argument (2019, pp. 36–8) can be summarized as follows:

(1) Anomalous events are events that cannot be explained through physical causes.
(2) If God intervened these would be anomalous events.

(3) Anomalous events would be 'patently obvious' to scientists.
(4) When scientists examine the world, they don't regularly find anomalous events (events that cannot be explained by physical causes).
(5) Therefore, God does not intervene.

There are queries to be raised with each of the (1–4) premises of this argument. First, since there is no theoretical reason why God might not intervene as a partial cause influencing but not determining an outcome, it seems that Goff needs to strengthen the first premise to the claim that anomalous events are events that cannot be exhaustively, or at least sufficiently, explained through physical causation. This would be inconsistent, however, because in the final chapter of the book, when discussing free will, Goff is entirely comfortable with the partial causal pressure of (rational) inclinations informing events (even, in theory at least, for non-human organisms and objects) (*ibid.*, pp. 204–5). If anomalous events are those for which we cannot *in practice* give a full and complete physical and causal explanation, then anomalies happen the majority of the time. As such, a term like 'anomalous' is probably misleading, since it carries connotations of 'rare' or 'out of place'. After all, if God exists and acts in creation, then we might expect this neither to be rare nor out of place.

Second, a theological compatibilist (arguably the majority of the Christian tradition historically) would reject premise 2. Theological compatibilists believe that, due to transcendence, God can act in and through physical causes, as well as sometimes apart from them if God wishes. Neo-Thomists, for example, argue that all the causation that we encounter and that scientists can describe is known as secondary causation. The world of secondary causes is created and sustained in every instance by God, who is the primary cause of all things. On this two-tiered schema, God could choose to act directly without a mediating secondary cause, which is how miracles are said to work, but God can also create or direct secondary causes in a way that would be undetectable to humans (as we ourselves are also secondary causes). There are certainly potential weaknesses of this position and I am not strongly advocating for it. But I do think it is sufficiently cogent to show that there is a rich variety of options for considering divine action on offer within Christian theology, not all of which presume incompatibilism. I think it is most likely that God acts in a wide variety of specific ways and that we shouldn't treat verbs like 'act' as closed concepts (Abraham, 2017).

As an (historically fraught) example, Goff describes how both William Paley and Charles Darwin agreed that the origin of complex organisms required an explanation (Goff, 2019, pp. 7–8). Paley opted for an intelligent designer and Darwin for natural selection. The way Goff describes this, it seems as if we have to choose between theism and evolution. The problem with this is that for the majority of Christians both Paley and Darwin were (at least partially) right. It is quite normal for Christians to claim that scientific descriptions of an event might be a description of how God, as creator, decided to achieve a certain outcome. Again, there are a wealth of compatible but distinct options to choose from here: God might set up natural laws in advance, God might gift certain powers and tendencies to specific species or individuals, God might input information some way along the evolutionary chain, God might lure or attract creatures toward a certain future, and there may be other options not listed here (see Kojonen, 2016).

This kind of compatibilism can also be seen in Christian ideas on the soul (*contra* Goff, 2019, p. 32). With regards to the origins of the soul, Christianity contains a persistent minority tradition of traducianism, which argues that each individual person's soul is not injected *ex nihilo* into foetuses but is inherited along with our bodies via sexual reproduction. There are various proposed mechanisms for this inheritance within the history of traducianism, most of which have some relevance to panpsychism. Some proponents propose that souls are complex entities that fission or partition to generate the child's soul. Others have suggested that the nutritive soul of the gametes can develop into the rational soul of a human or that the gametes carry more basic soul-stuff which becomes a human soul at syngamy. Others confess the precise mechanism of soul generation to be a mystery whilst still affirming that soul creationism is unnecessary (for discussion, see Swinburne, 1986; Crisp, 2006; and Leidenhag, forthcoming). In terms of affirming the importance and compatibilism between body and soul, a more central doctrine to point to, which all orthodox Christians (and many Jews and Muslims) affirm, is the resurrection of the body as an essential part of the afterlife. Indeed, for Christians it is the physicality of Jesus's resurrection, not a disembodied afterlife, that is the foundation of their faith and hope. At least within Christian theology, the premise that divine activity (or other immaterial phenomena) stands in opposition to physical causation is somewhat question-begging.

Interestingly, one could interpret Goff's panpsychism as a compatibilist version of 'dualism' (or duality at least) all the way down. Goff rejects naturalistic dualism, in part because he is doubtful that dualists will ever be able to solve the interaction problem. Instead, he argues that, on his panpsychism, consciousness is the intrinsic nature of the physical, which works in and through the physical. Whilst divinity is not (often) seen to be the intrinsic aspect of creation, traditional Christian theism also argues that God can work in and through created reality. We could even suggest that God is present within and sometimes influences the conscious experience of panpsychist entities at multiple levels of scale; comforting humans, rearranging broken legs, and moving water molecules to form a path through the Red Sea (Leidenhag, 2021, pp. 130–8). All of these things can be achieved in tandem with the psychophysical duality of the natural world. Here God would be making a difference in the world (so, we might say 'intervening'), but this could not meaningfully be described as anomalous because in a panpsychist world no event can be sufficiently explained in terms of physical causation alone. As such, we have no reason to suppose that such divine interventions would be patently obvious to scientists.

The third and fourth premises appear highly doubtful to me. There are numerous events for which we lack sufficient or complete physical explanations — particularly those related to medicine and neuroscience, which Goff mentions specifically (Goff, 2019, pp. 37–9). Whilst many believe that in theory there is a sufficient psychophysical causal explanation for every event, identifying all the causes involved in any particular real-world event is extremely difficult if not impossible. This does not guarantee that philosophical principles like reductionism or causal closure are certainly wrong, but only that we are not in a position to know for sure and so cannot rule out the existence of God (or free will, and other such phenomena) on the basis of an unfounded allegiance to such principles. The human body and brain are simply too complex and the real-world variables too numerous. The implication of this is that we should not expect scientists to have identified divine activity and we should be wary of any claims to that effect. Not because miracles do not happen — I believe that they do and with marvellous frequency — but because this is not the sort of thing that I expect scientists to be able to predict, corroborate, quantify, or control for experimentation.

Even within controlled environments, many scientific experiments contain anomalous results; so, it seems inaccurate to say that scientists (or anyone else) do not encounter anomalies that surprise us, contravene our expectations, and cannot be easily explained. As Thomas Kuhn famously argued, unless such anomalies accumulate with predictability and regularity, they will not lead to any kind of paradigm shift or force us to change our explanatory structures (Kuhn, 1962/2012, p. 82). A good example is that consciousness is too regular an 'anomaly' in a merely physical universe to be ignored. The almost universal testimony of conscious experience amongst humans demands a paradigm shift away from mere physicalist thinking, such as is offered by panpsychism. The experience of a miracle, by contrast, is an unpredictable free act by a personal agent, which might require a paradigm shift in our theological views, but does not require a revolution in the practices or theories of natural science. Miracles are not the sort of thing that the physical sciences can investigate, and we should not expect to find any 'gaps' in the causal chain to prove the existence of such anomalous events. In many ways, this appreciation for the proper limits of natural science is a similar argument to the one that Goff makes in several places for panpsychism against materialism (Goff, 2019, pp. 65–9, 75, 85).

But, as I've already hinted at above, there is a more interesting point to be made here. Rather than pairing up divine intervention and the dualist's interaction problem in the hopes that rejecting one will lead to the rejection of the other as well, panpsychism makes divine activity in the natural world more explicable. This is because panpsychism describes a world that is not as physically determined or mechanistically shallow as might be supposed, but one with internal depth; a world filled with intrinsic natures of consciousness. Panpsychism allows theologians to articulate how God's presence might be in the world, not in a spatial sense, but within the experience of creatures as a felt second-personal presence (Leidenhag, 2021, pp. 130–8). In the Christian tradition, notions of divine presence and divine action are often linked, if not sometimes treated as synonymous (Arcadi, 2018, pp. 96–101). To say that God is present within all things is also to say that God acts, either concurrently, persuasively, decisively, or in other ways. To be sure, panpsychism is not a *deus ex machina*. Panpsychism cannot solve all the various objections to divine action (many, for example, are really forms of the problem of evil, rather than anything particularly to do with action or causation discourse) and panpsychism does not help articulate all the very many ways God might act in the universe. However, it does give more resources to this area of philosophical theology and does so in a way that is consonant with philosophical, biblical, and liturgical expressions of God's intimate yet hidden presence (Leidenhag, 2021, pp. 129–30).

This experiential depth to all things that panpsychism posits is why panpsychism so often invites connection with mysticism, to which Goff tells us he is drawn. But rather than this resulting in a form of immamentist spirituality (that is, a spirituality that denies anything transcendent or beyond nature), this depth to nature might just as easily be viewed as a door to the supernatural, to a transcendent God that exists beyond space-time and who lovingly interacts with the intrinsic nature of every particle, piranha, or person.

The discussion above considered how Goff put (supernatural) theism in competition with panpsychism largely through some form of assumed incompatibilism. I have argued that there are established mainline theological positions within Christianity (and other theistic traditions) than undermine the grounds for this constructed opposition. But it might be argued that I have missed the heart of Goff's objection to theism. At several points in *Galileo's Error*, Goff appeals to the principle of explanatory simplicity and to Ockham's razor in order to make various arguments. Ockham's razor is 'the principle that, all things being equal, we should try to make our theories of reality as simple as possible' (Goff, 2019, pp. 48–9). Or, 'don't believe in more things if you can get away with fewer' (*ibid.*, p. 134). The appeals to various forms of theological compatibilism may seem particularly vulnerable to Ockham's blade.

However, the 'all things being equal' and 'if you can' are important. Such qualifications indicate that one can only use Ockham's razor when comparing two explanations with the same explanatory power for the same relevant phenomena. The desire for simplicity has to be subordinate to the desire for explanation. This is why panpsychism itself is not as vulnerable to Ockham's razor as many have previously supposed; materialism (and Galileo's legacy on natural science) cannot explain consciousness and so we need to posit something more. Goff's 'simplicity argument' is to posit consciousness within matter, so we still have one type of thing, psychophysical stuff, and do not have to resort to stronger forms of dualism. Goff then makes the same move with regard to explaining mystical experiences. He sets up a choice between seeing mystical experiences as an illusion or a supernatural dualism, and offers panpsychism as the simpler middle path: 'formless consciousness is the ultimate nature of *physical reality*' (*ibid.*, p. 207). It is this formless consciousness that he suggests the mystics may be encountering. This spirituality without transcendence or supernaturalism may have the advantage of simplicity (ontological monism), but at too high an explanatory cost. For the explanatory power of theism is not confined to explaining mystical experiences down the ages, but to explaining mystical experiences *and* providing an ultimate explanation for everything else.

3. Why a Panpsychist Should Also Be a Theist

So far, I have argued for why an opposition between panpsychism and supernatural theism is false. God could just have easily created a panpsychist universe as any other and, as I will conclude, I think we have good reasons for speculating that God would desire a panpsychist universe in particular. One could, then, stop here by acknowledging the neutrality between panpsychism and belief in God; a panpsychist is as welcome to be a theist as any substance dualist, hylomorphist, or (local) materialist.[4] However, I want to make a more positive case for why a panpsychist should embrace theism.

There are at least three different kinds of materialists, which I will call illusionists, reductionists, and non-reductive physicalists (emergentists). Illusionists deny there is anything called 'consciousness' or qualia that demands an explanation — it's an illusion. Reductionists think that conscious experience can be explained away by reducing it to something non-experiential, namely mere physical processes. Panpsychists reject both of these by affirming that consciousness exists and demands a non-reductive explanation.

Non-reductive physicalists (or, emergentists) such as Alex Moran in this journal issue, argue that wholly non-experiential matter can give rise to consciousness; it's an emergent feature of the physical world. For a while, philosophers of mind were optimistic that emergence theory was the way to go. After all, we are pretty confident that complex physical features of the world (like liquidity and life) are emergent, so why not consciousness too? However, one reason for the revival of panpsychism is that the argument for brute emergence is looking increasingly implausible. This is the so-called genetic argument for panpsychism, sometimes referred to as the argument from non-emergence or the argument from origination. Put simply, this

---

[4] I list these as common positions amongst theists, but they should not be considered exhaustive.

argument denies that mind or experience could ever emerge from wholly non-experiencing material stuff. Consciousness cannot suddenly appear in a world in which nothing conscious has previously existed, not even in a latent or potential form.

It was this argument that was operative in Thomas Nagel's famous essay 'Panpsychism', which has been so influential on the field. Nagel argued that 'unless we are prepared to accept the alternative that the appearance of mental properties in complex systems has *no causal explanation at all*, we must take the current epistemological emergence of the mental as the reason to believe that *constituents have properties of which we are not aware*, and which do necessitate these results' (Nagel, 1979, p. 187). That is, to have a metaphysically and scientifically plausible account of the emergence of consciousness, emergence theorists need to become panpsychists. Emergence might have an important part to play within a panpsychist account of consciousness (it may help us explain how we get complex human minds from a combination of simple atomic or bacterial minds), but it cannot explain how we get consciousness into the universe in the first place.

Galen Strawson makes this point through a comparison between 'brute emergence' and the creation of reality when he writes that if emergence is intelligible then 'it will be intelligible to suppose that existence can emerge from (come out of, develop out of) non-existence' (Strawson, 2008, p. 66). Similarly, J.P. Moreland writes:

> The emergence of consciousness seems to be a case of getting something from nothing. In general, physical-chemical reactions do not generate consciousness, not even one little bit, but they do in the brain, yet brains seem similar to other parts of organisms or bodies. How can like causes produce radically different effects? The appearance of mind is utterly unpredictable and inexplicable. This radical discontinuity seems like an inhomogeneous rupture in the natural world. (Moreland, 2003, p. 209)

Even though he subsequently rejects panpsychism, Moreland hints towards three of the main arguments for panpsychism: the argument from non-emergence ('getting something from nothing'), the argument from continuity ('brains seem to be similar with other parts of organisms or bodies'), and the evolutionary argument (a dissatisfaction with 'radical discontinuity' or an 'inhomogeneous rupture in the natural world'). A panpsychist might tie these problems together in the following way: if the universe is made up of the same stuff throughout, and some of this stuff so-arranged is definitely conscious and other parts appear not to be, then either we need to introduce inhomogeneous ruptures into the natural world — either by radical emergence or regular divine interjection — or we need to say that there is a continuum of consciousness throughout the universe. Panpsychists, like Goff and myself, opt for the latter.

It is surprising that Goff does not explicitly make this genetic argument for panpsychism in *Galileo's Error* and does not tackle emergence in any detail. To my mind it is an argument that contemporary panpsychists must make since emergence theory is often seen as their main philosophical rival. In making the genetic argument, panpsychists argue for the existence of minds of which we are not directly aware, cannot be empirically detected, and which they cannot arrive at through an argument from analogy;[5] and panpsychists make this argument on the basis of the principle that experience cannot come from non-experience — something cannot come from nothing.

In making this argument, the panpsychist is committing herself to at least one of two explanatory principles. She either rejects the radical novelty of the emergence of mind from matter because it violates the causal principle, according to which every contingent being has a cause of its existence and that the perfection of the effect is found in the cause. Or, she rejects the emergence of mind from matter because brute emergence cannot uphold the principle of sufficient reason which states that 'no fact can be real or existent, no statement true, unless there be a sufficient reason why it is so and not otherwise' (Leibniz, 1951, p. 527; *cf.* Nagel, 2012, p. 17). It is because the panpsychist cannot abide the abandonment of these metaphysical principles which undergird scientific and rationalist endeavours that, instead, she chooses to posit mind as fundamental to the universe.

One of the interesting facts about this argument for panpsychism is that the principle of sufficient reason and the causal principle form the backbone of the most common forms of the cosmological arguments

---

[5] The argument from analogy is a common answer to the problem of other minds. The problem of other minds, put simply, is to ask on what basis we believe that other people whom we interact with are conscious minds rather than mindless robots behaving as if they were conscious subjects. One popular answer is that we believe that other minds exist, that other people are conscious subjects, because we make an analogy from our own physiology and experience and, on the basis of sufficient similarity between myself and others, believe that others are conscious in a manner similar to myself. Since panpsychism affirms some measure of consciousness to entities extremely unlike myself, such as simple single-cell organisms, vast galaxies, and even inorganic particles, the argument from analogy seems unavailable to them.

for the existence of God: the Leibnizian cosmological argument argued from contingency, the Thomistic cosmological argument argued from sustained existence, and the Kalaām cosmological argument argued from a temporal beginning. These cosmological arguments, and others, are a family of *a posteriori* arguments that infer a necessary and creative being (God) from particular facts about the universe. Many of the critiques against the cosmological arguments, such as those put forward by David Hume and Immanuel Kant, attack either the causal principle or the principle of sufficient reason. However, these points of critique are not open to the panpsychist who has already endorsed these principles within the genetic argument for panpsychism.

Whereas the philosopher of religion asks, 'Why is there something rather than nothing?' and 'Why is that something like this?', the philosopher of mind asks, 'Why are there experiencing subjects rather than philosophical zombies?', and 'Why are these experiencing subjects as they are?'. Either there is an answer to these questions in terms of a reason, pre-existing conditions, or a cause (i.e. God/fundamental mentality), or there is no explanation at all and the universe/experiencing subjects just exist as a brute fact and have always done so. Since the panpsychist is dissatisfied with conscious organisms being a brute fact, it seems likely that she should find the brute fact of the universe's existence to be dissatisfying also.

The panpsychist posits consciousness as fundamental because the contingency of the human mind needs to be explained. But, in doing so, the panpsychist does not reduce the contingency of consciousness in general, she only ties the contingency of consciousness to the contingency of the universe as a whole. As such, the panpsychist uses the principle of sufficient reason to form a full, but not complete, explanation for human consciousness. This explanation is full because, whilst human consciousness is satisfactorily explained,[6] the puzzling aspect of consciousness has not disappeared entirely, it has just been moved to the fundamental level of reality (Pruss, 2006, p. 17; Swinburne, 2004, p. 78). The existence of consciousness becomes no more (and no less!) surprising than the existence of any other fundamental feature of the world; space-time, natural laws, quantum vacuums, or whatever. Whilst it is beyond the bounds of panpsychism

---

[6] Assuming panpsychists can eventually give an adequate account of mental combination or assuming cosmopsychists give an account of individuation/decombination.

as a position within philosophy of mind, these fundamental realities still need explaining.⁷ The most parsimonious and consistent way for this to be achieved would be to extend the same logic and explanatory principles that one applies within the universe to the universe at large. What I am suggesting is for the panpsychist to take her reasoning a step further and posit that there is a first cause or sufficient reason for the existence of the panpsychist universe, just as she posits that there must be a causal explanation or sufficient reason for the appearance of human minds within the cosmos.

Is it legitimate to demand an explanation of the universe based on the prior demand for an explanation of the mind? Even if human consciousness is a contingent fact demanding an explanation, perhaps we could still say — as Bertrand Russell did — that the universe 'just is' (Russell, 1948/1964, p. 175). Russell argued that theists are making a fallacy of composition in supposing that, just because the contents of the universe are contingent, so too is the universe itself as a whole contingent. A panpsychist might be inclined to champion Russell's argument here, since she is particularly alert to the fallacy of composition in defending panpsychism from the absurdity that all things that are composed of consciousness are themselves conscious (e.g. slippers, spoons, stars, etc). However, by making consciousness fundamental and intrinsic to the physical universe, the panpsychist blocks this Russellian critique of the cosmological argument. If the fundamental and intrinsic nature of everything in the universe ceased to exist, then the universe itself would cease to exist (Reichenbach, 1972, chapter 5). If fundamental consciousness is contingent, then so too is the universe as a whole. In which case, the panpsychist universe demands an explanation for its existence.

---

[7] Dasgupta (2014) explores a notion of the principle of sufficient reason whereby 'every substantive fact has an autonomous ground' (p. 384), and that autonomous grounds include essences. In Dasgupta's terminology and in the logic of grounding, I might be read as saying that God is an 'autonomous ground', because God is not apt for being grounded, whereas the universe is a substantive fact — apt for being grounded. In traditional Thomist metaphysics this is expressed by saying that, uniquely for God, God's essence is existence, thereby making God the only necessarily existing — and in Dasgupta's terminology the only autonomous — entity. Dasgupta does in fact acknowledge that something(s) with the essence of existence may be an implication of his proposal — only he thinks this is compatible with atheism-cum-naturalism; so here we would have to debate the definition of 'god' (pp. 396–9). However, I have chosen to stick with more traditional language of contingency and necessity, because I have some reservations about employing grounding to describe the God–world relation.

Another way to consider this argument is to debate the proper location of 'bruteness', or what is left unexplained within any account of reality. This is an issue Goff mentions a few times in *Galileo's Error*. In particular Goff writes, 'Everyone takes some facts as basic and unexplained. Some people take the laws of physics as an unexplained starting point; others the existence of God; others the laws of logic and mathematics. I take the reality of consciousness as a fundamental starting point' (Goff, 2019a, pp. 198–9). Expressed this way, what one takes as brute seems to be little more than a personal choice or matter of taste, which we can make no arguments for or against. But this is clearly not the case; the place of bruteness is important in evaluating competing theories. Despite their simplicity, the panpsychist rejects various forms of materialism because materialists take the explanandum as brute — either the appearance of consciousness is an illusion or the mind–body relation as brute emergence. This is why Goff concludes that 'contemporary materialism is not a solution but a stubborn refusal to face up to the problem' (*ibid.*, p. 96). The same could be said of Russell's statement that the universe just exists. If we consider the panpsychist universe as our explanandum and apply the same logic that the panpsychist applies against materialism, then we will need to posit a necessary being to stop the infinite regress of explanations in a non-arbitrary manner (*cf.* O'Connor, 2013, p. 42).

There is one further way for a panpsychist to attempt to avoid the conclusion of theism. Panpsychist philosophers D.S. Clarke and Freya Mathews have each suggested that this particular universe of fundamental consciousness is the necessarily existing entity at the end of the explanatory chain (Clarke, 2003, p. 120; Mathews, 2003, p. 61). This is to claim not only that a panpsychist universe does exist, but that it must exist. It might seem that, by positing a necessary consciousness, Clarke's and Mathews' panpsychism is starting to look quite a lot like some form of theism or pantheism, particularly if this consciousness is 'like us' or 'welcomes us' (e.g. Goff, 2019a, p. 217).[8] What should we make of such a claim? A theist might politely remind their interlocutor that necessity is a very strong claim, and not something to be appealed to lightly. In particular, to posit the actual

---

[8] Goff has also tentatively explored the idea that the universe is a conscious agent in order to explain cosmic fine-tuning (Goff, 2019a).

panpsychist universe as existing necessarily will create at least three difficulties for the panpsychist.

First, to claim that something exists of metaphysical necessity is to claim that it is self-caused or, to use Timothy O'Connor's phrase, necessary beings are 'absolutely invulnerable to nonexistence' (O'Connor, 2008, p. 70). If something is a necessary being then questions like 'Why does it exist?' are inappropriate questions, which is why Stephan Hawking's question 'Who created God?' is a philosophical non-starter (Hawking, 1988, p. 174; cf. Clarke, 2003, p. 120). If this were applied to this universe then it means abandoning our current cosmological theories with regards a beginning of this universe and its eventual heat death. It seems like a bad idea to set panpsychism up in conflict with current science, merely to avoid the conclusion of theism.

Second, if this (exact) panpsychist universe exists necessarily, then, unless the universe also has some contingent intentional will and power, what causally follows will also be of necessity and there is no contingency left in the universe at all, which as Bruce Reichenbach writes 'is a disquieting notion' to say the least (Reichenbach, 2021).

Third, if the panpsychist were to claim that the universe exists of logical necessity then they would be suggesting that, were we to properly understand the concept, it is inconceivable that it could be otherwise. Put another way, if it is conceivable that not-x, then x does not exist by necessity (but exists contingently). Logical necessity is the strongest sense of the claim of necessity, and some theists do not even want to posit the logical necessity of God (e.g. Swinburne, 2004, pp. 79, 148). Moreover, since the panpsychist typically follows David Chalmers in using conceivability arguments against physicalism, they should be hesitant about invoking logical necessity and thereby removing conceivability arguments from the playing field. If the universe's non-existence is conceivable (which I think it clearly is) then to also state that the universe is necessary is to sever any link between conceivability and possibility. However, if a panpsychist severs such a link then they can no longer argue from conceivability to the possibility of zombies, which many panpsychists are wont to do in order to ward off physicalism.[9]

---

[9] I owe a debt of thanks to Philip Goff for helping me clarify and articulate this argument (which is not to say that he agrees with me here).

Best return to the altogether more parsimonious and straightforward claim that this panpsychist universe is contingent and can be best explained by affirming a transcendent, creator God. In the words of Charles Taliaferro, 'Theism can thus provide an explanation for the existence of the panpsychistic cosmos as well as for the different levels of consciousness pervading it' (Taliaferro, 2018, p. 369).

4. Conclusion

Let me restate again how much I enjoyed *Galileo's Error*, and how much Goff and I agree on many of these arguments. In fact, the fundamental agreements between panpsychism and theism is the central point of this paper. It is not uncommon for philosophy of mind and philosophy of religion (and Christian theology) to bump up against one another and overlap. A common way to draw the line between philosophy of mind and theism is based on an analogy between an (immaterial) human mind and an immaterial divine mind, and the question of how these minds interact with material bodies. As Yujin Nagasawa (2020) has argued, if one makes this sort of move based on an analogy of immaterial substance (by replacing phenomenality with divinity) then panpsychism would seem parallel to polytheism rather than monotheism. If one adopts a more cosmopsychist view then the corresponding position in philosophy of religion is pantheism, although I have argued elsewhere that such a move is not as straightforward as it may seem at first (Leidenhag, 2019).

But, for a theologian who affirms the transcendence of God and the claim that God created all things out of nothing — as mainstream traditions within Judaism, Christianity, and Islam all affirm — then any move from mind to God on the basis of an analogy of substance should be suspect. Human minds are not a little bit like God, not even the minds of substance dualists. The ultimate divide, according to (supernatural) theism is not between immaterial things (Gods, angels, minds) and material things, but between God and everything that is not God, between Uncreated necessity and created contingency. Creation not only includes physical matter, but everything that God has created, such as consciousness. It is because of this transcendence that God's action is not in competition with (or incompatible with) creaturely action, physical causation, or scientific explanation.

Why posit such a transcendent being in the first place? As argued in the latter half of this paper, when the entire system demands an explanation then the simplest and most explanatorily satisfying solution is to posit something beyond that system. Here the extension of panpsychism into theism is not achieved on the basis of an analogy of immaterial substances, but on the consistent use of explanatory principles and epistemic virtues: the principle of sufficient reason, the causal principle, and the desire for simplicity but not at the cost of explanatory power. I can agree with Goff that panpsychism is the best explanation on offer within philosophy of mind, but it is not the best explanation on offer within philosophy of religion.

If we accept theism, particularly along the lines of the Christian faith, then the claim that God chose to create a panpsychist world makes good sense. If God values human minds, freedom, and relationship, then we should expect God to create a world rich with ubiquitous mentality. If God desired to create a world in which to become incarnate, in which to dwell as in the temple, to be present within and unite Godself with, then a panpsychist world of order, unity, simplicity, and spiritual depth seems to have been a good idea (*cf.* Page, 2020, pp. 351–3). Far from being mutually exclusive alternatives, I can't think of a worldview more explanatorily and personally satisfying than a combination of theism and panpsychism.

References

Abraham, W.J. (2017) *Divine Agency and Divine Action: Exploring and Evaluating the Debate*, vol. 1, Oxford: Oxford University Press.

Arcadi, J.M. (2018) *An Incarnational Model of the Eucharist*, Cambridge: Cambridge University Press.

Brooke, J.H. (1991) *Science and Religion: Some Historical Perspectives*, Cambridge: Cambridge University Press.

Clarke, D.S. (2003) *Panpsychism and the Religious Attitude*, New York: SUNY Press.

Crisp, O.D. (2006) Pulling traducianism out of the Shedd, *Ars Disputandi*, **6** (1), pp. 265–287.

Dasgupta, S. (2016) Metaphysical rationalism, *Noûs*, **50** (2), pp. 379–418.

Goff, P. (2019a) *Galileo's Error: Foundations for a New Science of Consciousness*, New York: Vintage Books.

Goff, P. (2019b) Did the universe design itself?, *International Journal for Philosophy of Religion*, **85**, pp. 99–122.

Hardin, J., Numbers, R.L. & Binzley, R.A. (eds.) (2018) *The Warfare between Science and Religion: The Idea that Wouldn't Die*, Baltimore, MD: Johns Hopkins University Press.

Harrison, P. (2015) *The Territories of Science and Religion*, Chicago, IL: University of Chicago Press.

Hawking, S. (1988) *A Brief History of Time*, New York: Bantam.

Kojonen, E.V.R. (2016) Salvaging the biological design argument in light of Darwinism?, *Theology and Science*, **14** (3), pp. 361–381.

Kuhn, T. (1962/2012) *The Structure of Scientific Revolutions*, Chicago, IL: University of Chicago Press.
Leibniz, G. (1951) The monadology, in Wiener, P. (ed.) *Leibniz Selections*, New York: Charles Scribner's Sons.
Leidenhag, J. (2019) Unity between God and mind? A study on the relationship between panpsychism and pantheism, *Sophia: International Journal of Philosophy and Traditions*, **58** (4), pp. 543–561.
Leidenhag, J. (2021) *Minding Creation: Theological Panpsychism and the Doctrine of Creation*, London: Bloomsbury/T&T Clark.
Leidenhag, J. (forthcoming) A panpsychist view: Souls as fundamental and combined, in Farris, J. & Leidenhag, J. (eds.) *Humans, Souls and Origins: A Conversation*, London/New York: Routledge.
Lightman, B. (ed.) (2019) *Rethinking History, Science and Religion: Exploring Complexity*, Pittsburgh, PA: Pittsburgh University Press.
Mathews, F. (2003) *For Love of Matter: A Contemporary Panpsychism*, New York: SUNY Press.
Moreland, J.P. (2003) The argument from consciousness, in Copan, P. & Moser, P. (eds.) *The Rationality of Theism*, London/New York: Routledge.
Nagasawa, Y. (2020) Panpsychism versus pantheism, polytheism, and cosmopsychism, in Seager, W. (ed.) *The Routledge Handbook of Panpsychism*, New York: Routledge.
Nagel, T. (1979) *Mortal Questions*, Oxford: Oxford University Press.
Nagel, T. (2012) *Mind and Cosmos: Why the Materialist Neo-Darwinian Conception of Nature is Almost Certainly False*, Oxford: Oxford University Press.
O'Connor, T. (2008) *Theism and Ultimate Explanation: The Necessary Shape of Contingency*, London: Wiley-Blackwell.
O'Connor, T. (2013) Could there be a complete explanation of *everything*?, in Goldschmidt, T. (ed.) *The Puzzle of Existence: Why is there Something Rather than Nothing?*, New York: Routledge.
Page, B. (2020) Arguing to theism from consciousness, *Faith and Philosophy: Journal of the Society of Christian Philosophers*, 37 (3), pp. 336–362.
Pruss, A. (2006) *The Principle of Sufficient Reason: A Reassessment*, Cambridge: Cambridge University Press.
Reichenbach, B. (1972) *The Cosmological Argument: A Reassessment*, Springfield, IL: Charles Thomas.
Reichenbach, B. (2021) Cosmological argument, in Zalta, E.N. (ed.) *The Stanford Encyclopedia of Philosophy*, [Online], https://plato.stanford.edu/archives/spr2021/entries/cosmological-argument/ [30 March 2021].
Russell, B. (1948/1964) Debate on the existence of God, reprinted in Hick, J. (ed.) *The Existence of God*, New York: Macmillan.
Strawson, G. (2008) *Real Materialism and Other Essays*, Oxford: Oxford University Press.
Swinburne, R. (1986) *The Evolution of the Soul*, Oxford: Oxford University Press.
Swinburne, R. (2004) *The Existence of God*, revised ed., Oxford: Oxford University Press.
Taliaferro, C. (2018) Dualism and panpsychism, in Brüntrup, G. & Jaskolla, L. (eds.) *Panpsychism: Contemporary Perspectives*, Oxford: Oxford University Press.

Sarah Lane Ritchie[1]

Panpsychism and Spiritual Flourishing

Constructive Engagement with the New Science of Psychedelics

Abstract: This article discusses the potential implications of panpsychism for the study and pursuit of spiritual flourishing, with a focus on emerging scientific and philosophical research on psychedelics. Psychedelics research (1) has the means to reliably and safely produce the conditions in which transformative experiences routinely occur, thereby allowing for robust neuroscientific and psychological research, and (2) has attracted a growing body of philosophical and theological work on the metaphysical and epistemological possibilities of such experiences. I begin with a discussion of recent scientific work on psychedelics. I then discuss the epistemic status of psychedelic experiences, where the metaphysics of panpsychism is particularly interesting. I suggest there exists a mutually reinforcing relationship between panpsychism and the metaphysical possibility of a veridical interpretation of psychedelic states, and that this conceptual congruence has important implications for research on spiritual flourishing. This flourishing need not be understood in a theological manner, although it is, I suggest, entirely consistent with at least some naturalistic theological frameworks: the main goal of the article is to map the conceptual terrain in which conversations about spiritual flourishing, psychedelics, and panpsychism might take place.

Correspondence:
Email: sarah.laneritchie@ed.ac.uk

---

[1] University of Edinburgh, Scotland, UK.

1. Introduction

In the final pages of *Galileo's Error*, Philip Goff acknowledges the tantalizing possibility that panpsychism might contribute to the development of a naturalized spirituality. Discussing the potential connections between panpsychism and the phenomenology of mystical experiences, Goff writes, 'I can't help being excited by the possibility that, in a panpsychist worldview, the yearnings of faith and the rationality of science might finally come into harmony' (Goff, 2019, p. 215). For Goff, such a project remains a fascinating possibility, but he admits that 'the appropriate attitude to mystical experiences for those who haven't had them is probably one of agnosticism, the withholding of belief either that mystical experiences provide genuine insight into the nature of reality or that they are delusions' (*ibid.*). Given the scope and nature of Goff's philosophical work, this agnosticism is perhaps appropriate, and it is for others to develop the contours of an interdisciplinary project integrating the insights of panpsychism, the 'yearnings of faith' (or what I will call 'spiritual flourishing'), and current scientific work in these areas.

My goal in this article is not to debate the finer points of panpsychism as a philosophical position, but to demonstrate the potential implications of panpsychism for the study and pursuit of spiritual flourishing, with a particular focus on emerging scientific and philosophical research on psychedelics. There are, of course, many ways in which an embrace of panpsychism might affect various aspects of human flourishing, and many dimensions even of spiritual flourishing in particular. While these will be touched on briefly in what follows, my focus here will be on what Goff describes as the 'yearnings of faith' and spiritual experiences. The interdisciplinary study of psychedelics lends itself particularly well to this discussion, in so far as such research (1) has the means to reliably and safely produce the conditions in which mystical experiences routinely occur, thereby allowing for more robust neuroscientific and psychological research on such phenomena, and (2) has attracted a growing body of philosophical and theological work on the metaphysical and epistemological possibilities of such experiences. Specifically, I argue that there exists a mutually reinforcing relationship between panpsychism and the possibility of a veridical interpretation of psychedelic states, and that this conceptual congruence has important implications for interdisciplinary research on spiritual flourishing. (By 'veridical', I do not mean a realist claim that all mystical or psychedelic experiences accurately and necessarily represent reality, but only that there *may* be metaphysical models allowing for the theoretical possibility of real knowledge arising from some such states, in some cases.) This spiritual flourishing need not be understood in a theological manner, although it is, I suggest, entirely consistent with at least some naturalistic theological frameworks.

I begin with a discussion of recent scientific work on psychedelics, with a particular emphasis on psychedelic experiences' significant impact on human flourishing. I then move to a discussion of the epistemic status of psychedelic experiences, where we come to a metaphysical crossroads of sorts. On one hand, one might recognize the existential value of psychedelic experiences, but interpret as metaphysically illusory the mystical phenomenology often associated with these experiences. On the other hand, there are philosophical resources that permit the interpretation of psychedelic experiences as metaphysically illuminative in some way. Panpsychism here proves particularly compelling: if one is persuaded by interdisciplinary work on both panpsychism and recent research on the therapeutic and existential value of some psychedelic experiences, the spiritual content of some psychedelic experiences need not be interpreted only as an existentially useful fiction. My intention here is not to argue for a particular metaphysical framework (naturalistic or theological), but rather to map the conceptual terrain in which conversations about spiritual flourishing, psychedelics, and panpsychism might take place. What is evident, I suggest, is that the natural (but not necessary) compatibility between panpsychism and research on psychedelics suggests truly exciting possibilities for spiritual flourishing.

2. Spiritual Flourishing

In a recent work on what the authors term 'neuroexistentialism', Owen Flanagan and Gregg Caruso write that the 'really hard problem' facing humans today is this: 'How — given that we are natural beings living in a material world and given that consciousness is a natural phenomenon — does human life mean anything? What significance, if any, does living our kind of conscious life have' (Caruso and Flanagan, 2018, p. 10). In response to this existential anxiety, scholars from across disciplines have begun to develop frameworks of meaning, of spirituality, that take seriously the reality that humans are wholly natural, embodied beings, and these embodied beings are in principle (if not yet in practice) fully explicable and subject to empirical analysis. Of course, scholars like Flanagan and Goff will disagree on questions within the philosophy of mind, but they share in common a naturalistic, monistic impulse: whatever consciousness is, it is part and parcel of nature as a whole. In short: I do not have an immaterial soul that will live on after I die, there is nothing ontologically unique about me as a biological organism among other biological organisms, and many find good reason to doubt that the traditional omnipotent God of classical theism offers the sure metaphysical foundation for spirituality that we might wish for.

Given this stark depiction of reality, what are we to do with the very real, visceral 'yearnings of faith' that lead us to seek out transcendence and connection with Ultimate Reality? Are we fated to exist only in a disenchanted world? For those committed to a dichotomous, flatly reductionistic worldview, the responsible move here might be to do away with childish fancies and resign oneself to the cold, hard facts of a materialist universe. Others, though, embrace contemporary science and a broad naturalism, but still remain open to the possibility of 'something more'.

For the sake of brevity, I will here describe this search as the search for *spiritual flourishing*, by which I mean *an holistic sense of wellbeing or wholeness that, while involving the physical, mental, and relational components of human flourishing, is experienced by the individual as somehow going beyond those components*. For the sake of this particular discussion, I posit three markers of spiritual flourishing. First, spiritual flourishing is marked by *transcendence* — the experience of being part of something larger than oneself, caught up in a reality that is felt to be beyond one's 'normal' constrained reality. Second, spiritual flourishing will involve *connectedness* with Ultimate Reality (however naturalistically or theologically defined), the rest of the natural world, and/or other humans. Third, spiritual flourishing will often involve a sense that the world is marked by *enchantment*, or imbued with potential and even a 'magic' that seems to be inadequately conveyed by scientific descriptions of the world. A human life marked by a pervasive sense of transcendence, connectedness, and enchantment might align well with fourteenth-century English mystic Julian of Norwich's statement that 'all shall be well, and all shall be well, and all manner of thing shall be well' (Julian of Norwich, 2015, p. 174). Note that these three markers of spiritual flourishing are not exhaustive, technical, or dependent on particular metaphysical commitments. They are intended only as useful heuristics and 'pointers' in discussions of spiritual flourishing, and could well be supplemented or replaced by other such markers. In other words, this is a partial list of phenomenological qualities often attending spiritual flourishing, which may or may not be contextualized in theological or religious frameworks. Most importantly for our discussion here, transcendence, connectedness, and enchantment all feature prominently in contemporary research on psychedelics, to which we now turn.

3. The New Science of Psychedelics

Recent years have seen something of a renaissance in scientific research on psychedelics. Dedicated research centres and programmes are now active at leading academic institutions such as Imperial College London, Johns Hopkins University, and Massachusetts General Hospital, among others. An ever-growing list of peer-reviewed journal articles and research studies contribute to a now impressive body of scholarship devoted to understanding the mechanisms and effects of psychoactive compounds on human beings. Biologically, classic psychedelics (namely, LSD and psilocybin) are remarkably safe and physically well-tolerated (Nichols, 2016). They are not addictive and, indeed, have been demonstrated to be strikingly effective at *treating* addiction in controlled, therapeutic settings (Garcia-Romeu, Griffiths and Johnson, 2014; Bogenschutz *et al.*, 2015). Much of the current neuroscientific research on psychedelics has examined the therapeutic potential of these substances in treating a variety of mental health conditions, including major depression, anxiety, and PTSD (Carhart-Harris *et al.*, 2016; Grob *et al*, 2011). The effects have been admittedly dramatic, with psychiatrists, therapists, and neuroscientists highlighting how much more effective psychedelics are than traditional pharmaceutical and therapeutic interventions and treatments (Johnson, *et al.*, 2014). Of course, there is much work to be done on the potential therapeutic use of psychedelics, but existing and emerging research suggests that, when contextualized within therapeutic, controlled settings that prioritize preparation and the long-term work of integration, psychedelic experiences can be remarkably effective healing modalities.

What is currently under-researched are the effects of psychedelics on clinically healthy individuals, who may not suffer from depression or PTSD, but who experience deep existential distress and a lack of spiritual flourishing. Yes, psychedelics can be effective at mitigating certain conditions, but do they contribute to human flourishing, even spiritual flourishing? A growing body of scholars and researchers argue that this is, indeed, a realistic possibility. It is striking to note that some of the most interesting and suggestive studies on psychedelics involve what might be termed 'existential concerns', even if studied in a context of more traditionally clinical diagnoses. For example, the most well-researched therapeutic application of psychedelics involves end-of-life anxiety in terminally ill patients. Numerous studies have now demonstrated that psychedelics have the potential to significantly reduce the fear of death in these patients, sometimes dramatically so (e.g. Grob *et al.*, 2011; Gasser *et al.*, 2014). As Johns Hopkins psychologist William A. Richards writes:

> It is of interest that cancer patients and others who find such profound mystical experiences in their memory banks are not necessarily convinced of personal immortality... Rather, they tend to report a conviction that Eternity, or Infinity, a state of consciousness outside of time, is so unquestionably real to them that it does not much matter one way or another whether the everyday personality survives when the body stops functioning and decomposing. (Richards, 2015, p. 48)

In other words, the therapeutic effects of some psychedelic experiences can be viewed as secondary effects of a more holistic and comprehensive existential encounter. This bodes well for the application of psychedelics for the ends of spiritual flourishing.

At this point in the discussion, we can begin to focus more specifically on the phenomenology of the psychedelic experience itself. It is becoming increasingly clear that the quality and intensity of psychedelic experiences play a determinative role in their measurable long-term effects. More specifically, studies have repeatedly demonstrated that psychedelic experiences deemed 'mystical' are reliable indicators of the experiences' lasting psychological benefits (e.g. MacLean, Johnson and Griffiths, 2011). As Chris Letheby put it, 'there is mounting evidence that classic psychedelics can durably improve quality of life by inducing mystical or spiritual experiences, bolstering the case for them as a potential remedy to neuroexistential anxiety and disenchantment' (Letheby, 2017, p. 629). Of course, the correlation of mystical experiences and psychological benefits does not in itself necessitate a veridical interpretation of the psychedelic experience. It is entirely possible that a mystical psychedelic experience could be metaphysically misleading while still contributing to physical, mental, and spiritual flourishing. What is notable here is the empirical observation that it is the experiential quality and intensity of the psychedelic experience that, at least in part, determines the long-term effects desired by clinicians (and, presumably, the patient or participant herself). If one is interested in applying psychedelics to the pursuit of psychological or spiritual flourishing, then it is of strategic and pragmatic value to pay very close attention to the mystical experience itself.

Mystical experiences, of course, have long been a topic of scholarly discussion and empirical research, and will not be explored in depth here. It is, however, helpful to note the qualities that often attend experiences described as 'mystical', as well as the sorts of insights that tend to result from such experiences. In his groundbreaking *The Varieties of Religious Experience: A Study in Human Nature* (comprised of his 1901–1902 Gifford Lectures at the University of Edinburgh), psychologist William James describes four qualities that may function as general markers of religious experience: ineffability, a noetic quality, transiency, and passivity. Mystical experiences are marked by *ineffability*, with the individual finding herself at an utter loss when it comes to adequately describing what she has experienced. As James poetically describes this, 'One must have musical ears to know the value of a symphony; one must have been in love oneself to understand a lover's state of mind' (James, 1985, p. 302). Second, and perhaps most importantly, there is a *noetic quality* attending mystical states, and 'mystical states seem to those who experience them to be also states of knowledge' (*ibid.*). Those who have had mystical psychedelic experiences often report that these states are far more objective and 'real' than what is usually considered 'normal' consciousness experience. Perhaps less importantly, mystical states are marked by *transiency* — they are time limited in duration, though can have powerful lasting effects, and they are marked by *passivity*, with the individual feeling as though she is being encountered in some way, or is 'along for the ride'. James's markers of mystical experiences are not the only framework which can be applied to these states, of course, but they comprise a useful lens for our discussion here.

Beyond the markers of mystical states more broadly, those having psychedelic experiences often report similar themes in terms of qualitative content. In analysing the 'intuitive knowledge' gained through psychedelic experiences, Richards highlights the centrality of six core insights: (1) God, (2) immortality, (3) interrelationships, (4) love, (5) beauty, and (6) emerging wisdom (Richards, 2015, p. 41). By 'God' Richards does not mean the classical theistic God of Western Christianity, but something more akin to 'Ultimate Reality'. One might name this Ultimate Reality 'Allah', 'Brahman', 'Nothingness', or 'The Ground of Being', but 'a fascinating discovery in speaking with mystics from different world religions is how secure they appear in their certainty of the reality of a sacred dimension of consciousness and how little they care about what words one may choose to describe it' (*ibid.*, p. 41). This sense of transcendent connection is consistent with the spiritual flourishing described above, whether contextualized in purely naturalistic or theological frameworks. Richards' second theme, immortality, is similarly fluid in terms of metaphysical content: by immortality Richards means that 'intuitive conviction that the eternal realms of consciousness are indestructible and not subject to time' (*ibid.*, p. 46). This is not necessarily a conviction that one will live forever, but rather an awareness that one's core self is in some way part of fundamental reality. Similarly, psychedelic experiences often result in an enduring sense of interconnectedness with other humans, animals, and the natural world more broadly: here again the implications for spiritual flourishing are readily apparent. And finally, a profound sense of love, an awareness of beauty, and an emerging wisdom are enduring features of mystical psychedelic experiences, resulting in what James called 'fruits for life' as psychedelic experiences are integrated into one's life and practice in the months and years after the experience (James, 1985, p. 327).

One might be forgiven for ending the discussion here, content with the therapeutic and pragmatic effects of psychedelics. After all, the growing body of research on psychedelics supports the claim that psychedelics are not only safe and beneficial for therapeutic purposes, but hold significant potential for applications in quality of life and emotional and relational well-being. Further, the phenomenological qualities and enduring effects of psychedelic experiences suggest that careful, holistic use of psychedelics within appropriately controlled settings might well have the potential to advance spiritual flourishing: this spiritual flourishing is marked, I have suggested, by connectedness, enchantment, and transcendence, whether understood in naturalistic or theological terms. These tangible benefits are compelling, of course, but they do not address the pressing metaphysical questions that arise in discussion of psychedelics: is there any epistemic value to be gained in psychedelic experiences? Do such experiences actually give us access to a deeper knowledge of reality, or are they merely useful, beneficial illusions? It is this question to which we now turn.

4. Panpsychism and the Epistemic Status of Psychedelic Experiences

It is one thing to acknowledge that psychedelic experiences might be illusory but beneficial for one's spiritual flourishing, and something very different to suggest that psychedelic experiences are 'windows through which the mind looks out upon a more extensive and inclusive world' (James, 1985, p. 339), as James describes the epistemological possibility of mystical states. The possibility of the in-principle *potential* of veridical interpretations of psychedelic experiences looms large in interdisciplinary discussions, with philosophers and theologians in particular beginning to engage with the relevant scientific research in meaningful ways (e.g. Letheby, 2015; Cole-Turner, 2014). What is particularly interesting is the important role of philosophy of mind in these discussions — and, in particular, the centrality of panpsychism in both naturalistic and theological framings of psychedelic experiences. In short, there are natural — but not necessary — connections between panpsychism and the metaphysical possibility of veridical interpretations of psychedelic states, and these panpsychism-informed interpretations are compatible with both naturalistic and theological metaphysical commitments.

Of course, one must adopt extreme caution in applying the term 'veridical' to any experience: mundane, mystical, psychedelic, or otherwise. I do not here suggest that all psychedelic or mystical experiences should be treated as corresponding to actual reality. Clearly this cannot be the case (given the diverse content of such experiences), nor would I advocate a naïve realism about psychedelic experiences in particular, especially in cases where psychiatric pathology is present. As James notes in his own discussion of mystical experiences, 'What comes [from the mystical experience] must be sifted and tested, and run the gauntlet of confrontation with the total context of experience, just like what comes from the outer world of sense. Its value must be ascertained by empirical methods, so long as we are not mystics ourselves' (James, 1985, p. 338). *In discussing the potential for veridical interpretations of psychedelic experiences, I refer only to the hypothesis that one's philosophy of mind and metaphysical framework might allow for the structural possibility that psychedelic states may have epistemological value.* Perhaps it is helpful here to name the specific challenges that arise when defending the claim that a psychedelic experience can give one access to a deeper knowledge of reality. First, there is what may be called the 'authenticity objection'. This objection stems from an intuitive sense (often rooted in theological commitments) that a mystical experience is only 'real' if an individual plays no intentional role in its induction. In other words, psychedelic experiences are not 'real' encounters with Ultimate Reality because the subject has intentionally manipulated her neurobiological state. As this concern is not directly relevant to the argument here, and I have addressed it more fully elsewhere, I will say only that the intentional curation of embodied spiritual practice and experience has been a necessary part of human religion and culture for thousands of years (Ritchie, 2021). Theological systems, too, have the conceptual resources with which to accommodate and encourage self-directed spiritual formation and even fairly intense experiences for theological ends.

More important for our discussion here is the subtler and more insidious influence of a totalizing physicalism that reduces extraordinary conscious experiences to neurological events. There is an influential physicalist ontology that is often operative in discussions about psychedelic states, and this physicalist ontology necessitates that interlocutors reject the epistemic possibilities of psychedelic states. By 'physicalist ontology', I do not mean the *broadly* physicalist affirmation that to be an organism with the sorts of brains and bodies that humans have, in the sorts of environments conducive to human life, just is what it is to be conscious. This broadly physicalist commitment is one to which I happily subscribe: it is not a reductionist commitment but an anti-dualist commitment. To be broadly physicalist, in this sense, is to affirm that, whatever the ontology one ascribes to the fundamental 'stuff' of the physical world, no further 'stuff' needs to be added for consciousness to be a possibility. (This is very similar to a simple monistic impulse, which both panpsychists and reductive physicalists share.)

Rather, the totalizing physicalist ontology at issue here is one that necessitates a dismissal of the epistemological potential of psychedelic states, on the assumption (perhaps unwitting) that consciousness is *reducible* to neural activity in the brain. Or, to put this differently: one can accept that consciousness is fully natural and a 'non-spooky' feature of the physical world, without also accepting that consciousness is reducible to brain activity. If consciousness were reducible to brain activity in relatively complex organisms, then it would indeed be difficult to ascribe any epistemological value to psychedelic experiences. Such experiences would be fully explainable in neurological terms and, because consciousness would be confined to a relatively small number of biological organisms, would have no obvious mechanism by which to illuminate the nature of reality. Psychedelic experiences might, in the reductive physicalist view, be extraordinarily useful and beneficial, but they would have little chance of revealing anything of metaphysical import. Such experiences might even be seen as *spiritually* beneficial, but the metaphysical insights that often accompany them would have to be deemed illusory. In sum: for those interested in the potential epistemic value of psychedelic experiences, the challenge is to frame such experiences in a model where human consciousness 'in here' (i.e. individual consciousness experiences) is ontologically connected to reality 'out there' (i.e. reality as a whole, including the physical world).

Panpsychism, here, becomes evidently relevant. While it is perhaps true that one should not adopt a philosophical position because of its relative compatibility with insights arising from psychedelic experiences, it is not obviously true that transformative experiences should *not* impact one's philosophical position. And, of course, there are persuasive reasons why one might find panpsychism appealing on its own terms, regardless of its conceptual usefulness when debating psychedelic experiences. For the purposes of this discussion, though, panpsychism suggests itself as a promising view due to (1) the emphasis on mind as fundamental to physical reality, and (2) the potential for affirming ontological connections between human minds and Ultimate Reality.

4.1. Mind and nature

First, panpsychists are unapologetic in their claim that mind is fundamental to reality, and, in particular, *physical reality*. Naturally, scholars differ on the details of exactly *how* mind is to be viewed as fundamental to the natural world, but they are in general agreement that 'consciousness is a fundamental and ubiquitous feature of the physical world' itself (Goff, 2019, p. 23). One might join Goff in affirming a Russell-Eddington approach in which 'consciousness is not separate from the physical world; rather consciousness is located in the intrinsic nature of the world' (*ibid.*, p. 174). Or, one might be persuaded by a Whiteheadian panexperientialism in which each actual entity is a centre of real experience, agreeing with Matthew Segall that 'physics has moved beyond the substantialist view of matter' (Segall, 2020, p. 123) implied by talk of intrinsic natures. Still others may be persuaded of the merits of an emergentist variety of panpsychism, perhaps subscribing to Philip Clayton's 'gradualist panpsychism' in which 'increasing complexity across biological evolution brings more and more complex awareness, with human consciousness being the most advanced form of embodied awareness that we have yet discovered' (Clayton, 2020, p. 194). The differences between these varieties of panpsychism are important. For the purposes of the discussion here, however, what is most salient is the basic panpsychist recognition of the ubiquity (even a 'proto' ubiquity; e.g. Chalmers, 2017) of mind in the physical world. If mind (in some form) is viewed as a part of fundamental reality, rather than a mysterious ontological add-on or a radically emergent phenomenon, then it begins to seem at least possible that even our more exceptional conscious experiences are epistemologically illuminating. If consciousness and experience are *normalized* across the physical world, then even psychedelic experiences (and mystical experiences more broadly) could be considered 'part and parcel' of the natural world. If psychedelic experiences are seen as a subset of conscious experience, which is in some sense inherent to the fundamental nature of physical reality, then it becomes at least plausible that such exceptional experiences may be informative about reality. It is not as though the 'real' physical world is somehow 'out there' and ultimately disconnected from mystical experiences, but rather that the 'real' physical world necessarily includes mind (in some form) all the way down.

In this way, panpsychism offers suggestive possibilities for philosophical engagement with research on psychedelics. Panpsychists reject any rigid boundary between consciousness and reality as a whole. As Goff writes, 'on the panpsychist view, the universe is *like us*; we *belong* in it' (Goff, 2019, p. 217). And if the universe itself is fundamentally like us in some meaningful, mind-full way, then one might argue that mystical experiences should also be viewed through this monistic lens. Of course, the possibility of veridical interpretations of psychedelic experiences has long been of interest to philosophically-inclined scholars. In contemporary work, for example, both Alfred North Whitehead and Baruch Spinoza have proved immensely valuable for philosophical engagement with both panpsychism and psychedelics (e.g. Sjöstedt-Hughes, 2015; Skrbina, 2017). Detailed discussion of these various philosophical options will not be undertaken here, but it is important to highlight the salient critique — sometimes explicit — running through such scholars' work. Namely, this: the rejection of the mere veridical *possibility* of mystical states (including, but not limited to, psychedelic experiences) often betrays an implicit (if not explicit) materialist ontology. In one sense, this implicit ontology is forgivable: modern science has accomplished remarkable things by paying attention only to the physical stuff of the world and ignoring subjective experience. But as Goff argues, 'Galileo's error was to commit us to a theory of nature which entailed that consciousness was essentially and inevitably mysterious' (Goff, 2019, pp. 21–2). If Goff is correct, the modern scientific project, at least in its current form, is not equipped to deal with the ontology of conscious experience, precisely because 'Galileo decided that science was not in the business of dealing with consciousness' (*ibid.*, p. 23). Those of us who consider ourselves to be hard-nosed empiricists thus quite naturally adopt an unwitting resistance to ontological claims about conscious experience (psychedelic or otherwise), as our inherited philosophy of science does not afford us the requisite tools to make sense of it. We may acknowledge that our conscious experience *represents* physical reality to some extent, but rarely do we think of our conscious selves as being wholly *like* the rest of the physical world. It is entirely possible that a comprehensive understanding of exceptional conscious experiences is a scientific possibility — but this is a possibility that is likely to be realized only through an expansion of the current scientific paradigm. Godehard Brüntrup and Ludwig Jaskolla describe this as the idea that 'physical structure as described in the formalized language of physics cannot by itself provide the ultimate grounding of reality but rather needs to be complemented by nonstructural intrinsic facts which escape the vocabulary of physics' (Brüntrup and Jaskolla, 2017, p. 1). Or, more simply, 'we must somehow find a way of making consciousness, once again, the business of science' (Goff, 2019, p. 23). Panpsychism reframes the ontology of nature itself, normalizing mind — including extraordinary conscious experiences — as of a piece with fundamental reality.

4.2. Mind and Ultimate Reality

Second, panpsychism is a constructive position for those interested in veridical interpretations of psychedelic experiences in so far as it provides a philosophical structure for affirming ontological connections between humans and Ultimate Reality. Again, by 'Ultimate Reality' I do not refer necessarily to a theistic version of God. Ultimate Reality may be framed in terms of God, a higher power, the Ground of Being, Brahman, or some other purely naturalistic conception (for example, 'The Nameless') (Richards, 2015, p. 42). Those having mystical psychedelic experiences routinely report being encountered by Ultimate Reality in a profound, transformative, truly ineffable manner. Further, these experiences are often described as 'unitive experiences', involving a weakening or dissolution of the rigid boundaries between the self, others, nature, and Ultimate Reality (e.g. Pollan, 2018, chapter 6). More specifically, mystical psychedelic experiences often involve the experience of being immersed in, or even one with, Ultimate Reality (e.g. Richards, 2015, chapter 5; Griffiths *et al.*, 2006). This is an uncomfortable truth for those of us who are naturalistically inclined, as it seems to link the transformative spiritual and therapeutic effects of psychedelics with a supernaturalist metaphysic that we are reluctant to affirm. As Letheby laments, 'some evidence suggests that the existential reenchantment occasioned by psychedelics depends crucially on the induction of mystical experiences involving apparent encounters with transcendent non-natural levels of reality… If this is so, then psychedelics would seem less a means to making peace with a naturalistic worldview than a means to becoming persuaded of its falsity' (Letheby, 2017, p. 624). Of course, many will have no issues accepting a religious, transcendent depiction of reality, but others seek naturalistic ways of affirming the epistemological value of psychedelic experiences with Ultimate Reality. As Richards describes this, 'a reasonably well-integrated, stable person with articulate verbal skills who recalls such a profound unitive state of consciousness, were he to speak at all, might own the ultimate insight that in the final analysis the energy that makes up his life is "God"' (Richards, 2015, p. 45). It is evident that 'God', used in this sense, can be framed within theological or naturalistic models. What is necessary, then, are metaphysical frameworks that make sense of the experiential encounter with Ultimate Reality occasioned by psychedelics, without dismissing the experience as illusory on one hand, or inherently supernatural on the other.

One way to do this is to argue that the neurochemical mechanisms of psychedelics temporarily *inhibit* the illusory effects of normal cognitive processes, thereby giving subjects increased epistemic access to reality. This is the position put forward by Letheby, whose aim within the philosophy of psychedelics is to develop a fully naturalistic framework for understanding psychedelic experiences as epistemically valuable. He affirms that 'there are key elements of psychedelic spirituality that can be practiced by a naturalist in intellectual good faith, and that represent a promising path to reenchanting the naturalistic world on its own terms' (Letheby, 2017,

p. 624). Letheby recognizes the centrality of mystical encounters with Ultimate Reality in the most transformative psychedelic experiences, but argues that these experiences reveal something about our own cognitive fictions, rather than the existence of a supernatural God. Specifically, Letheby highlights research suggesting that psychedelics act (at least in part) by decreasing activity in the default mode network (DMN), a neural network that seems to be involved in generating a sense of self and cognitive rigidity (in addition to other activities that seem to contribute to human misery in varying degrees) (Letheby, 2015). Robin Carhart-Harris and his colleagues suggest that down-regulation of the DMN is linked to the ego-dissolution and mind-expansion so germane to mystical psychedelic experiences (Carhart-Harris *et al.*, 2014). And if Thomas Metzinger is correct that the 'self' is essentially a useful fiction (an entirely arguable but plausible hypothesis), then a psychedelic experience in which one feels connected with all reality, even Ultimate Reality, might actually represent the diminution of a cognitive distortion (Metzinger, 2003). As Letheby writes, subjects 'gain experiential knowledge of the contingency of their own sense of self by experiencing its temporary subtraction from their phenomenal space' (Letheby, 2015, p. 187). In this way, then, a psychedelic experience involving a sense of one's connectedness with Ultimate Reality might represent a more accurate metaphysical picture than that afforded us by 'normal' everyday consciousness.

This is where engagement with panpsychism becomes particularly constructive. Mystical psychedelic experiences often involve a sense of encountering Ultimate Reality, and this encounter is, in turn, often experienced as a weakening or dissolution of the rigid boundaries between oneself and the rest of reality. This phenomenological reality is suggestive, in so far as it entails potential overlap with central features of panpsychism. Panpsychism, in its many forms, entails the claim that, at the very least, human minds are ontologically connected to all of nature: in the words of Arthur Eddington, 'the stuff of the world is mind-stuff' (Callaway, 2014, p. 274). This raises the vital question of whether human minds represent distinct entities (selves), or whether it is more accurate to say that all conscious selves are really part of, or a manifestation of, a single universal mind. David Skrbina poses this question thus: 'The central issue here is whether we speak of such mind as "mind of single universal" (God, the Absolute, the World Soul, and so on) or of mind as attributable to each thing in itself (of each object's possessing its own unique, individual mind).

The former view would be a monist conception of mind, the latter a pluralist concept' (Skrbina, 2005, p. 21). I will not address the details of this debate here, but suggest that either metaphysical possibility is in-principle compatible with the claim that mystical psychedelic experiences *may* yield knowledge about the way things 'really are'. For example, a monist version of panpsychism is remarkably consistent with unitive psychedelic experiences that include ego-dissolution and a sense of oneness with Ultimate Reality. Alternatively, a pluralist approach to panpsychism allows for an easy framing of mystical experiences in which one experiences being in intimate connection with Ultimate Reality, without being identified with Ultimate Reality.

Both options are compatible, in principle, with purely naturalistic as well as theological metaphysical views. For example, naturalist panpsychists might draw on the radical monism of Baruch Spinoza, for whom 'all of reality consists of a single substance, called "God" or "Nature" depending on the context and circumstances... God was not a transcendent being, not a personal being, not a moral being, but simply the totality of existence' (Skrbina, 2005, p. 106). Within a Spinozan framework, unitive psychedelic experiences might well be interpreted as veridical: conscious experience is intrinsically 'of a piece' with Ultimate Reality, and 'normal' perceptions to the contrary are illusory. Other naturalists may prefer a radically pluralist panpsychist account that preserves the integrity of distinct conscious selves. Whitehead's insistence on the existence of a great many 'actual entities' (AEs) is compelling here. Clayton explains that Whitehead's approach to AEs 'requires us to think of each such moment of creative becoming as a separate entity or occasion, existing on its own. Of course, one can be a radical pluralist in this way and still hold that AEs are so interdependent that they are internally related. This would mean a radical pluralism of psyches' (Clayton, 2020, p. 193). A Whiteheadian naturalist, then, can affirm the panpsychist commitment to mind-as-fundamental, while still prioritizing the ontological status of individual conscious entities, or moments of creative becoming. This, in turn, is entirely consistent with psychedelic experiences in which one is encountered by Ultimate Reality, without being dissolved into that Ultimate Reality. Segall summarizes this tension well in his description of Whitehead's panpsychism: 'There is a universal soul, a psyche of the cosmos, a primordial actuality or God of this world, *and* there are countless creatures creating in concert with it' (Segall, 2020, pp. 127–8). My intention here is not to adjudicate between radically pluralist and radically monist approaches to panpsychism, but to demonstrate that there are naturalistic metaphysical options for those wishing to take seriously the epistemic possibilities of research into psychedelic experiences.

More explicitly theological possibilities are also worth mentioning. There are a variety of theistic naturalisms that prioritize empirical enquiry, motivated by a naturalistic impulse and seeking what Clayton has called 'maximum traction' between theological ideas and scientific research (Clayton, 2008, pp. 54f.). For example, *panentheism* has emerged as a position that is remarkably resonant with both panpsychism and the possibility of veridical interpretations of psychedelic experiences. Panentheism is a metaphysical position stating that the entire natural world exists within God, but also that God is, in some sense, more than the natural world. What panentheists share is what Niels Henrik Gregersen describes as '*the intuition of a living two-way relation between God and world, within the inclusive reality of God*' (Gregersen, 2004, p. 22). For the panentheist, a complete description of nature will necessarily include the *truly* natural relationship between nature and Ultimate Reality, and 'there is no "place outside" the infinite God in which what is created could exist' (Peacocke, 2007, p. 22).

It is this enhanced naturalism that makes panentheism such a rich framework for engaging with both panpsychism and psychedelic experiences. In fact, the potential consonance between panpsychism and panentheism is so suggestive that at least one edited volume has published on the subject, in addition to special issues and individual journal articles (e.g. Brüntrup, Göcke and Jaskolla, 2020). Brüntrup, Göcke and Jaskolla describe this suggestive overlap thus:

> Although panpsychism and panentheism prima facie refer to different areas of subject matter, they exhibit astonishing structural similarities: Both panpsychistic and panentheistic approaches generally mediate between dualistic and monistic theories by avoiding a complete ontological separation of God and the world, or mind and matter. The rejection of reductionism and the legacy of unity metaphysics can also be seen as common ground. God and world, as well as mind and matter, are regarded as different but nevertheless intrinsically related to each other. (*ibid.*, p. 1)

Of course, all this is rather vague, and Clayton cautions that, in this discussion, 'one immediately recognizes that the relationships between panpsychism and panentheism are rather more complex than one might have thought. There are no simple entailments' (Clayton,

2020, p. 94). Quite a lot hinges on what it means for an entity, or for Ultimate Reality, to be mental or mind-full. I will not explore the nuances of this debate here, but simply highlight that, at the very least, there are rich theological possibilities for the panpsychist who is so inclined. These theological possibilities include the potential for veridical interpretations of psychedelic experiences. If it is at least possible that God, or Ultimate Reality, is ontologically connected to all of nature in mind-full way, then mental experiences (including psychedelic experiences) of that Ultimate Reality can at least be plausibly framed as theologically sound and epistemically valuable. Panentheism bridges the metaphysical chasm between God and nature, and panpsychism affirms that mind is fundamental to nature in some way. While one may have reasons to reject panentheism, panpsychism, and/or a veridical interpretation of psychedelic experiences, there are at least conceptual resources available for those wishing to affirm that such mystical experiences truly do provide 'windows through which the mind looks out upon a more extensive and inclusive world' (James, 1985, p. 339).

5. Conclusion

Often marked by a sense of transcendence, connectedness, and enchantment, spiritual flourishing is an holistic pursuit that escapes easy definition and, importantly, need not be framed in religious terms (though, of course, it *can* be). I have here examined the potential of psychedelic experiences to contribute to spiritual flourishing, asking whether there are ways to interpret such experiences as veridical: yielding ontological insights that lead to new knowledge and new ways of existing within the world. Panpsychism here proves particularly suggestive, for the panpsychist insists that even the most extraordinary conscious experiences are fundamentally connected to all of reality. All conscious experiencing, including psychedelic experiencing, shares its ontology with the whole of the natural world. This being the case, it is at least plausible to suggest that mystical psychedelic experiences are able to yield insights that *may* be rooted in fundamental reality, even Ultimate Reality. This is a discussion of metaphysical possibilities, but not necessities.

The real work of spiritual flourishing lies ahead: a 'eudaimonic programme' would develop what Flanagan describes as 'a framework for thinking in a unified way about philosophical psychology, moral and political philosophy, neuroethics, neuroeconomics, and positive psychology, as well as about transformative mindfulness practices that have their original home in non-theistic spiritual traditions' (Flanagan, 2007, p. 4). But, if successful, an open and interdisciplinary exploration in this area might well lead to the development of (and access to) embodied, transformative pathways of spiritual flourishing in an increasingly disenchanted world. This, I suggest, is exactly the sort of flourishing described by Goff in the closing pages of his book: 'Panpsychism offers a way of "re-enchanting" the universe... the universe is *like us;* we *belong* in it. We need not live exclusively in the human realm, ever more diluted by globalization and consumerist capitalism. We can live in nature, in the universe. We can let go of nation and tribe, happy in the knowledge that there is a universe that welcomes us' (Goff, 2019, p. 217).

References

Bogenschutz, M.P., Forcehimes, A.A., Pommy, J.A., Wilcox, C.E., Barbosa, P.C.R. & Strassman, R.J. (2015) Psilocybin-assisted treatment for alcohol dependence: A proof-of-concept study, *Journal of Psychopharmacology*, **29**, pp. 289–299.

Brüntrup, G. & Jaskolla, L. (2017) Introduction, in Brüntrup, G. & Jaskolla, L. (eds.) *Panpsychism: Contemporary Perspectives*, pp. 1–16, New York: Oxford University Press.

Brüntrup, G., Göcke, B.P. & Jaskolla, L. (2020) *Panentheism and Panpsychism: Philosophy of Religion Meets Philosophy of Mind*, Leiden: Brill.

Callaway, H.G. (2014) *Arthur S. Eddington, The Nature of the Physical World: Gifford Lectures of 1927, An Annotated Edition*, Newcastle-upon-Tyne: Cambridge Scholars Publisher.

Carhart-Harris, R.L., et al. (2014) The entropic brain: A theory of conscious states informed by neuroimaging research with psychedelic drugs, *Frontiers in Human Neuroscience*, **8**, pp. 1–22.

Carhart-Harris, R.L., et al. (2016) Psilocybin with psychological support for treatment-resistant depression: An open-label feasibility study, *Lancet Psychiatry*, **3**, pp. 619–627.

Caruso, G.D. & Flanagan, O. (2018) Neuroexistentialism: Third-wave existentialism, in Caruso, G.D. & Flanagan, O. (eds.) *Neuroexistentialism: Meaning, Morals, & Purpose in the Age of Neuroscience*, pp. 1–22, New York: Oxford University Press.

Chalmers, D.J. (2017) Panpsychism and panprotopsychism, in Brüntrup, G. & Jaskolla, L. (eds.) *Panpsychism: Contemporary Perspectives*, pp. 19–47, New York: Oxford University Press.

Clayton, P. (2008) *Adventures in the Spirit: God, World, Divine Action*, Simpson, Z. (ed.), Minneapolis, MN: Fortress Press.

Clayton, P. (2020) Varieties of panpsychism, in Brüntrup, G., Göcke, B.P. & Jaskolla, L. (eds.) *Panentheism and Panpsychism*, pp. 191–203, Leiden: Brill.

Cole-Turner, R. (2014) Entheogens, mysticism, and neuroscience, *Zygon*, **49**, pp. 642–651.

Flanagan, O. (2007) *The Really Hard Problem: Meaning in a Material World*, Cambridge, MA: MIT Press.
Garcia-Romeu, A., Griffiths, R.R. & Johnson, M.W. (2014) Psilocybin-occasioned mystical experiences in the treatment of tobacco addiction, *Current Drug Abuse Reviews*, **7**, pp. 157–164.
Gasser, P., Holstein, D., Michel, Y., Doblin, R., Yazar-Klosinski, B., Passie, T. & Brenneisen, R. (2014) Safety and efficacy of lysergic acid diethylamide-assisted psychotherapy for anxiety associated with life-threatening diseases, *Journal of Nervous and Mental Disease*, **202**, pp. 513–520.
Goff, P. (2019) *Galileo's Error: Foundations for a New Science of Consciousness*, New York: Vintage.
Gregersen, N.H. (2004) Three varieties of panentheism, in Clayton, P. & Peacocke, A. (eds.) *In Whom We Live and Move and Have Our Being: Panentheistic Reflections on God's Presence in a Scientific World*, Grand Rapids, MI: Eerdmans.
Griffiths, R.R., Richards, W.A., McCann, U. & Jesse, R. (2006) Psilocybin can occasion mystical-type experiences having substantial and sustained personal meaning and spiritual significance, *Psychopharmacology*, **187**, pp. 268–283.
Grob, C.S., Danforth, A.L., Chopra, G.S., Hagerty, M., McKay, C.R., Halberstadt, A.L. & Greer, G.R. (2011) Pilot study of psilocybin treatment for anxiety in patients with advanced-state cancer, *Archives of General Psychiatry*, **68**, pp. 71–78.
James, W. (1985) *The Varieties of Religious Experience*, The Works of William James, Vol. 13, Cambridge, MA: Harvard University Press.
Johnson, M.W., Garcia-Romeu, A., Cosimano, M.P. & Griffiths, R.R. (2014) Pilot study of the 5-HT(2A)R agonist psilosybin in the treatment of tobacco addiction, *Journal of Psychopharmacology*, **28**, pp. 983–992.
Julian of Norwich (2015) *Revelations of Divine Love*, Windeatt, B. (trans.), New York: Oxford University Press.
Letheby, C. (2015) The philosophy of psychedelic transformation, *Journal of Consciousness Studies*, **22** (9–10), pp. 170–193.
Letheby, C. (2017) Naturalizing psychedelic spirituality, *Zygon*, **52**, pp. 623–642.
MacLean, K.A, Johnson, M.W. & Griffiths, R.R. (2011) Mystical experiences occasioned by the hallucinogen psilocybin lead to increases in the personality domain of openness, *Journal of Psychopharmacology*, **25**, pp. 1453–1461.
Metzinger, T. (2003) *Being No-One: The Self-Model Theory of Subjectivity*, Cambridge, MA: MIT Press.
Nichols, D.E. (2016) Psychedelics, *Pharmacological Reviews*, **68**, pp. 264–355.
Peacocke, A. (2007) *All That Is: A Naturalistic Faith for the 21st Century*, Minneapolis, MN: Fortress Press.
Pollan, M. (2018) *How to Change Your Mind: The New Science of Psychedelics*, London: Penguin Press.
Richards, W.A. (2015) *Sacred Knowledge: Psychedelics and Religious Experiences*, New York: Columbia University Press.
Ritchie, S.L. (2021) Integrated physicality and the absence of God: Spiritual technologies in theological context, *Modern Theology*, **37**, pp. 296–315.
Segall, M. (2020) The varieties of physicalist ontology: A study in Whitehead's process-relational alternative, *Philosophy, Theology, and the Sciences*, **7**, pp. 105–131.

Sjöstedt-Hughes, P. (2015) *Noumenautics*: *Metaphysics — Meta-ethics — Psychedelics*, London: Psychedelic Press.
Skrbina, D. (2005) *Panpsychism in the West*, Cambridge, MA: MIT Press.

Philip Goff[1]

Putting Consciousness First

Replies to Critics

I am honoured and humbled that these 21 incredible scientists and thinkers have responded to my work.[2] These essays have both challenged and stimulated me. There is much disagreement, as there should be in these matters about which there is little consensus. But there are also many of us who are journeying to the same location, albeit via different paths.

This also gives me an opportunity to sketch in more detail what a post-Galilean science of consciousness, one in which consciousness is taken to be a fundamental feature of reality, might look like. It's time non-reductionists about consciousness stopped spending all their time justifying their existence and got on with building an interdisciplinary research programme to rival the dominant materialist paradigm. In *Galileo's Error*, I argue that the problem of consciousness is rooted in the philosophical foundations of science. It follows that if we want to solve the problem, we need to rethink what science is. This will involve the joined-up efforts of physicists, philosophers, neuroscientists, and many others. In bringing together non-reductionists about consciousness from so many different fields, this special issue takes an important step towards that goal. It's time to build the post-Galilean paradigm.

Correspondence:
Email: philip.a.goff@durham.ac.uk

---

[1] Durham University, UK.
[2] Unfortunately it was not possible to write a reply to Delafield-Butt's fascinating essay, due to last minute submission.

1. Replies to Scientists

1.1. Reply to Carlo Rovelli

Carlo Rovelli's proposal is ingenious. However, I'm not persuaded it provides an adequate solution to the hard problem of consciousness. There are two aspects of consciousness that give rise to a hard problem: qualitivity and subjectivity. Qualitivity consists in the fact that experiences involve *qualities*: the smell of coffee, the taste of mint, the deep red you experience as you watch the setting sun. Subjectivity consists in the fact that these experiences are *for* someone: there is something that it's like *for me* to experience that deep red.

These two aspects of consciousness give rise to two 'hard problems'. Either would, in my view, be sufficient to refute materialism. But the hard problem of qualitivity is more pronounced, which is why I tend to focus on it. Physical science tells us a purely quantitative story of causal structures. In that kind of vocabulary, you simply can't articulate the qualities of our experience. That's an *expressive* limitation of the language of physics. But I think it entails an *explanatory* limitation. Because if I wanted to reductively explain the redness of a red experience in purely physical terms, my theory would have to articulate the redness and then explicate it terms of more basic physical structures. If physical theory can't even articulate the redness, then it can't reductively explain it. (Of course, we can capture in quantitative terms the location of redness in the similarity space of colour along the dimensions of hue, lightness, and saturation, but that kind of purely structural information doesn't fully convey the redness of a red experience.)

The hard problem of subjectivity is a very similar challenge, but more subtle. You can't articulate in the language of causal structures the hypothesis that a certain physical system has *experience*. Many materialists have tried to get around this by articulating in the language of causal structures something that *sounds a bit like* experience. We can talk, for example, about how a certain physical system is 'sensitive' to certain features of the environment, say, the red colour of objects, in the sense that it's causally set up to track red objects or respond in some way to the fact that things are red. That sounds a bit like saying the system is experiencing red. But of course it's not the same thing at all. The fact that a system causally-mechanically responds to a certain feature of reality doesn't entail that it has any experience of that feature. Another popular option has been to talk about the capacity of system to monitor its internal states, which can obviously be captured in causal-structural terms. That sounds a bit like the system is 'aware' of its internal states. Well, it depends what you mean by 'awareness'. If you just mean 'causally responds to', then, yes, the system is aware. But if you mean that the system 'experiences' its internal states, then this is no way implied by its causal responsiveness to its inner states.

Rovelli's position is a highly novel form of this old strategy of describing something that sounds a bit like experience, and then suggesting that this solves — or at least helps with — the hard problem. As Rovelli says, his favoured interpretation of quantum mechanics is essentially 'perspectival'. Having a 'perspective' seems to imply having experience, or something on the way towards it. But this is just a reflection of ambiguity in the word 'perspective'. We sometimes say that something 'has a perspective' to mean that there is something that it's like to be it, that it *experiences*. But when in relational quantum physics we say that a system 'has a perspective', we mean something quite different: that the theory cannot be applied to systems in isolation but only in relation to each other. Accepting that physical systems 'have perspectives' in the relational quantum physics sense is totally consistent with physical systems not 'having perspectives' in the experiential sense.

I don't think Rovelli would deny this. After all, he's not saying his view is literally a form of panpsychism. But if we haven't managed to articulate, and thereby explain, subjectivity in the terms of physical science, then we haven't addressed the hard problem. And even if relational quantum mechanics could close the 'subjectivity gap' — the gap between the processes of physical science and the having of experience — this would still leave the 'qualitivity gap' — the gap between the quantitative features of physical science and the qualities of experience — as wide as ever.

I'm not sure Rovelli would disagree with this either, as much of his article is devoted to rejecting the idea that there is a hard problem in the first place, arguing that 'what we can "conceive" depends on the conceptual structure we have, and this keeps changing and includes a great deal of presuppositions, sometimes wrong'. I am totally on board with this when it comes to anything other than consciousness. The unique thing about consciousness science is that we have a fundamental explanandum that does not come from public observation and experiments, but from the immediate awareness each of us has of the qualities of our experience (as well as the fact that we are experiencing those qualities). Introspection is fallible in all sorts of ways. However, the basic concepts we use to articulate the fact that we experience, and that our experience involves qualities, are not subject to scientific revision in the way the concepts of time, space, and solidity are. Why not? Because we know that we have experience, and that our experience involves qualities, with a greater certainty than we know any empirical fact.

Of course, one could reject all this, and follow Daniel Dennett in holding that there is no first-person explanandum for a science of consciousness, that the only task for any kind of science is to explain the facts of public observation and experiment. But in that case, there is no hard problem, and hence no need to employ relational mechanics to address the hard problem. At the end of the day, I don't see stable middle ground between Dennett and myself.

1.2. Reply to Sean Carroll

Sean Carroll worries that any view according to which consciousness is a fundamental feature of reality will end up requiring modifications to a well-confirmed scientific theory, namely the Core Theory. However, panpsychists comes in two varieties — weak and strong emergentist — and the former kind agree with Carroll that *biological* consciousness is weakly emergent (from more fundamental forms of consciousness). I will argue that both weak and emergentist panpsychist avoid the challenges Carroll raises.

1.2.1. Weak emergentist panpsychism

The Core Theory describes the causal dynamics of fields and particles, as determined by their physical properties, such as mass, spin, and charge. According to Russellian panpsychism, those physical properties are forms of consciousness. Hence, according to panpsychism, the Core Theory describes the causal dynamics of fields and particles *as determined by the forms of consciousness they instantiate*. Rather than physics not leaving room for consciousness to have an impact, the entire story of physics is the story of what consciousness does. Of course, when you're doing physics, you don't know that's what you're studying. But that's just because physics is only concerned about causal dynamics and abstracts away from the nature of the things underlying those dynamical structures. Doing physics is like playing chess when you don't know what the pieces are made of.

Carroll says panpsychists 'analogize' consciousness to electric charge. That's not quite right. According to panpsychism, charge *is* a form of consciousness. Carroll objects that 'electric charge is a paradigmatic example of a property with dynamical consequences; placed in an electric field, particles with opposite charges move in opposite directions'. The fact that electric charge has dynamical consequences is totally consistent with its being a form of consciousness; it just follows that that form of consciousness has dynamical consequences, it does stuff. Carroll also worries that the charge of an elementary particle is unchanging throughout its entire existence. But the experience of a human being is constantly changing due to the constantly changing processes of the brain; there is nothing incoherent in the idea of a very simple mind with unchanging experience.

Carroll then raises two possibilities: either there is a one-to-one correspondence between physical properties and elementary forms of consciousness or there isn't. The Russellian panpsychist will certainly opt for the former, given that physical properties just are forms of consciousness. But Carroll now worries that the panpsychist is proposing the following causal chain:

> Physical property *causes* mental property *causes* behavioural impact.

Or

> Physical property (upon which mental property supervenes) *causes* behavioural impact.

In either case, Carroll claims, this is 'functionally equivalent to' and '[f]or all intents and purposes… equivalent to…'

> Physical property causes behavioural impact.

The true panpsychist view is captured by neither of these. There is no causal relation between the physical property and the mental property, as they are one and the same thing. And if anything, the supervenience relation goes in the other direction: a certain form of consciousness is correctly classed as 'charge' in virtue of its dynamical consequences. Carroll is right that these descriptions are 'functionally equivalent', if that just means that, from the perspective of a science that is only interested in behaviour, they come out the same. But to assume that such a science of behaviour captures the full story of what is going on is to beg the question against panpsychism. According to panpsychism, physics is so useful because it just focuses on behaviour, abstracting away from the real nature of the properties being studied.

1.2.2. Strong emergentist panpsychism

Strong emergentist panpsychism, as the name suggests, combines panpsychism with the kind of strong emergence Carroll discusses in his contribution. As strong emergentists, they believe that certain complex systems, such as conscious brains, have novel causal capacities that could not be predicted from knowledge of their basic components. Consider a superintelligence, of the kind imagined by Simone Pierre Laplace, who has total knowledge of the particles and fields covered by the Core Theory at time t, and tries to work out the state of my brain at t+1, solely on the basis of the Core Theory. If strong emergence is true, that superintelligence will make some false predictions about the locations of the particles in my brain at t+1, as it is relying entirely on the Core Theory and is ignorant regarding the contribution of the emergent causal capacities of my brain.

Crucially, however, this does not entail that strong emergentists are modifying the Core Theory. They can instead take the Core Theory to be complete on its own terms, i.e. as a complete theory of the inherent causal capacities of particles and fields. However, if there are strongly emergent wholes, these wholes themselves have irreducible capacities, capacities which *complement* the basic causal capacities of particles and fields. When these strongly emergent wholes come into being, they *co-determine* the evolution of the universe, in conjunction with the basic causal capacities of particles and fields (where the latter are perfectly captured by the Core Theory).

Carroll is right that the strong emergentist is obliged to do some serious theoretical work. But this theoretical work need not be conceived of as modifying the Core Theory, but rather as explaining how the causal capacities of strongly emergent wholes interact with the causal capacities of particles/fields to co-determine what will happen. Understanding strong emergence in this way gives us a response to Carroll's novel argument that 'based on purely physical grounds rather than consciousness-based motivations, our expectation that the laws of quantum field theory might break down in biological organisms would be very low indeed'. Maybe so, but we should think of strong emergence not as quantum field theory breaking down but as a new neurobiological theory kicking in. And the place to look when emergent neurobiological principles kick in is not physics but neurobiology.

It seems to me an open empirical question whether or not strong emergentism is true. Some philosophers (Papineau, 2001) have put forth an inductive argument against strong emergence: if there was strong emergence in the brain, we probably would have found it by now; therefore, it probably doesn't exist. In my early work, I assumed the soundness of this kind of argument, and used it as part of a defence of weak emergentist panpsychism. However, the more I talk to neuroscientists, the more I'm inclined to think the first premise of this inductive argument is false, as we simply don't know enough about the workings of the brain. We know a great deal about the basic chemistry of the brain: neurotransmitters, actions potentials, calcium chambers, etc. And we know a fair bit about the large-scale functions of the brain. However, we know very little about how these large-scale functions are realized at the cellular level (Cobb, 2020), i.e. about how the brain *works*. Because we have little clue about how the functions of the brain are realized, we have little clue about whether they are entirely realized by the basic electrochemical processes we understand so well, or whether at some point new causal principles kick in which, in conjunction with the electrochemistry, realize the brain's functions.

People get very excited about brain scans, but in fact they are very low resolution. Each pixel of an fMRI image corresponds to 5.5 million neurons, between 2.2 and 5.5×10^{10} synapses, 22 km of dendrites, and 220 km of axons (Logothetis, 2008). We are only 70% of the way through putting together a complete connectome of a maggot's brain, with its 10,000 neurons (Cobb, 2020, p. 257). The idea that we know enough about the workings of the human brain with its 86 billion neurons to know whether or not its workings involve strong emergence is not credible.

As we uncover more about the workings of the brain, if there does turn out to be strong emergence, this would provide a crucial way of making progress on identifying the neural correlates of consciousness. Matthias Michel (2019) has argued that for 150 years the science of consciousness has been having the same debates about where to locate consciousness, without significantly narrowing down the options. In my view, the root of these difficulties is that consciousness is not a publicly observable phenomenon. We are totally reliant on the external signs of consciousness, such as report. But the 'detection procedures' which underlie inference from external markers of consciousness to consciousness itself are controversial, and arguably impossible to justify empirically. For example, there is much debate over so-called 'overflow': the thesis that there are conscious experiences we are unable to attend to. If the overflow thesis is true, then it does not follow from the fact that somebody reports not having a given experience that they did not in fact have that experience. One manifestation of these difficulties is the perennial dispute (Odegaard, Knight and Lau, 2017; Boly *et al.*, 2017) over whether consciousness is located at the back or the front of the brain.

But if there is strong emergence, then it will show up empirically. There will be functions of the brain that are not realized solely in the underlying electrochemistry. To oversimplify: suppose it turned out that these irreducible functions were found in the front but not the back of the brain. This would provide powerful evidence that the seat of consciousness is located in the front of the brain.

It is notable that there are scientists exploring non-reductionist models. The neuroscientist Kevin Mitchell's (2019) working assumption is that there are strongly emergent dynamics in the brain, and he seeks ways of modelling these emergent dynamics. And Martin Picard's Mitochondrial PsychoBiology Laboratory model the behaviour of mitochondria in the brain as social interactions rather than attempting to reduce them to underlying chemistry. These are the kinds of research programmes non-reductionists about consciousness can build on in shaping the post-Galilean alternative to the materialist paradigm.

1.3. Reply to Marina Cortês, Lee Smolin, and Clelia Verde

Much of my reply to Carroll above doubles as a reply to Cortês, Smolin and Verde. Like Carroll, Marina Cortês, Lee Smolin, and Clelia Verde interpret Russellian panpsychism as holding that consciousness is 'not involved in the dynamical laws', and hence 'it can't matter one whit whether they are there are not'. But, as I responded to Carroll, Russellian panpsychists do not hold that the physical properties do all the work whilst micro-consciousness sits along for the ride. The Russellian view is that micro-consciousness is *all that exists* at the fundamental level, and hence micro-consciousness is doing the work if anything is. In referring to 'mass', 'spin', and 'charge', physics is — unbeknownst to itself — referring to kinds of micro-consciousness. Furthermore, as I also press in my response to Carroll, even if biological consciousness does make a fundamental causal impact on the universe, that doesn't entail modifying fundamental physics.

Having said all that, this is a fascinating proposal. Cortês, Smolin and Verde are doing precisely what Carroll suggests is required of a strong realist about consciousness: modifying physics in order to make space for consciousness to play a fundamental role in the causal evolution of reality. I would love to see the empirical implications spelt out in more detail, especially those pertaining to the neural correlates of consciousness. I discussed in my reply to Carroll how strong emergence may allow us to make progress on, say, the debate as to whether consciousness is in the back or the front of the brain. It's an intriguing prospect that Cortês, Smolin and Verde's proposal may be another way, e.g. if there turn out to be more events without precedent in the front of the brain. This is exactly the kind of interdisciplinary research programme we non-reductionists need to be formulating in building the post-Galilean paradigm. It's also exciting that the post-Galilean approach may lead to new insights in the ongoing attempts to understand quantum mechanics, or to reconcile quantum mechanics with general relativity (*cf.* my reply to Aleksiev below).

The other fascinating aspect of Cortês, Smolin and Verde's essay is the discussion of time. They worry about reducing temporal flow and causation to static, eternal truths. I'm a bit more agnostic on this issue. The reality of 'real time' is not an undeniable datum in the way the reality of consciousness is. I tentatively sketched an argument against four-dimensionalism (the view that all moments of time exist equally and hence time doesn't really pass) in my academic book *Consciousness and Fundamental Reality* (2017). In the universe as the four-dimensionalist conceives of it, there is nothing which (a) persists through time, and (b) has the kind of conscious experience we pre-theoretically associate with humans and other animals. According to four-dimensionalism, each person is associated with a four-dimensional 'space-time worm' stretching out over their entire life history. But if those space-time worms are conscious, they have a very weird kind of consciousness consisting of a unified experience of a whole life. The momentary stages of the space-time worms are the things that have 'ordinary' conscious experience, but these things are forever located at one temporal location and do not persist through time. This argument pressures the four-dimensionalist to deny a lot of common sense, but I suspect many of them would be happy to do so (maybe not Carroll, who is keen to preserve the Manifest Image).

1.4. Reply to Anil Seth

I'm so glad Anil Seth was able to contribute at the last minute. He gives a rigorous and compelling defence of a popular way of defending materialism, and this gives me a chance to clarify my opposition to materialism in response.

Consciousness is not a normal scientific phenomenon. In general, the aim of science is to account for the data of public observation and experiment. But in the case of consciousness, the phenomenon we are trying to account for is not publicly observable. You can't look into someone's head and see their feelings and experiences. We know that consciousness exists not from observation or experiment but from our immediate awareness of our own feelings and experiences. Of course, science is used to dealing with unobservables but, in all other cases, we postulate unobservables in order to explain what can be observed. In the unique case of consciousness, *the thing we are trying to explain* is not publicly observable.[3]

Moreover, the qualities we know about via immediate awareness of our experience cannot be described in the purely quantitative language of physical science. You can't convey to a colour-blind neuroscientist what it's like to see red. Do I just mean you can't get a colour-blind neuroscientist to have a red experience by reading neuroscience? No, it's more than that. There is *information* that we get from attending to our experience, information that cannot be conveyed in the language of physical science. That descriptive limitation entails an explanatory limitation, as I explain in my reply to Rovelli above.

I therefore totally reject Seth's characterization of the hard problem as simply the challenge of explaining how consciousness exists in the first place, as though everyone agrees that we have good materialist explanations of the qualitative character of specific experiences. No: in order to give a materialist account of the qualitative character of red experiences, you'd have to be able to convey that qualitative character in the language of physical science, and that simply can't be done (see my reply to Rovelli). If we assume a functionalist notion of representation, then predictive processing can help explain the representa-

---

[3] Matthias Michel (in our discussion on the *Mind Chat* podcast) suggested to me that this is no different to dinosaurs, who can't be publicly observed because they no longer exist. However, our grounds for believing dinosaurs existed consist entirely of publicly observable data. The crucial point in the case of consciousness is that *there are data that aren't publicly observable data*.

tional content of experiences. But, unless we can capture in the quantitative language of cognitive functions and predictive processing the qualities of our experience, a reductive explanation of those qualities is impossible.

Am I suggesting that we can't deal with consciousness experimentally? Not at all. Although we can't publicly observe consciousness, we can ask someone what they're feeling and experienceing, or observe external markers of their consciousness. If we do this while scanning their brain, we can establish correlations between the neural happenings which we can observe and the private experiences we cannot. This is important data, but it's not a complete theory of consciousness. What we ultimately want is to explain *why* brain activity goes along with conscious experience. Because consciousness is not publicly observable, this is not a question one can answer with an experiment (all experiments can do is establish more correlations). At this point we must turn to philosophy, examining the various proposals philosophers have offered to account for the fact that brain activity is correlated with experience: materialism, dualism, panpsychism, and idealism are the standard options. (Or rather, we turn to what is currently philosophy but I hope it will one day be established science: philosophy is what exists when the rules of the game aren't agreed upon.)

Seth of course disputes this, holding that explanatory bridges can be built from publicly observable mechanisms to private experiences. But when you look into the details, it's the old 'bait-and-switch' trick Chalmers (1995) exposed twenty-five years ago. Seth starts off talking about the qualitative character of the experience, and then subtly moves to a purely functionalist notion of content (e.g. predictive processing), giving the impression they're the same thing.[4] Low and behold, a purely functionalist notion of content can be accounted for in functionalist terms!

The whole allure of the materialist line is rooted in the idea 'But look how well science has done!'. The analogy to life is supposed to support this. But the data relevant to life are all publicly accessible. Explaining privately accessible qualities is a totally different explanatory enterprise to anything else science deals with; that's why Galileo set it outside of the domain of science, allowing scientists to focus on

---

[4] To be clear, I'm not saying anything nefarious is going on here. These are difficult questions, and I have no doubt that Seth is as intent on getting to the truth as I am.

what can be observed and quantified. We're now going through a phase of history in which physical science has gone so well that many are inclined to think it's the road to all truth, including a complete theory of consciousness. The irony is that physical science has gone so well precisely because Galileo put consciousness outside of its scope of enquiry. The fact that physical science has been able to reductively explain so much since it set aside the private qualities of experience gives us no grounds for thinking it will be able to reductively explain the qualities of experience themselves.

Seth thinks I'm setting the bar too high. Why think scientific explanations should be intuitive? It's nothing to do with what is or isn't 'intuitive'. The point is just that we need an explanation of the phenomenon. Either the qualities of experience are fundamental, or we can deduce them from more fundamental facts. You can't deduce the qualities of experience from the physical facts; therefore, we can't explain the qualities of consciousness in terms of the physical facts; therefore, materialism is false. Seth hasn't given us any grounds for doubting either of the premises of this argument.

Seth says that panpsychism isn't testable. But nor is materialism. Both panpsychism and materialism are philosophical theories that go beyond the correlations established by the experimental science of consciousness. The advantage of going for panpsychism is that you swap the hard problem for the combination problem, and the latter looks more tractable. We already have worked out theories (Mørch, 2018; Goff, 2021) that are both empirically adequate and eliminate explanatory gaps. Still, Seth exhorts us to 'wait and see' whether a method designed to explain quantifiable, publicly observable data will somehow explain privately accessible qualities, even though it's never had any success in that explanatory project and we have other explanatory paradigms that are already yielding insights.

Towards the start of his article, Seth says that panpsychism is unpopular among many academic researchers on consciousness, citing a survey at the ASSC. This is potentially misleading, as it implicitly identifies 'academic researchers of consciousness' with 'academic researchers working on the *experimental science* of consciousness'. Are philosophers not academic researchers? What Seth and I are arguing about here is the nature of the hard problem, and whether it is philosophically coherent to propose a materialist answer. This is a philosophical question and, among academic philosophers working on that question, panpsychism is a well-respected view.[5] It's true that among philosophers and scientists specifically working on the *experimental* science of consciousness, materialism dominates. There's a job of work to be done communicating to the broader community that, among those working on these more theoretical questions in philosophy of consciousness, materialism no longer dominates in the way it once did.

Finally, like most of the scientists in this volume, Seth assumes that panpsychism requires rewriting physics. I've explained in my responses to Carroll and Cortês, Smolin and Verde why this misunderstands the view.

It's not either/or. We need the experimental science of consciousness. But, because consciousness is not publicly observable, we also need philosophical theorizing. The latter without the former is lame but the former without the latter is blind. I'm sure this debate will go on, and I'm looking forward to talking at length with Anil when Keith Frankish and I host him on the *Mind Chat* podcast next month.

1.5. Reply to Christof Koch

I have learnt a great deal from my interactions with Christof Koch, and I take us to be comrades-in-arms in the same struggle. However, there are also important disagreements between us.

Koch questions the view of Galileo I present in *Galileo's Error*. This understanding of Galileo is not something I came up with, but one I have learnt from historians of science. Filip Buyse, for example, says the following:

> The corporeal domain was limited by Galileo to the domain of primary properties, which could be described in mathematical terms alone. Man, however, could not be described in mathematical terms — or, at least, not completely. For Galileo, the being of man also included the domains of color, odor, doubt, joy, etc. Man was thus understood to distinguish himself from other beings precisely by his capacity to have these sorts of affections. Man, one could say, is a bundle of secondary qualities. (Buyse, 2015, p. 34)

Given that I am assuming a pretty standard view in the history of science, I don't think it's correct to say I'm 'retrofitting [my] ideas of

---

[5] We'll have more precise data on this when the new *PhilPapers* survey results are published this October.

the mind onto Galileo'. However, Koch's article did press me to think harder about whether this familiar historical interpretation is correct.

Koch seems to imply that the Lockean distinction between primary and secondary properties has nothing to do with Galileo. However, Galileo makes a big deal of distinguishing the properties of 'material or corporeal substance' — size, shape, location, motion, number — from properties such as 'white or red, bitter or sweet, noisy or silent, and of sweet or foul odour', which reside not in corporeal substance but in the 'sensible body' (Galilei, 1623/2008, p. 185). So we certainly get something that looks very much like the Lockean distinction.

What did Galileo mean by 'sensible body' ('*corpo sensitivo*')? Koch suggests we should understand this in opposition to the Aristotelian notion of the sensitive soul ('*l'amina sensitive*'). However, Galileo also uses the term '*l'amina sensitive*' to describe the sensitive body, and the fact that Galileo says the qualities of consciousness are *not* in the corporeal body but *are* in the sensible body would seem to suggest a kind of Aristotelian dualism. I therefore find plausible Buyse's interpretation that, 'by the use of the terms "sensible body", Galileo suggests that, besides the body, there is an Aristotelian, immaterial sensitive soul, which is linked with the body and which functions as a principle of distinction between human, animal, vegetative, and unliving bodies' (Buyse, 2015, p. 33). Koch provides no counter-evidence to this interpretation of Galileo as an Aristotelian dualist.

Combining this with Galileo's famous declaration that mathematics should be the language of natural philosophy, whose subject matter is corporeal substance, we can surmise that Galileo thought:

(1) The qualities of consciousness do not reside in matter but only in the ensouled body.
(2) Mathematics is the appropriate language for describing the properties of corporeal substance, the proper subject matter of natural philosophy.

If Galileo did think the qualities of consciousness could be captured in mathematics, why did he deny that they resided in corporeal matter?

In any case, my main aim in appealing to Galileo is to resist the common argument: 'Physical science has done so well, of course it's going to explain consciousness.' Regardless of what Galileo thought, those of us who do think the qualities of consciousness resist mathematical analysis can argue that physical science has gone so well since Galileo showed physical scientists how to ignore these qualities. The fact that physical science has gone so well when it ignores the qualities of consciousness doesn't give us grounds for thinking it will do well when it comes to dealing with the qualities themselves.

Despite our common interest in panpsychism, Koch worries that philosophy ultimately gets us nowhere and the only way we're going to make progress on consciousness is by getting on with the experimental science. The subtitle of *Galileo's Error* is 'Foundations for a New Science of Consciousness'. I am not trying to do science, in this book, but rather to lay the foundations for a systematic methodology for investigating consciousness. I have explained in my reply to Seth above why I think dealing with consciousness essentially involves philosophical theorizing as well as experiments and won't repeat those points here.

On the face of it, Koch seems to be proposing a *materialist* theory of consciousness: he suggests that conscious experience is identical with maximal Φ, which is a purely quantitative, physically realized property. In *Galileo's Error*, I reject such materialist accounts, because I don't think you can account for the *qualities* of consciousness in the purely *quantitative* language of mathematical physical science (see my responses to Rovelli and Seth above for further detail). Koch responds that IIT provides a counter-example to this claim:

> Haun and Tononi... published a detailed, mathematical account of how the phenomenology of two-dimensional space, say an empty canvas, can be fully accounted for in terms of intrinsic causal powers of the associated physical substrate, here a very simple, grid-like neural network.

I have no doubt that we can in principle map out the quantitative structure of visual experience in mathematical language. This is important work. But such a mathematical description cannot fully capture the qualities that fill out that structure. If it could, we could use the mathematical description to explain to a colour-blind neuroscientist what it's like to see colour, which I think Koch would agree is absurd. As I argue above in my reply to Rovelli, this descriptive limitation of purely quantitative language entails an explanatory limitation. If a purely quantitative theory can't even convey the qualities of experience, then it certainly can't reductively account for them.

There's a certain irony in Koch saying we need to stop philosophizing and get on with the science. If Dennett was saying this, I could understand. But Koch defends integrated information theory

(IIT) and IIT is a highly philosophical theory. It employs contentious philosophical notions, such as 'intrinsic existence' and 'intrinsic causal power'; and it justifies itself not only through empirical investigation but on the basis of five axioms, supposedly known on the basis of introspection, which are then translated into five postulates the theory claims are necessary and sufficient for consciousness. I'm not objecting to any of this; I think that we need to bring science and philosophy together to deal with consciousness. But I think Koch needs to accept that the theory he's defending is knee deep in philosophical assumptions and arguments. What's the difference between my claim that we can know through introspection that consciousness has qualities that can't be captured in purely quantitative terms, and Koch's claim that we can know through introspection that consciousness exists 'for itself'?

It might sound like Koch and I have strongly opposed views. In fact, this is the rivalry of small differences (likewise, I find I have bigger fights with friends on the same wing of politics than I do with friends on the other wing). Whilst I disagree with much of Tononi's philosophical framing of IIT,[6] I'm increasingly thinking the empirical core of IIT may be the only extant philosophically coherent theory of consciousness. This is because it is the only theory of consciousness able to provide *non-vague* physical correlates of consciousness, 'non-vague' in the technical sense of not admitting of borderline cases. I will finish by briefly outlining this thought.

There are fuzzy borderline cases between being 'tall' and being 'not tall'; one of the things I believe we know about consciousness in virtue of our direct awareness of it is that there cannot be fuzzy borderline cases between consciousness and non-consciousness. Something is either experiencing or it's not. IIT can account for such sharp boundaries: a system becomes conscious at the precise moment it comes to embody more integrated information than its parts. Contrast with global workspace theory (GWT), according to which a system is conscious just in case it 'broadcasts' information to a wide variety of its cognitive systems, if it embodies 'fame in the brain', to use Dennett's memorable phrase. How famous does the information need to be? Whatever level is deemed sufficient, there will inevitably be some fuzzy borderline cases between being famous enough and not famous enough, and hence some fuzzy borderline cases whereby

---

[6] I prefer Mørch's (2018) and my own (2021) philosophical reworkings of IIT.

there's no fact of the matter as to whether the system is conscious. Therefore, IIT passes, whereas GWT fails, a crucial test of a theory of consciousness.

The post-Galilean research programme needs to develop and institute these kinds of introspection-based tests of theories of consciousness. They may be crucial in making progress where purely third-person experimental methods break down due to the fact that consciousness is not publicly observable. The post-Galilean research programme will start to be noticed when it starts getting results.

1.6. Reply to Robert Prentner (and Donald Hoffman et al.)

I have a lot of time for the interface theory of consciousness — or 'idealism' as we philosophers have been calling it for a couple of hundred years — which Robert Prentner has defended, together with Donald Hoffman, Chetan Prakash, Manish Singh, and Chris Fields (also a contributor to this special issue). Both panpsychists and interface theorists share something crucial in common: rather than starting with the physical world and trying to account for consciousness in terms of it, we start with consciousness and build up from there. I'm delighted Prentner is on board with the aspiration to build a post-Galilean science of consciousness. It's early days in building this new paradigm, and I'm glad different people are developing different non-reductionist options to see where they lead. Having said that, in my personal view, a panpsychist approach is more promising.

Prentner raises two criticisms of my panpsychist approach. Firstly, he claims that, in contrast to the interface theory, panpsychism remains a kind of dualism. Secondly, Prentner suggests that my approach in particular is too hostile to mathematics. The first criticism involves a misunderstanding of the theory. As I explain in my response to Carroll above, Russellian panpsychists don't think that particles have physical properties and consciousness properties; the claim is rather that the physical properties of the particles *are* forms of consciousness. To be fair to Prentner, he accurately reports that we Russellian panpsychists distinguish between the dispositional properties that physics tracks and the intrinsic nature that underlies them. But these aren't really two different kinds of property. All that really exists, according to Russellian panpsychism, are various forms of consciousness. The 'dispositional properties' are just a way of talking about what those forms of consciousness do. Matter is what consciousness does.

As for the accusation that Russellian panpsychists are hostile to mathematics, Prentner never substantiates this claim. The only claim we make about the limitations of mathematics is that you can't fully capture the qualities of consciousness in the purely quantitative vocabulary of mathematics. But Prenter agrees with that claim! It's true that there's not much mathematics in *Galileo's Error*. But that's because this is a book that aims at *laying the foundations* for a science of consciousness rather than itself engaging in first-order science. Panpsychists can employ mathematics as much as interface theorists, as can be seen by the highly mathematical work of integrated information theorists.

I turn now to concerns I have with the interface theory, as defended by Prentner, Hoffman, *et al*. These objections are intended in the spirit of collegial challenge. As I understand the interface theory, it commits to networks of conscious agents underlying the physical world. To make good on this commitment, interface theorists will have to postulate structures of conscious agents that facilitate not only the emergence of the structures of fundamental physics, but also the emergence of subjects like us who have the experience of interacting with a virtual world that can be accurately predicted by the equations of physics. This is a Herculean task. It's been hard enough to get the equations of physics we already have. Coming up with a whole new level of mathematical structure underneath that, which yields precisely the same predictions, is rather a large challenge. Moreover, I can't see the motivation for taking on that challenge. Rather than postulating a *new* level of mathematical structure *underneath* physics, why not locate the networks of conscious agents *inside* the mathematical structure of physics itself? This is essentially the Russellian panpsychist approach.

Taking the panpsychist approach also provides a wonderfully simple account of why it is that we emergent subjects experience a world that is accurately described by the equations of physics. According to Russellian panpsychism, there really is a world that is accurately described the equations of physics — networks of conscious agents realizing the mathematical structure of physics — and we perceive that world. In contrast, on the interface theory, we emergent subjects are interacting with networks of conscious subjects that are *not* accurately described by physics, in such a way that we seem to experience a world that *is* accurately described by physics. Dr Prentner (Hoffman *et al.*): cut out the middle man! Much simpler to suppose that we seem to experience the world physics describes because we *do* experience the world physics describes.

Donald Hoffman has argued that it is highly unlikely that we experience the world we seem to experience, given that our senses have evolved for survival fitness rather than for truth. This is supposedly proved by the 'fitness beats truth theorem' (Prakash *et al.*, 2017). My objection to this argument — as I put to Hoffman during our three-hour discussion for Annaka Harris's forthcoming audio series — is that it overgeneralizes in a problematic way. If we ought to doubt the testimony of our senses, then we ought similarly to doubt the testimony of our evolved capacity for forming judgments concerning the mental states of others. We are hardwired to judge the emotions of others on the basis of their behaviour and facial expressions. But if this hardwired capacity was evolved for truth rather than survival, and if this is sufficient for us to reject the deliverances of our sensory perception, then we ought likewise to reject the deliverances of our judgments about other minds. We ought to think others are zombies, or at least have no faith in our judgments that crying indicates sadness. I consider this a *reductio ad absurdum* of the 'fitness beats truth' argument.

1.7. Reply to Chris Fields

As with Prentner, I am inclined to think Chris Fields and I are basically on the same path. The goal for both of us is a robust, empirically adequate scientific theory in which the reality of consciousness is fundamental. This is what I mean by the 'post-Galilean science of consciousness'.

However, like Prentner, Fields mistakenly takes me to be hostile to mathematics. But, again, my only claim is that the qualities of experience cannot be reductively accounted for in mathematical terms, and — like Prentner — I think Fields agrees with this. It's true that I don't employ much mathematics myself, but that's because I'm a philosopher trying to lay the groundwork for a science of consciousness, rather than a scientist formulating a first-order theory.

I've suggested we're basically on the same path, but Fields' main line of critique involves a scepticism about some of my fundamental aims. Fields questions the need for a theory of qualia *per se*, i.e. a theory that explains, for any given experience, why the experience has the qualitative character it does; and he diagnoses my desire for a theory as arising from two sources: (a) an assumption that qualia are compositional, (b) a yearning for human specialness. I reject both of these diagnoses. Strong emergentist panpsychists may reject compositionality about qualia. And whilst Chapter 5 of *Galileo's Error* explores the implications for human existence of the Russellian panpsychism that is defended in earlier chapters, the arguments in earlier chapters are certainly not rooted in a desire for a world consonant with human happiness. I'm interested not in the view that I'd like to be true but in the view that's most likely to be true.

At least this is what I aspire to, and I believe I succeed fairly well at separating my theoretical work from my personal desires. As far as I can judge it, I really am obsessed with truth, with trying to have my best guess as to what reality is like. Maybe I'm totally deluded in this regard, and my fundamental motivation is yearning for the Cosmic Daddy of my Catholic upbringing. But any theorist could be psychoanalysed for the 'real' motivations underlying their apparent desire for truth. The Churchland/Dennett-esque idea that science as we currently conceive it is just obviously the path to all truth is terribly comforting.

My basic assumption here is that reality is intelligible. We have to start with some basic, unexplained facts, but there should in principle be an intelligible story as to how non-fundamental facts emerge from fundamental facts. Human qualia are real. And therefore we need an explanation of *what is going on in reality* to account for their existence. Perhaps, as the dualist or strong emergentist panpsychist supposes, human qualia are fundamental features of reality. Perhaps, as the materialist or the reductive panpsychist supposes, human qualia can be reductively explained in terms of more fundamental features of reality (just because *some* consciousness is fundamental, it doesn't follow that *human* consciousness is fundamental). Either way, the fundamental drive of science and philosophy is to explain.

Fields wonders why it is 'satisfying' to postulate consciousness as the intrinsic nature of matter, again suspecting that I'm looking for a view that 'satisfies' in some emotional sense. I take it that what we are looking for is the most parsimonious account of the data. As explained above, Russellian panpsychism offers a radically non-dualistic, and therefore radically parsimonious, theory. Rather than postulating physical properties *and* consciousness properties, we identify the two. Physical properties just are consciousness properties characterized in terms of how they behave. Like most of the scientists in this special issue, Fields misinterprets Russellian panpsychism as the view that consciousness, as the intrinsic nature of matter, doesn't do anything.

Quite the contrary, if consciousness is all there is, either consciousness does something or nothing does anything.

2. Replies to Philosophers

2.1. Reply to Luke Roelofs

I don't have too much to say in response to Luke Roelofs' excellent essay, simply because I am entirely in agreement with most of its contents. Roelofs does an amazing job debunking the various charges that panpsychism is 'anti-science'.

Having said that, there are some slight differences between the two of us. If panpsychism is a middle way between dualism and materialism, I am slightly closer to the dualist pole and Roelofs to the materialist. Roelofs once told me that if they weren't a panpsychist they'd be a materialist, where I would certainly embrace dualism before materialism. As Roelofs says, I'm less convinced of micro-reductionism than they are. Roelofs says micro-reductionism fits better with the trajectory of science. But whether or not everything that happens is totally determined at the level of fundamental physics is an empirical question, and I don't think we're yet in a position to judge empirically whether it's true (see my reply to Carroll). Until such a time as we can assess the view empirically, I'm not sure it's appropriate to draw inferences about the nature of reality from the hunches and methodological assumptions of working scientists.

Overall, I'm much more persuaded of the clash between materialism and consciousness realism than I am of the clash between dualism and the empirical facts. And given my greater affinity with dualism than materialism, I'm not totally ready to sign up to the 'Monist United Front' Luke proposes (opposition to dualism is a tentative third clause in the post-Galilean manifesto). Maybe if I set up the 'Anti-Materialist United Front' in the building next door, Roelofs and I can run a tea tent in the space between the two organizations.

2.2. Reply to Annaka Harris

Annaka Harris's solution to the combination problem is intriguing and something I will continue to reflect on. I'm not yet convinced, however, that it resolves all of the challenges that go under that banner. Harris proposes that all physical systems are associated with some form of conscious experience, presumably both at the micro and the macro level. But if consciousness exists both at the neurophysiological level and at the level of fundamental physics, a crucial theoretical question arises:

> *The reduction question*: Can the consciousness of the brain be reductively explained in terms of the consciousness of the particles making up the brain, in something like the way the liquidity of water can be reductively explained in terms of the underlying chemistry?

If we want to answer 'yes' to the reduction question, then the combination problem returns as the challenge of how to make sense of such a reduction. But if the answer is 'no', the spectre of epiphenomenalism looms again. If Roelofs (this issue) and Carroll (this issue) are right that all of my behaviour can be causally explained in terms of micro-level goings-on that are the proper subject of matter of physics, then the consciousness that exists at the macro-level of the brain — what we think of as 'Philip Goff's consciousness'[7] — (if irreducible) has nothing left to do, no role to play in governing behaviour. I know from her work more generally that Harris doesn't believe in free will, but it seems a more extreme step if the contents of my mind have absolutely no impact on the physical world, e.g. if the ideas I am thinking of in composing this reply play no role in determining what words I'm typing on the page.

In my recent work (Goff, 2021), I have developed a form of 'hybrid panpsychism', which tries to address this dilemma by distinguishing sharply between subjects and their experiences, holding that the former are 'strongly emergent' (i.e. they can't be reductively explained) whilst the latter are 'weakly emergent' (i.e. they can be reductively explained, in terms of consciousness at the level of physics). My hope is that this combination of strong and weak emergence can allow me to avoid *both* the combination problem (as I'm not trying to reductively explain subjects of experience) *and* the threat of epiphenomenalism (as human experience can be reduced to, and thereby derives its causal power from, the level of fundamental physics). Of course, this strategy involves precisely the commitment to subjects that Harris rejects. But it seems to me that this allows me a

---

[7] I am just pointing to a certain form of experience and do not mean to commit to a self/subject that has that experience.

way of avoiding problems that I can't see how Harris's form of panpsychism can avoid.[8]

2.3. Reply to Damian Aleksiev

Whilst most discussion of panpsychism focuses on the combination problem, Damian Aleksiev is pioneering work on a quite different challenge: what he calls 'the missing entities problem'. I will here sketch some tentative responses to this new set of challenges, which I hope to give a fuller response to in future work. Aleksiev's concerns seem to be instances of the *structural mismatch problem*: the challenge for the panpsychist of explaining how the structure of human consciousness relates to the structure of consciousness at the fundamental level (which, for the panpsychist, is the level of basic physics). To take a very simple form of this: the physical structure of the parts of my brain which support my consciousness are highly complex, if we take it down to the detail of every individual quark and electron, whereas that enormous complexity doesn't seem to be present in *my* consciousness.

Wrestling with the structural mismatch problem has been my main motivation for returning to a 'cosmopsychist' form of panpsychism, according to which the fundamental forms of consciousness are the intrinsic natures of universe-wide fields, forms of consciousness which are borne by the universe as a whole. Some of that very complex experience borne by the universe is located in my head. But, on the hybrid cosmopsychism theory I have developed (Goff, 2021), that's not *my* experience. Rather, my experience consists of a 'thinned out' version of the very busy experience borne by the universe inside my head. To make this clearer, consider an analogy. Imagine someone experiencing a countryside scene of a lake, trees and birds. Now suppose we take out some of the detail, so that the person is only experiencing the lake without the trees and birds. The latter experience would be a 'thinned out' version of the former. Similarly, my experience is a thinned out version of the very rich experience inside my head: it's what results when you take the very busy experience inside my head (corresponding to the very complex physical structure at the fundamental level of the brain) and take out a lot of detail.

---

[8] I should say that Harris responded to Goff (2021) in an earlier version of her contribution, but I advised as editor focusing on laying out her own position, given the tight word limit.

It's not clear to me why this same account could not be used to address at least the space-time gap and the quantum gravity gap. Perhaps experience at the fundamental level is timeless and has an esoteric geometrical structure. But the process of thinning out that structure to form my experience will result in experience with a quite different structure. It may very well lack the metric of fundamental space-time, and the very selective cutting and pasting involved in putting together my experience may be precisely what yields its temporal character.[9]

The quantum state gap is more troubling, as it's not clear to me that we could get from the structure of the quantum state to the structure of human consciousness via thinning out. To make this clear consider, for the sake of simplicity, a situation we never actually find in the real world. Suppose all of the particles in the universe were in an utterly determinate state. This corresponds to all the amplitude of the wave function occupying a single location in $3 \times N$ space. Interpreted as a form of panpsychism, that amplitude is a form of consciousness. But it's just a blob lacking any kind of spatial extension or internal structure. It's clear that we can't get from a simple structureless blob to the complexities of human consciousness by taking away some detail.

The single most important claim of *Galileo's Error* is that we need to take the reality of consciousness, as we know it 'from the inside', as a fundamental scientific datum in its own right. If that datum rules out certain interpretations of quantum physics, e.g. those in which fundamental reality is entirely constituted by the wave function, then we should not hesitate to reject those interpretations on that basis. Indeed, this could be one important way of deciding between the many interpretations of quantum mechanics, and an important aspect of the post-Galilean research programme (*cf.* Cortês, Smolin and Verde's contribution to this special issue). I have explored this problem in detail in my paper 'Quantum Mechanics and the Consciousness Constraint' (Goff, forthcoming).

---

[9] It's true that my most recent version of panpsychism is not fully the constitutive version Aleksiev is critiquing in his paper. However, although I have departed from fully constitutive panpsychism, I have done so in order to address the combination problem not the structural mismatch problem. The basic idea I now refer to as 'thinning out' was present in the version of constitutive panpsychism I defended in *Consciousness and Fundamental Reality* (2017). Hence, I'm not persuaded that the issues Aleksiev raises here give us grounds for dropping constitutive panpsychism.

2.4. Reply to Alyssa Ney

Alyssa Ney raises some great challenges to my views. In responding, I'd like to start with one small but important point which I think is a misunderstanding of my view. My view is not that we cannot use the tools of physical science to investigate consciousness, but rather that the tools of physical science *alone* cannot give a full account of consciousness. Neuroscience can provide us with an account of the physical correlates of conscious experience, but cannot, in my view, give a fully satisfactory explanation of *why* conscious experience is correlated with brain activity. I'm proposing *adding* to the toolbox of the science consciousness, not taking away from it.

Ney draws attention to the ambiguity in *Galileo's Error* regarding whether I'm subscribing to scientific instrumentalism or structural realism. This is partly the result of the necessity of being less precise in a book aimed at a general audience. It's much clearer (I hope!) in my academic book, *Consciousness and Fundamental Reality*, that I advocate structural realism. I appreciate that the line in *Galileo's Error* Alyssa points to — '...physical science doesn't even tell us what matter does...' — might seem to suggest otherwise. But whilst I think physical science doesn't strictly speaking tell us what anything does — that would require a non-circular specification of the manifestation of some physical property — I do think it can accurately capture the abstract causal structure of reality.

One might get the impression reading Ney's article that I'm wildly at odds with orthodoxy in philosophy of science. However, as Ney says, structural realism is orthodoxy in philosophy of science, and to that extent I am completely in line with orthodoxy. It is true, as Ney says, that most philosophers of science would go for *ontic* structural realism — all that exists is structure — rather than *epistemic* structural realism — all science can tell us about is structure. But that's because most philosophers of science aren't thinking about how to accommodate consciousness into their view, and, to the extent that they do, they're likely to adopt materialism. I don't think I've kept hidden that I'm proposing radical change — the subtitle of *Galileo's Error* calls for a 'New Science of Consciousness' — and to that extent I'm challenging orthodoxy. Whether that radical change is called for depends on whether the arguments against materialism about consciousness are successful (I briefly outline these arguments above in my replies to Rovelli and Seth), and Ney doesn't in this article deal with those arguments.

On free will, I agree with Ney that a compatibilist would be unmotivated to consider the 'pan-agentialist' position I explore in the final chapter. My argument here is conditional: *if* you're a libertarian about free will, you should be a pan-agentialist; but I don't offer an argument for the antecedent of that conditional. Again, this might not have been completely clear due to the necessity in a popular book of not defining everything totally precisely. The argument is (I hope!) laid out in a more rigorous fashion in my article 'Panpsychism and Free Will' (2020).

On the foundations of morality, I agree that, without some way of accounting for prudential oughts, the account I sketch wouldn't provide an objective basis for morality. But it may perhaps reveal some false presuppositions in egoism; and if you combine the view I outline with internalism about reasons — roughly, the view that one has reason to do what one desires — we may have an objective basis for demonstrating that those with self-interested desires ought to have altruistic desires.

I also agree with Ney that, if analytic naturalism were plausible, Sam Harris would have a way of avoiding the concerns I raise with his view; and perhaps that is Harris's view. Having said that, I don't think analytic naturalism is a very plausible view. For one thing both ethical egoism — the view that I only have reason to do what is good for me — and non-egoistic ethical realism — the view that I have reason to do both what is good for others as well as myself — seem quite clearly non-contradictory, which leaves analytic naturalism with no resources to say which of these mutually contradictory views is correct.

Ney points out that there are other options for naturalistic reductions of ethics, which is of course correct. For what it's worth, I favour pretty hardcore non-reductionist realism about value, but I certainly don't take myself have to have defended that in *Galileo's Error*. With both the discussions of free will and morality, there are a hundred different positions, and if I'd spent time precisely articulating what I took myself to have shown, then this chapter would have been a lot less accessible. One useful thing about this journal issue is that I can further clarify some of my views, and I'm grateful to Ney for pushing me to do so.

On the big picture meaning of life stuff, my core concern with materialism is that it is incompatible with the reality of conscious experience, and hence our official worldview is incompatible with the thing of which we are most certain and the thing which gives life meaning and value. I appreciate that what I've just said assumes the falsity of materialism, which Ney disputes. At the end of the day, everything will hang on the arguments against materialism about consciousness.

2.5. Reply to Keith Frankish

Is Galileo's Real Error believing that qualities exist anywhere? Reading Keith Frankish's ingenious argument, I can almost go along with him. I have learnt so much from thinking my way inside Frankish's worldview, occupying his world for thirty minutes or so (no longer, in case I never get back...). But I can't help thinking it doesn't quite work out.

Let's suppose that Frankish is right that the sensory qualities — colour, sounds, smells, tastes — don't exist anywhere, either in the mind or the external world. Focusing on the case of colour, Chalmers uses the term 'Edenic' for colours, sounds, smells, tastes, etc. as we naïvely take them to be — sensory qualities as they were in the Garden of Eden. We can thus define Frankish's 'Edenic illusionism' as follows:

(1) The mind represents physical objects to have Edenic qualities.
(2) Edenic qualities don't really exist, either in the mind or the external world.

It's clear that the mind can represent things that don't exist, e.g. if I hallucinate a pink elephant. Similarly, just because the mind represents objects to have Edenic qualities, it doesn't mean that they really do.

Does the hard problem of consciousness go away if we accept Edenic illusionism? My case against materialism is built upon the claim that we can't account for the qualities of consciousness in the purely quantitative language of physical science. But if these qualities don't exist at all — either in the mind or the external world — then haven't we entirely removed all the troubling phenomena that threaten the materialist worldview?

I don't think so. Even if my mind doesn't contain Edenic qualities, it nonetheless has the property of *representing* Edenic qualities. And the property of representing Edenic qualities is just as irreducible as Edenic qualities themselves. The argument against materialist reduction of Edenic qualities is roughly as follows (see my reply to Rovelli above):

(1) Edenic qualities can be reductively explained in the terms of physical science only if Edenic qualities can be fully captured in the terms of physical science.
(2) Edenic qualities cannot be fully captured in the terms of physical science.
(3) Therefore, Edenic qualities cannot be reductively explained in the terms of physical science.

The same form of argument applies to the property of representing Edenic qualities:

(1) The property of representing Edenic qualities can be reductively explained in the terms of physical science only if the property of representing Edenic qualities can be fully captured in the terms of physical science.
(2) The property of representing Edenic qualities cannot be fully captured in the terms of physical science.
(3) Therefore, the property of representing Edenic qualities cannot be reductively explained in the terms of physical science.

In order to fully describe the property of representing Edenic red, you'd have to fully describe Edenic red. Thus, if you can't do the latter in the purely quantitative vocabulary of the physical science, you can't do the former in the purely quantitative vocabulary of physical science.

I often speak of the core challenge for the materialism as that of accounting for the 'qualities' of consciousness. And I am indeed inclined to the Galilean view that the sensory qualities that seem to be in the external world are really in the mind (see my reply to Liu below). But even if, strictly speaking, there are no qualities in consciousness, we still face the challenge of accounting for the property of consciously representing qualities. Either way, I don't think Frankish's strategy can save materialism.

2.6. Reply to Michelle Liu

Michelle Liu argues makes a powerful case for the polar opposite position to Frankish, for the real existence of Edenic qualities. Colours as we naïvely take them to be really do exist out there on the surfaces of objects. Call this view 'Edenic realism'.

The first thing I would note is that the existence of Edenic properties is subject to doubt in the way that the qualities I directly apprehend in my experience are not.[10] I can doubt that the rose in front of me is really red — I can even doubt whether it exists — but I cannot doubt that my experience of the rose has a reddish qualitative character.

Moreover, I am not persuaded by Liu's theoretical argument for Edenic properties: that if we deny their existence, it is extremely hard to explain how we come to perceptually represent such non-existent properties. As David Hume observed, we 'gild and stain' the world with our sentiments. When I see something nauseating, I experience the 'disgustingness' as an intrinsic property of the object of my disgust, even though, of course, there isn't really any such intrinsic property as 'disgustingness'. I don't see why something similar couldn't be going on in the case of colour experience: I misrepresent an intrinsic quality of my experience of a rose as an intrinsic quality of the rose itself. I agree with Liu that the experience of colour is much more robust than the experience of disgustingness; the former as opposed to the latter is not dependent on past experience and is 'just as perceptually salient and persistent as that of shape and size'. But that is presumably because the way we experience colour is hardwired into us from birth, whereas the experience of disgustingness is partly shaped by our life experience. This difference seems to me perfectly compatible with the commonality that they both involve misrepresenting an intrinsic quality of experience as an intrinsic quality of the object of experience.

I'm more persuaded by Liu's argument that a world without Edenic properties is of significantly less value than the Edenic world, due to the absence of the beautiful colours that seem to fill our world. This is a consideration I hadn't properly appreciated before. Of course, one could retort: 'Why think reality is likely to be of value?' But any kind of theorizing about reality depends on anti-sceptical assumptions, e.g. that we're not in the Matrix, that other people have consciousness. Perhaps Liu can make a case that assuming Edenic realism is justified by a similar kind of anti-sceptical assumption.

However, ultimately I think there are more powerful considerations on the other side of the ledger. My main motivation for rejecting Edenic realism is due to issues in the philosophy of perception, which

---

[10] Or at least the fact that I consciously represent those qualities is not subject to doubt (see my reply to Frankish).

aren't Liu's focus here. Edenic realism goes naturally with a 'naïve realist' view of perception, according to which perception involves a *direct* connection with things in the world: when I'm looking at a rose, the red quality in my experience just is the red quality of the rose. Worries arise when we try to square this with the existence of hallucinations. Clearly, when I'm hallucinating a red rose, I'm not in direct contact with the Edenic redness of a physical object, as there's no red object there. In this case, then, the redness must be in my head. So it seems that we must say that when I'm veridically perceiving a red rose, what I'm in contact with is redness out there in the world, whereas when I'm hallucinating a red rose, the redness is in my head.

I think there are a number of problems with this position, which I hope to develop in future work. Just to focus on one issue: the mental redness in the hallucination has the same character as the Edenic redness in the veridical experience. How does the cognitive system manage to locate a mental property to use in the hallucination with the same character as the Edenic property of the veridical case? And why have we evolved to create a deceptive experience in hallucination? If the cognitive system somehow 'knows' it's hallucinating (which it presumably must do as it activates the mental redness when and only when there is an hallucination of red), why not flag up that this is an hallucination rather than encouraging the person to think it's a true experience of red by activating a mental quality that resembles Edenic redness? If these issues could be addressed, I would be willing to look again at Edenic realism, but for the moment I remain unpersuaded.

2.7. Reply to Alex Moran

Like Liu, Alex Moran is inclined to believe in Edenic qualities, and hence worries that my panpsychist view is at best an incomplete theory of reality: even if it can account for the qualities of the mind, it is unable to account for the qualities of the external world: the colours, sounds, smells, and tastes. He then offers a compelling presentation of two alternative views with the potential to do both. My reply to Liu above, in which I have argued against the existence of Edenic properties, doubles as a reply to Moran: if there are no Edenic properties, then there is no additional explanatory obligation for the panpsychist over and above accounting for the qualities of the mind. I think Moran is right that other forms of Russellian monism work better if we're trying to account for both experiential qualities and

Edenic properties. But, given my doubts about Edenic properties, I'm not yet motivated to embrace these alternatives to panpsychism.

The other fascinating aspect of Moran's proposal is his articulation and defence of *grounding physicalism*, which is an emerging position in the philosophical literature more generally. I certainly agree with Moran that there should be room for a physicalist position according to which the relationship between mental properties and physical properties is looser than identity (in *Galileo's Error*, but not in my academic work, I deliberately oversimplify by construing materialism as an identity theory). Philosophers tend to call this looser relationship 'grounding' whilst scientists call it 'emergence'. However, I struggle to make sense of a grounding relation that involves no explanatory connection between the fundamental and the non-fundamental. Surely physicalists must *explain* the existence of consciousness in terms of the physical properties of the brain, in such a way that the explanatory gap is closed.

Moran resists this by comparing grounding to causation, which many philosophers now take to be compatible with the existence of explanatory gaps: what causes what is discovered through empirical investigation not rational deduction. But, in that case, I don't understand the difference between 'grounding' and 'causation'. Chalmers is associated with the property dualist position according to which consciousness is causally brought into being by the physical properties of the brain, in accordance with fundamental psychophysical laws of nature. I can't see the substantial difference between that position and the position Moran and others are calling 'grounding physicalism'.

2.8. Reply to Ralph Weir

Ralph Weir presents a powerful argument that the post-Galilean science of consciousness should embrace substance dualism. I think he's right that one can't accept the logical possibility of a zombie (a body without a mind) whilst denying the logical possibility of a ghost (a mind without a body). However, the real-world implications of a possible ghost are less straightforward than the real-world implications of a possible zombie. Crucially, I'm not persuaded that the mere possibility of my ghost twin entails that *I* could exist in a disembodied state.

I think the issue hangs on the metaphysics of substance. Roughly speaking, there are two rival views: bundle theory and substance-attribute theory. According to bundle theory, a physical object is just a bundle of its properties: an electron, for example, is nothing more than its mass, its negative charge, and so on, 'bundled together' (however that is cashed out). According to substance-attribute theory, in contrast, there aren't merely the properties of the electron, but also the thing that *has* the property, also known as the electron's 'substratum'.

I think Weir's argument is sound if we assume bundle theory. If my conscious mind is nothing more than a bundle of its conscious states, and that bundle of conscious states could exist in the absence of anything else, it follows that my conscious mind could exist in the absence of anything else. However, now suppose my conscious mind is a substratum that *has* its conscious states. Is that substratum *essentially* tied to any of its properties? Four possibilities suggest themselves:

(1) **My substratum is essentially tied to experiential properties and only to experiential properties:** My substratum can exist only in so far as it is bearing experiential properties, but it could exist without bearing any other kinds of property.
(2) **My substratum is essentially tied to experiential properties but also to certain physical properties (for the purposes of this discussion let's suppose — contrary to panpsychism — that physical properties are wholly non-experiential):** My substratum can exist only in so far as it is bearing experiential properties and certain physical properties, but could (and perhaps does) exist without any other forms of being.
(3) **My substratum is essentially tied to certain physical properties but not to experiential properties:** My substratum could exist without bearing experiential properties but could not exist without bearing certain physical properties.
(4) **My substratum is not essentially tied to any of its properties:** On this view, I could exist as pretty much anything: an abstract object, a non-experiencing boiled egg, whatever.

I don't think we can settle on introspective grounds which of these options is correct. Through introspection, I know about the essential nature of my conscious states, and I know that *those states* could exist without any other properties. I arguably also know that there is something that bears my conscious states: my substratum. But I have no way of knowing on grounds of introspection whether, in conceiving of my substratum as *a bearer of conscious states*, I thereby grasp its complete, essential nature. This seems to me essentially the flaw in

Descartes' argument for dualism, and the worry remains, I think, for Weir's argument.

To be fair, Weir does acknowledge something like this response, referring to it as belief in 'transcendent egos', saying in a footnote that these might be conceived of as substrata. However, that might suggest a certain 'add on' to a theory of mind. However, the choice between bundle theory and substance-attribute theory is a fundamental theoretical choice — a point we face in deciding, not merely on a theory of mind, but on a theory of reality itself. Which way we go with regards to that fundamental choice-point will determine whether or not Weir's argument is sound, or so it seems to me.

2.9. Reply to Galen Strawson

Galen Strawson and I have the same view about the nature of reality. We both believe that consciousness can't be accounted for in the terms of physical science, but that that's okay because physical science doesn't tell us the nature of matter. Consciousness is totally physical, but it's an aspect of the physical that goes beyond what physical science can teach us about. We're also both inclined to think that consciousness exhausts the nature of physical reality. In terms of living philosophers, Strawson is the father of the post-Galilean revolution.[11]

Despite our agreement on issues of substance, we've had a fifteen-year long dispute about what to call the view we both accept. Strawson is very passionate about calling it 'physicalism' and/or 'materialism'. I think it's better to use those words to describe the view we're opposed to, namely that the nature of consciousness can be entirely accounted for in the terms of physical science.

It doesn't seem to me that there's any fact of the matter we're arguing over here. As Humpty Dumpty wisely observed, we can use words how we wish. Strawson is better versed than I in historical matters, so I won't dispute his claim to be using the word 'materialism' in a way that fits better with its historical usage. However, I think we should decide how to use the words 'materialism' and 'physicalism' on pragmatic grounds. Specifically, I think we should go with the meanings of these terms that are most useful for having the debates we

---

[11] I will always be deeply indebted to Galen for inspiring me with the confidence to defend a whacky view for no other reason than that it's probably true.

want to have. There's a big and important debate between those who do and those who don't think consciousness can be accounted for in the terms of physical science, and, to have that debate, it's incredibly useful to have words that identify people in terms of which side of that debate they're on. We can't say 'dualists' and 'anti-dualists', as people on our side of the debate include panpsychists and neutral monists. I think it makes sense to use the terms 'physicalists/materialists' and 'anti-physicalists/materialists'.

Part of what's going on here is that Strawson has very little time for the view of our opponents — the group I call 'physicalists' and he calls 'physicSalists'. Strawson (2018) argues that physicSalism amounts to the denial of the reality of consciousness, which he has famously called 'the Denial' and the 'silliest claim ever made'. Of course, I agree with Strawson that we can't fully account for consciousness in the terms of physical science, and to that extent I agree that physicSalism does commit one to the denial of the reality of consciousness. Where I think we disagree is that I don't think that deep philosophical truth is *obvious*, such that anybody who opposes it is in denial or deluded. Many of the physicSalists in this volume — Ney, Carroll, Rovelli, Seth — fully believe in the reality of consciousness — feelings, experiences, sensations — whilst also believing that physical science can in principle give a full explanation of consciousness. I think they're wrong but I don't think they're stupid for thinking this. Philosophy is hard!

Having said this, there are the physicSalists who explicitly *do* deny the reality of phenomenal consciousness, such as contributor to this special issue Keith Frankish. I agree with Strawson that this view is very implausible. But, again, I think philosophy is hard, and I can understand how honest and earnest philosophical reasoning can get you to that position. Of course, in some sense I think that reasoning is flawed; that follows from the fact that I reject that philosophical view. But there's plenty of honest and sincere disagreement in philosophy.

Moreover, I think Strawson and I should welcome the growth of illusionism about consciousness. The more it becomes accepted that the choice is between illusionism and anti-physicSalism, the more I believe society will come to reject the physicSalism we both find so implausible. Indeed, the growth of Dennett/Frankish-style illusionism as the dominant materialist position may ultimately be the sociological cause of the post-Galilean science of consciousness moving from fringe to mainstream.

3. Replies to Theologians

3.1. Reply to Joanna Leidenhag

Joanna Leidenhag does a great job of describing more nuanced ways of understanding the relationship between God and the world. But I think maybe she's reading a little too much into my references to God in *Galileo's Error*. These discussions weren't really intended to engage with the question of miracles or God's existence, but rather to illustrate the challenges to dualism, which I tried to illustrate by imagining a God who intervenes *very regularly*, and hence makes Her presence very well known. As I say in the book, '...if God does intervene in the world, She doesn't act enough to make Her presence *obvious*' (Goff, 2019, p. 37). My fundamental objection to traditional theism is not to do with the possibility of miracles but is rather rooted in the traditional problem of evil and suffering. I think it's highly implausible that a loving God who could do anything would create and sustain a world with so much pain. Obviously, this is a huge debate I can't get properly into here.

My disagreements with Leidenhag begin in the latter half of the article, where she presents an intriguing argument that the case for panpsychism should also lead one to theism. Strawson (another contributor to this special issue) argues for panpsychism on the grounds that we can't intelligibly get consciousness from non-consciousness, but I've never supported that argument. Whilst I don't think we can intelligibly get from the purely *quantitative* facts of physical science to the *qualitative* facts of conscious experience, I'm not sure we can rule out that there are some special kinds of properties of matter which are not themselves forms of experience, but which somehow combine in various ways to produce experience; philosophers call these 'protophenomenal properties' (Goff, 2006; 2009).

My argument for panpsychism is rather based on considerations of simplicity. I believe that we *can* account for human consciousness in terms of more fundamental forms of consciousness, and that such a panpsychist theory is the most parsimonious theory alternative to materialism (which I think should be rejected on the grounds that it can't account for consciousness). Panpsychism is the most parsimonious theory able to account for both the reality of consciousness and the data of public observation.

Leidenhag is correct that my case for panpsychism relies on the idea that there must be an intelligible connection between fundamental and non-fundamental facts. In my academic work, I call (a more detailed version of) this principle 'minimal rationalism'. We could roughly define it as follows:

> *Minimal rationalism (MR for short)*: For any non-fundamental fact F, it must be possible in principle to derive F from the complete fundamental story of reality.

However, I think MR is a much more modest principle than the principle of sufficient reason (PSR). It's one thing to say that there must be an intelligible connection between the fundamental and non-fundamental facts, quite another to say that there must be an intelligible story about why the fundamental facts exist in the first place. Therefore, I think the panpsychist can quite consistently defend her position whilst rejecting theism by defending MR and rejecting PSR.

Having said that, I am rather sympathetic to PSR. I'll say more on that presently, but let's first think about where the PSR leads us. An infinite regress of universes is no good, as then you don't have an explanation of why the infinite regress of universes has always existed (as opposed to, say, an infinite regress of ghosts, or nothing at all). The only way there could be a complete explanation of everything is if there is something that explains its own existence; PSR leads inexorably to the postulation of a *self-explainer*.

How can we get a grip on the idea of a self-explainer? Perhaps by considering its polar opposite: an impossible being. An impossible being is a being whose nature — or rather the nature it would have if it existed — explains its non-existence. We have a good grip on many impossible beings; for example, square circles. Once you grasp the nature of a square circle, you just see that it can't possibly exist. By analogy, a self-explainer would be a being such that, once you grasped its nature, you'd see that it *must* exist. We have no positive understanding of such a being, but that doesn't mean there isn't such a thing. Dogs can't do mathematics; maybe forming a positive conception of a self-explainer is similarly beyond us. The limits of human comprehension are not the limits of reality.

If we accept PSR, we have to postulate a self-explainer. But why accept PSR? Quite simply, it seems to me that, all things being equal, a theory of reality in which there are no brute facts is superior to a theory of reality in which there are brute facts. This principle leads us to postulate a self-explainer — in order to avoid brute facts — unless there is some comparable theoretical reason to avoid postulating a self-explainer.

Could Ockham's razor give us grounds for avoiding postulating a self-explainer? Not if we think of Ockham's razor, as I do, as a principle telling us what kinds of explanation we should seek — the most parsimonious ones — rather than a principle telling us what should or shouldn't be explained in the first place. I suppose someone might say that we shouldn't postulate something that does explanatory work when we don't understand how that explanatory work is done. But that seems to me false: we can have very good reason to suppose that something had a cause, even if we don't understand how that cause did its causing. Thus, I can't see any good reason *not* to postulate a self-explainer and, in the absence of some such reason, the theoretical attraction of avoiding brute facts ought to lead us to postulate a self-explainer.

For these reasons, I think there probably is a self-explainer. However, I don't see why the self-explainer can't be the universe itself. Leidenhag raises three objections to this view, which I lay out below together with my responses:

(1) Objection: The universe began to exist, and hence doesn't exist necessarily. Response: It's true that the *temporal phase* of the universe began to exist, but it could be that, in the absence of time, the universe exists in a timeless form. Leidenhag may object that we have no reason to think the universe could exist in a timeless form. I would respond: PSR entails we must choose between two hypotheses: a self-explaining universe which can exist in a timeless form, or a supernatural self-explainer distinct from the physical universe. The former hypothesis is more parsimonious, and therefore the one we ought to go for.

(2) Objection: If the universe exists necessarily, nothing is contingent. Response: I don't think that's quite right. If there is quantum indeterminacy, then this will introduce an element of contingency. And the necessity of the universe's existence is consistent with the emergence of libertarian free will.

(3) Objection: It is conceivable that the universe doesn't exist, and my commitment to the zombie argument involves a commitment to conceivability entailing possibility. Putting these together, I ought to believe in the possible non-existence of the universe. Response: It's also conceivable that God doesn't exist. Or at least it's conceivable that an all-knowing, all-powerful, and perfectly good being doesn't exist. Of course, we define 'God' as existing necessarily, but I could equally define a new term, 'super-

universe', which refers to the universe but stipulates that it exists necessarily. I think that if there is a God who exists necessarily, it must be in virtue of some aspect of Her nature beyond our comprehension. Why not take the more parsimonious option of ascribing that nature to the universe itself?

3.2. Reply to Sarah Lane Ritchie

Sarah Lane Ritchie is here arguing for a conclusion that is often dismissed as 'new age fluffy thinking'. But she argues with such rigour, clarity, and nuance that such an accusation would be ridiculous, rooted only in ignorance and prejudice.

It's important to note that many philosophers and scientists defending panpsychism are completely atheistic secularists. Roelofs (a contributor to this special issue) and Chalmers are cases in point. They don't believe in a transcendent reality, but they do believe in feelings and experiences, and hence want an account of how those things fit in to our overall theory of reality. I'm sure Ritchie would agree with all of this.

Having said that, I would say that panpsychism removes some reasons *not to believe* that mystical experiences are veridical. Suppose one has a mystical experience with the following content: *there is a higher form of consciousness underlying all things*. If you're a materialist, then you have to think this experience is a delusion, as what this theory suggests concerning fundamental reality is inconsistent with the story about fundamental reality we get from physics. But if you *already* think that fundamental reality is constituted of forms of consciousness, then it's not too much of a leap to suppose that the higher form of consciousness you seem to be aware of in having the mystical experience is also part of the fundamental story of reality.

But just because you have no reason *not* to believe X doesn't mean you have a reason *to believe* X. Richard Dawkins doesn't buy the problem of evil as a reason to doubt the existence of God, but he nonetheless ascribes a very small probability to God's existence because he thinks there's no evidence for it. Many people think there's no rational basis for the mystic to trust what her experience seems to be telling her about the nature of reality, as it could simply be a delusion caused by unusual brain activity. However, I'm sympathetic to William James's line (1902/1985, chapter XVII) that this common viewpoint betrays a double standard. It's true that the mystic can't prove that their experience corresponds to an external reality. But nor can any of us prove that our sensory experiences correspond to an external reality. All empirical knowledge begins with trusting what experience seems to be telling us. It's not clear how one could consistently tell the mystic it is irrational for her to trust her mystical experience without also denying the rationality of trusting one's own sensory experiences.

I've never had a mystical experience. But in some of my deepest experiences — certain moral experiences, engagements with nature, deep meditation, or watching the light of early morning or late evening — I have a fleeting sense of a greater reality at the root of things. I wouldn't go so far as to say I *believe* that those experience correspond to an external reality. But I engage them, I trust them, I work with them in my spiritual practice. Maybe I'm a fool for taking these epistemological risks. But as John Locke wisely observed, 'He that in the ordinary affairs of life, would admit of nothing but direct plain demonstration, would be sure of nothing in this world but of perishing quickly' (Locke, 1689/2008, Book IV, section 10).

References

Boly, M., Massimini, M., Tsuchiya, N., Postle, B.R., Koch, C. & Tononi, G. (2017) Are the neural correlates of consciousness in the front or in the back of the cerebral cortex? Clinical and neuroimaging evidence, *Journal of Neuroscience*, **37** (40), pp. 9603–9613.

Buyse, F. (2015) The distinction between primary properties and secondary qualities in Galileo Galilei's natural philosophy, *Cahiers du Séminaire québécois en philosophie moderne/Working Papers of the Quebec Seminar in Early Modern Philosophy*, **1**, pp. 20–43.

Chalmers, D.J. (1995) Facing up to the problem of consciousness, *Journal of Consciousness Studies*, **2** (3), pp. 200–219.

Cobb, M. (2020) *The Idea of the Brain*, London: Profile Books.

Galilei, G. (1623/2008) *The Assayer*, in Finocchiaro, M.A. (trans.) *The Essential Galileo*, pp. 179–189, Indianapolis, IN: Hackett Publishing Company.

Goff, P. (2006) Experiences don't sum, *Journal of Consciousness Studies*, **13** (10–11), pp. 53–61.

Goff, P. (2009) Why panpsychism doesn't help us explain consciousness, *Dialectica*, **63** (3), pp. 289–311.

Goff, P. (2017) *Consciousness and Fundamental Reality*, Oxford: Oxford University Press.

Goff, P. (2019) *Galileo's Error: Foundations for a New Science of Consciousness*, London: Pantheon/Rider.

Goff, P. (2020) Panpsychism and free will: A case study in liberal naturalism, *Proceedings of the Aristotelian Society*, **120** (2), pp. 123–144.

Goff, P. (2021) How exactly does panpsychism explain consciousness?, [Online], https://www.philipgoffphilosophy.com/.

Goff, P. (forthcoming) Quantum mechanics and the consciousness constraint, in Gao, S. (ed.) *Quantum Mechanics and Consciousness*, Oxford: Oxford University Press.

James, W. (1902/1985) *The Varieties of Religious Experience, The Works of William James*, Vol. 13, Cambridge, MA: Harvard University Press.

Locke, J. (1689/2008) *An Essay Concerning Human Understanding*, Phemister, P. (ed.), Oxford: Oxford University Press.

Logothetis, N.K. (2008) What we can do and what we cannot do with fMRI, *Nature*, **453**, pp. 869–878.

Michel, M. (2019) Consciousness science underdetermined: A short history of endless debates, *Ergo*, **6**, art. 28.

Mitchell, K. (2019) Beyond reductionism — systems biology gets dynamic, *Wiring the Brain*, Blog, 14 September, [Online], http://www.wiringthebrain.com/2019/09/beyond-reductionism-systems-biology.html.

Mørch, H.H. (2018) Is the integrated information theory of consciousness compatible with Russellian panpsychism?, *Erkenntnis*, **84** (5), pp. 1065–1085.

Odegaard, B., Knight, R.T. & Lau, H. (2017) Should a few null findings falsify prefrontal theories of conscious perception?, *Journal of Neuroscience*, **37** (40), pp. 9593–9602.

Papineau, D. (2001) The rise of physicalism, in Gillet, C. & Loewer, B.M. (eds.) *Physicalism and its Discontents*, Cambridge: Cambridge University Press.

Prakash, C., Stephens, K., Hoffman, D.D., Singh, M. & Fields, C. (2017) Fitness beats truth in the evolution of perception, *Acta Biotheoretica*, **69**, pp. 319–341.

Strawson, G. (2018) The consciousness deniers, *The New York Review*, 13 March, https://www.nybooks.com/daily/2018/03/13/the-consciousness-deniers/.

www.ingramcontent.com/pod-product-compliance
Lightning Source LLC
Chambersburg PA
CBHW071231230426
43668CB00011B/1386